Closed Nuclear Fuel Cycle with Fast Reactors

Closed Nuclear Fuel Cycle with Fast Reactors

Handbook of Russian Nuclear Power

Evgeny O. Adamov
Honored Scientist of the Russian Federation, Chairman of the Scientific and Technical Council of Rosatom State Corporation, Russia

ELSEVIER

ACADEMIC PRESS
An imprint of Elsevier
elsevier.com/books-and-journals

Academic Press is an imprint of Elsevier

125 London Wall, London EC2Y 5AS, United Kingdom
525 B Street, Suite 1650, San Diego, CA 92101, United States
50 Hampshire Street, 5th Floor, Cambridge, MA 02139, United States
The Boulevard, Langford Lane, Kidlington, Oxford OX5 1GB, United Kingdom

ISBN: 978-0-323-99308-1

For Information on all Academic Press publications visit our website at https://www.elsevier.com/books-and-journals

Publisher: Charlotte Cockle
Acquisitions Editor: Charlotte Cockle
Editorial Project Manager: Maria Elaine D
Production Project Manager: R.Vijay Bharath
Cover Designer: Christian J. Bilbow

Typeset by Aptara, New Delhi, India

Contents

Part IV
Advanced reactor technologies and the nuclear power engineering infrastructure

Preface

In 2000, the Russian Government approved the Strategy for Nuclear Power Development in Russia in the First Half of the 21st Century which was the first long-term program defining the main milestones and key objectives of nuclear power generation not only in the country but worldwide. It can be said that the objectives of the past two decades since that time have been mainly achieved:

1. The installed capacity of nuclear power reached the objective of 30 GW.
2. The capacity factor of NPPs exceeded 80% as early as in 2008.
3. The power units with high and intermediate degree of readiness were completed.
4. New SNF storage facilities were built at the Mining and Chemical Plant.
5. Decommissioning of the first generation power units and their replacement with third generation power units commenced at Novovoronezh NPP and Leningrad NPP.
6. A power unit with a BN-800 reactor plant was put into operation.
7. The scope of export contracts of the Russian nuclear industry for construction of NPPs, supply of nuclear fuel and services within the nearest 10 years exceeded 130 bln USD.

The Federal Target Program "New Generation Nuclear Power Technologies for the Period of 2010-2015 with the Extension up to 2020" approved in 2010 and completed in 2018 became the basis for the Integrated Nuclear Science, Engineering and Technology Development Program for the Period up to 2024. Both programs describe the practical implementation stages for the Initiative of V.V. Putin, President of the Russian Federation, for energy support of the sustainable humankind development, comprehensive solution of any problems related to nonproliferation of nuclear weapons, and ecological recovery of the Earth proposed at the Millennium Summit in 2000.

In 2013, the work was partially consolidated within the program "New Generation Nuclear Power Technologies for the Period of 2010-2015 with the Extension up to 2020" under the common title of Proryv Project Framework, and unique results were obtained due to their completion within a relatively short period of time.

The technology of new solid mixed nitride uranium-plutonium (MNIT) fuel optimal for fast reactors was developed. Experimental fuel rods with this fuel demonstrated their operability both in research reactors and BN-600 reactor plants with the maximum burnup of up to 7.5%. Construction of a pilot MNIT fuel fabrication facility (the fabrication/refabrication module – FRM) with the capacity of 14 tons per a year will be completed in 2022.

The design of a pilot and demonstration power unit with the capacity of 300 MW with MNIT fuel and lead coolant (BREST-OD-300) obtained a positive opinion of the State Expert Review Panel. Its commissioning for pilot operation is planned in 2026. The possibility to create a core with the reactivity margin in the course of power operation at the power level of more than 30% not sufficient for uncontrolled runaway of a prompt neutron reactor was demonstrated through calculations and experiments (on the upgraded BFS test facility). The R&D and design study of a power unit with a BN-1200 reactor are near completion. Subject to achievement of the established competitive ability, it can become the first high-capacity fast reactor marking the beginning of transition to two-component nuclear power.

The concept of the pyrochemical SNF processing technology providing for short external fuel cycles and nonseparation of uranium and plutonium for technology enhancement of the nonproliferation regime was defined and substantiated through experiments in the laboratory environment. Usage of borophosphate glass (more resistant to radiation than aluminophosphate glass currently used at Mayak Production Plant) for solidification of chlorine-containing waste as well as usage of a "cold" firepot as a melter for vitrification of high-level radioactive waste was substantiated. Basic designs of main and auxiliary process equipment for the hydrometallurgical process of the combined processing scheme for MNIT and MOX FR SNF were developed.

As far as fractionation of high-level radioactive waste is concerned, the technologies for extraction and separation of rare-earth and trans-plutonium elements, Am and Cm extraction and separation were developed and checked with real SNF. Engineering achievability of the actinide extraction level necessary for radiation-equivalent disposal of radioactive waste (more than 99.9%) was demonstrated through experiments. Upon transition to the chromatographic process for Am-Cm separation the amount of secondary radioactive waste was reduced more than hundredfold. The possibility for homogeneous Am burn-out was confirmed through experiments. The heterogeneous option is also considered as a redundant one.

New generation computer codes for reactor plants, NPPs, fuel cycle closure facilities and safety analysis as good as the best world-class products and in some cases leaving them behind were developed. All codes underwent verification and the majority of them have already passed certification.

Integration of the results obtained from R&D, design, engineering and technology development within the integrated project and the building information model (PDEC - Pilot Demonstrational Power Complex) supporting both its construction and subsequent operation and decommissioning is carried out within the framework of Proryv Project Framework for the first time in the country.

The purpose of Proryv Project Framework is to form the grounds for large-scale nuclear power with the NFC closure on the basis of fast reactors. Experimental verification of all closed NFC components will be carried out at the PDEC Complex including the FRM, the power unit with a BREST-OD-300 reactor, and the module for SNF processing and RW treatment for radiation-equivalent disposal. The PDEC Complex is being constructed at the site of the Siberian Chemical Plant.

Further improvement of NPPs with VVER reactor plants, particularly development of a reactor design with spectral regulation, is provided within the framework of the Integrated Nuclear Science, Engineering and Technology Development Program for the Period up to 2024. Feasibility of transition to supercritical parameters of coolant in order to increase the NPP efficiency factor is considered.

All these achievements formed a basis of the Strategy for Nuclear Power Development of the Russian Federation up to 2050 and Further Extension up to 2100 approved by the General Presidium of the Scientific and Technical Council of Rosatom in December 2018. The White Book was published for substantiation of its updated version as for compilation of the Strategy for Nuclear Power Development in Russia in the First Half of the 21st Century in the early zeros.

A.M. Lokshin, First Deputy CEO of Rosatom

Introduction

Development of nuclear power for more than half of a century has not yet resulted in any nuclear technology ready to compete with the traditional power technology on the scale of global power engineering. But this objective can be achieved within the nearest decades based on the extensive practical experience of its first stage.

In comparison with other power technologies nuclear power has important fundamental peculiarities:

- Nuclear fuel has millionfold concentration of energy and sustainable resources;
- The share of feedstock costs in nuclear power is considerably less than the cost of fossil fuel for fossil-fuel generation, and consequently the net cost of energy is less dependent on volatility of the fossil fuel market than under the current conditions;
- The nuclear power waste are of relatively small volume and can be localized reliably, and the most hazardous ones can be "incinerated" in nuclear reactors;
- Zero-carbon nuclear power is capable of becoming the top-priority environment-friendly generation subject to prevention of severe accidents.

All these aspects open up conceptually new opportunities and prospects, particularly for implementation of the fuel cycle enabling to obtain the required amount of energy to satisfy the energy needs of the humankind in case of any expected civilization development scenario from limited natural resources of fertile material for thousands of years. Nuclear power technically enables to implement the closed process cycle considerably reducing its environmental impact in comparison with the impact of other traditional power technologies. The potential of nuclear power for development of power grids in remote regions and arrangement of power installations for large transport facilities is also high. Nuclear power opens up great opportunities for usage of nuclear fuel to replace fossil fuel which may be efficiently used for other purposes: transport, pharmaceutics, perfumery, fertilizers, production of synthetic materials, foodstuff, paints, varnishes, plastics, etc., unlike the former.

Thus, nuclear power potentially has all necessary features for gradual replacement of the significant fossil-fuel power generation section and establishment as the predominant power technology.

Formation of the required preconditions for development of large-scale nuclear power and implementation of its fundamental peculiarities constitute the main body of the nuclear power development strategy.

The demand for the fundamental peculiarities of nuclear power will mean the demand for large-scale nuclear power.

Significance of nuclear technology development in Russia is determined by its national interests:

- Nuclear technologies remain the basis of the Russian defense capacity within the period under consideration;
- Without any limitation on the part of cheap fuel nuclear power opens up new opportunities for the Russian economy development;
- Large-scale nuclear power shifts the center of influence in energy production from fuel-mining industries and fuel transportation to state-of-the-art knowledge-intensive nuclear and auxiliary nonnuclear technologies, and from fertile materials to the products of these technologies in export, thus giving new momentum to social and cultural development of Russia;
- Developing nuclear power will enable to avoid the risks related to depletion of fossil fuel and international conflicts over its sources, thus facilitating stabilization of the political situation;
- Well-developed nuclear power eliminating the need for large-scale fossil fuel mining and transportation will ensure national energy security in the time of expected global climatic changes with increasing frequency and intensity of extreme natural phenomena;
- Engagement of plutonium from reduced nuclear warheads and nuclear fuel into the balanced closed fuel cycle of fast reactors will support the non-proliferation regime;
- Nuclear power will contribute considerably to increase of life duration and quality for the Russian citizens by facilitating safe economic and social development and habitat preservation.

The first long-term Strategy for Nuclear Power Development in Russia in the first half of the 21st century [9] was approved by the Russian Government in 2000. It was based on the first White Book of Nuclear Power issued in 1998 by NIKIET with engagement of the industry experts, the Russian Academy of Sciences and higher education institutions. The second expanded edition of the White Book was published in 2001 and had not been updated ever since [23]. Deployment of the works in accordance with the Federal Special Program "New Technology Platform for Nuclear Power" and the R&D, design, technology and engineering works performed within the framework of this program provided the material for the present edition.

Establishment of nuclear power

The possibility for arrangement of nuclear power was demonstrated in 1951 by start-up of the first EBR-1 reactor (Experimental Breeder Reactor I) which generated thermal energy used to produce electric power. It was sufficient to light four lamps of 200 W each. It is interesting to note that this reactor was already operated with fast neutrons as it was intended to confirm the theory of Fermi on the possibility for expanded breeding of fissile materials. The reactor had remained experimental within the entire lifecycle and was connected to the local grid of Arco (Idaho) only in 1955; so, although the plate "The World's First NPP" is installed on the building it is not consistent with the reality.

The research objectives were also pursued on the pilot NRX reactor - the predecessor of the well-known series of NPPs with CANDU reactors - which had reached criticality even earlier (in 1947).

The possibility for peaceful nuclear power in the USSR was demonstrated in 1954 by connection of the first nuclear power plant in Obninsk to the power grid in the USSR [67]. The first NPP capacity was only 5 MW (e), but the first installations were followed with construction of more powerful NPPs all around the world.

Construction of the NPP in Obninsk was a political action, as the USSR government did not see any prospects of nuclear power for the country due to significant resources of fossil fuel. Siberian NPP with the initial capacity of 100 and then 600 MW generating electric power in parallel with production of weapon-grade plutonium was commissioned only in 1958. Many NPPs had been built worldwide by 1964 when the first power units of Beloyarsk NPP and then Novovoronezh NPP were commissioned.

By 1980s, there were approximately 300 operating nuclear reactors [23] with the total installed capacity of about 200 GW (e) around the world. Nuclear power produced about 10% of the global amount of electric power. So, the capacity of nuclear power increased from 5 to 200,000 MW just in a quarter of a century. It is difficult to find any example of such fast introduction of a new power generating technology into the social life in the history.

Two circumstances had a profound effect on establishment of nuclear power in the country:

1) High growth rates in the traditional power engineering within the post-war decades - 5–7% worldwide, up to 12% per a year in the USSR, placing the same high demands on development of nuclear power;
2) Government investments into the reactor facilities and nuclear fuel cycle developed for military purposes.

On the basis of favorable operating experience of the first generation civil nuclear power installations the concept of two-stage development of nuclear power had established in the world:

1) Power engineering based on thermal reactors (TNR) and accumulation of plutonium in them in order to start fast reactors (FNR);
2) Development of large-scale nuclear power on the basis of FNRs gradually replacing the traditional power engineering based on fossil organic fuel.

The strategic aim is to acquire sustainable resources of cheap fuel - uranium and, probably, thorium - on the basis of nuclear breeding.

The tactic aim is to use ^{235}U-based thermal reactors (mastered for production of weapon-grade materials - plutonium and tritium - and for nuclear-powered submarines) in order to generate power and radioisotopes for the national economy and accumulate civil-grade plutonium for fast reactors.

It was considered that large-scale nuclear power gradually replacing the traditional one and having no further limitations on the side of cheap fuel resources could be developed on the basis of fast reactors as low-grade deposits are economically feasible for the reactors with the breeding ratio of more than one completely using natural uranium or thorium.

Success of nuclear technologies in the military area and the first NPPs gave rise to ambitious intentions to develop nuclear power at the scale of several thousand GW by the end of the XX century.

Potential of nuclear power

As of 2017 the total amount of reasonably assured uranium resources with the recoverable at a cost of not exceeding 130 USD/kg of uranium including the potential of mining through the use of underground leaching was 8.122 mln. tons, and with the extraction cost not exceeding 260 USD/kg of uranium - 10.652 mln. tons [191].

Let us evaluate **the potential of nuclear power based on thermal reactors with the open NFC** proceeding from the nominal resource of 10 mln. tons of uranium [45].

A TNR with the capacity of 1 GW (e) consumes ~150 tons of natural uranium within a year of continuous operation, produces ~8.76 TWh of electric power and generates 15 tons of irradiated nuclear fuel (SNF).

A total of 10 mln. tons of cheap uranium enable to generate about 60,000 GW (e) of electric power per a year at NPPs with TNRs, which is equivalent to continuous operation of a NPP with the capacity of 600 GW (e) for 100 years.

Then nuclear power based on NPPs with TNRs will come to an end (Fig. B1) leaving behind about 1 mln. tons of SNF and 9 mln. tons of waste uranium.

Let us assess **the potential of nuclear power based on fast reactors with the closed NFC.**

A total of 10 tons of ^{235}U (produced from 2000 tons of natural uranium) and 50 tons of natural uranium will be necessary for start-up loading of a FNR with the capacity of 1 GW (e) based on enriched uranium. Thus 2050 tons of

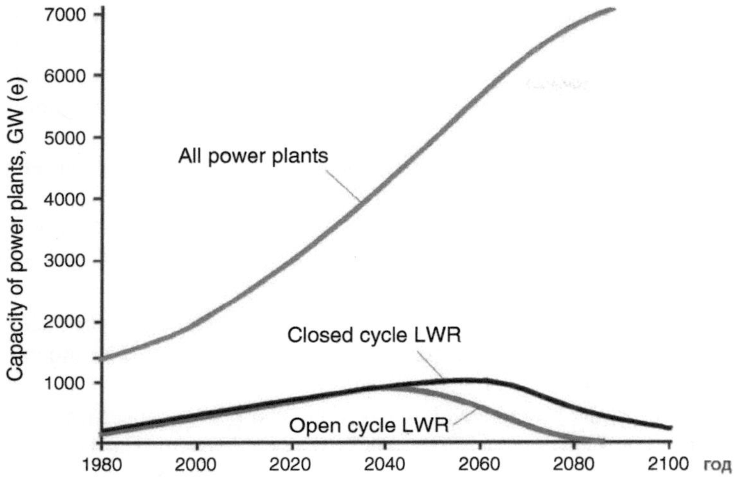

FIGURE B1 Approximate growth scenario for nuclear capacities without fast reactors (with the potential uranium reserves of ~10 mln. tons).

natural uranium are needed for start-up of 1 GW (e) based on enriched uranium, including 60 tons of nuclear fuel and 1990 tons of waste uranium.

A FNR with the capacity of 1 GW (e) consumes ~1 ton of natural uranium within a year of continuous operation, produces ~8.76 TWh of electric power and generates 1 ton of radioactive waste - RW (fission products). A FNR with the capacity of 1 GW (e) can operate on its own (1990 t) waste uranium for 1990 years.

A total of 10 mln. tons of cheap uranium enable to start approximately 5000 GW (e) of FNRs capable of operating for 1990 years on their own waste uranium with generation of about 10,000,000 GW(e) of electric power per a year which is equivalent to operation of a NPP with the capacity of about 10,000 GW (e) for 1000 years. Then nuclear power with FNR-based NPPs will come to an end (on cheap uranium) leaving behind about 10 mln. tons of radioactive waste in the form of fission products.

It is obvious that large-scale nuclear power is possible only with FNR-based NPPs with the closed nuclear fuel cycle (CNFC). And two strategies of FNR introduction into nuclear power are possible here.

The first strategy of FNR introduction to nuclear power is **to start FNRs based on plutonium from TNRs.**

A 1 GW (e) TNR with high burnup consumes about 10,000 tons of U_{nat} and produces about 7 tons of fissile Pu within the period of 50 years.

About 10 tons of Pu (7 +2~3 prior to onset of the self-sustainment mode) are needed to start a 1 GW (e) FNR based on Pu from a TNR, which is due to consumption of 14,000 tons of U_{nat} (within the period of 50 years) in order to provide nuclear fuel for the TNR with the capacity of 1 GW (e).

So, 714 GW (e) of FNRs can be started on plutonium from TNRs subsequent to running of all 10 mln. tons of U_{nat} through TNRs (not earlier than in 50 years).

Further rapid growth of FNR-based nuclear power capacities is possible only due to breeding of nuclear fuel in FNRs and requires higher breeding ration (BR) of FNRs.

The first strategy of FNR introduction to nuclear power is **to start FNRs based on enriched uranium.**

A total of 10 tons of ^{235}U are needed to start a 1 GW (e) FNR based on enriched uranium, which is due to consumption of 2000 tons of natural uranium for nuclear fuel of a FNR with the capacity of 1 GW (e).

So, 5000 GW (e) of FNRs can be started on enriched uranium produced from 10 mln. tons of natural uranium that is seven times as many (and much faster) than in the first case.

Further moderate growth of FNR-based nuclear power capacities is possible due to breeding of nuclear fuel in the FNRs with the breeding ratio of little more than one.

So, a FNR with the breeding ratio of more than one can provide for almost complete burning of the mined uranium, not approximately 1%. Hundredfold increase of energy output of the nuclear fuel as compared with TNRs enables to supply 4000 GW (e) of fast reactors with cheap uranium for 2500 years with low fuel (raw material) cost component, and the power generation of 10,000 GW (e) - for about 1000 years. Uranium from low-grade deposits with the resources exceeding the reserves of cheap uranium by a factor of hundreds and even thousands is also acceptable for FNRs (Fig. B2).

Current state of global nuclear power

In accordance with the IAEA data [170] 453 nuclear power units with the total installed power capacity (net) of approximately 399 GW(e) were operated worldwide, and 55 nuclear power units (56.6 GW(e)) were under construction as of February 2019. In 2017, the NPPs worldwide generated 2487 bln. kW·h of electric power in aggregate[1] (breakdown by countries - see Fig. B3) [170] or 10% (Fig. B4) [222] of the total global generation (the peak was 18% in 1996), while more than 65% were generated by thermal power plants based on fossil fuel. Eleven out of 55 nuclear power units being under construction at present are accounted for by China, 6 - by Russia and 7 - by India. The average age of NPPs operated worldwide is 30 years and 6 months as of February 2019 (particularly in Russia).

Global power consumption continues to grow in spite of great uncertainty in the prospects for further economic development, mainly due to developing countries, and this trend will remain, at least some decades ahead, given that almost a quarter of the planet population still has no access to any industrial energy sources.

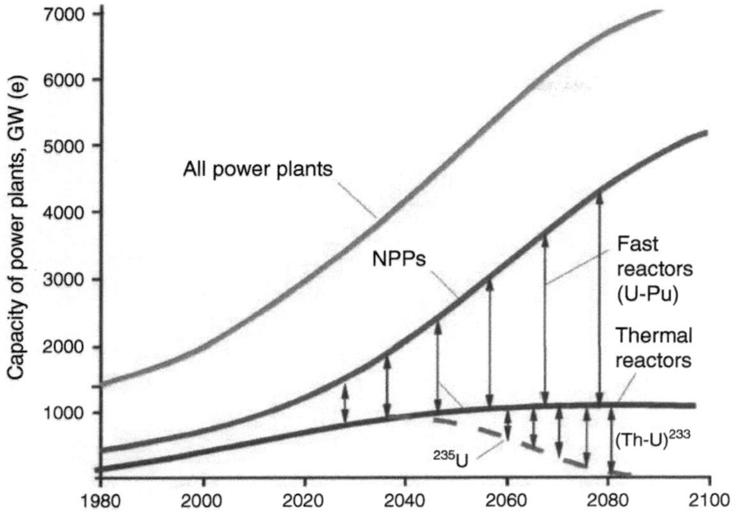

FIGURE B2 Possibilities for power generation growth in the world due to nuclear capacities including fast reactors.

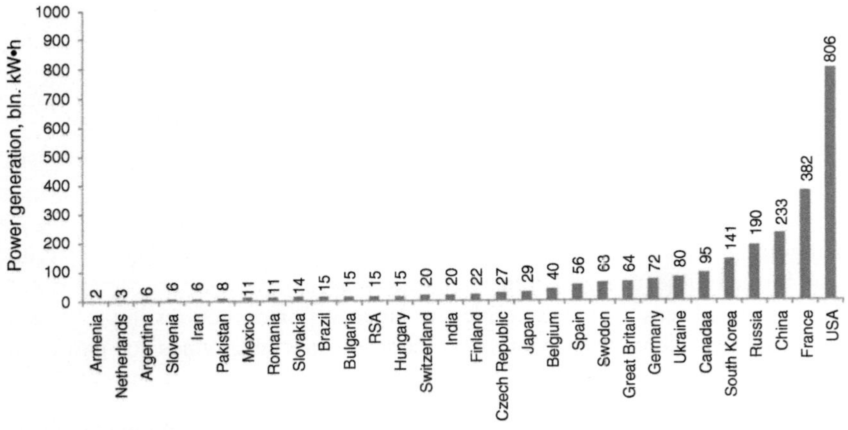

FIGURE B3 Electricity generation at NPPs worldwide in 2017 (according to the IAEA data).

Problems of nuclear power

Ambitious development programs for large-scale nuclear power turned out to be both economically nondemanded and technically unprepared.

Despite the significant role of nuclear power its certain crisis may be mentioned at this present time. The emerging trend for reduction of its share in the global power generation, curtailment of nuclear programs and developments in the developed Western countries are evidences of this. Besides, nuclear power is

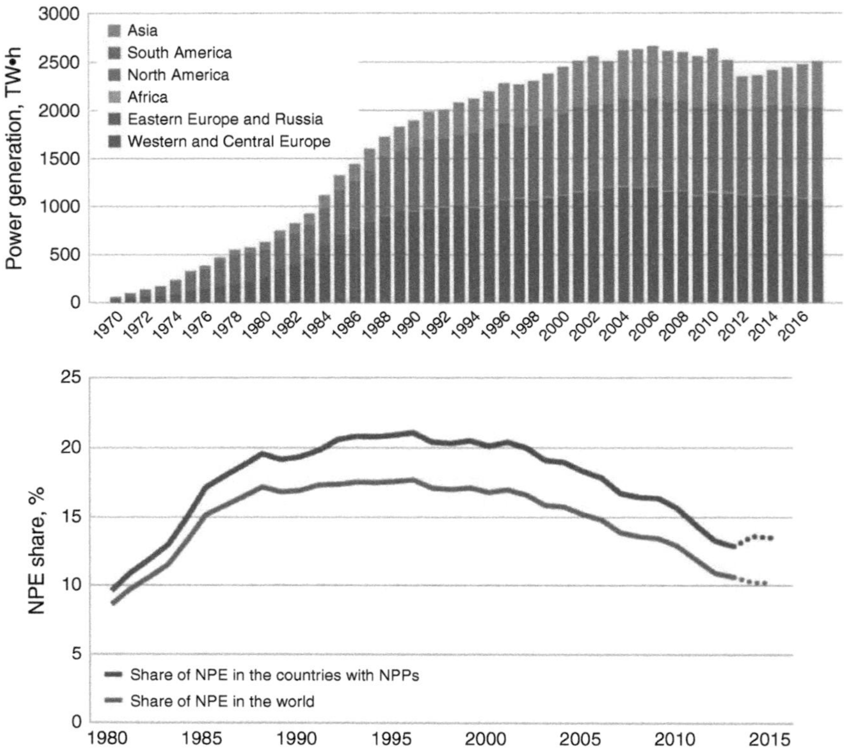

FIGURE B4 Nuclear power electricity generation worldwide (a) and share of NPPs (b) in the total global power generation (according to the data of World Nuclear Association).

being criticized up to the demand for its complete closure. Although this criticism is often subjective, and even completely biased, it should be recognized that there are valid reasons for the criticism. Nuclear power requires improvement like any other technology. Moreover, there are special grounds for particular attention toward it:

- Potential hazard of accidents with high environmental and economic damage (reality of this hazard is confirmed by a number of accidents);
- Accumulation of high-level and long-lived waste;
- Relation of nuclear power to the risk of nuclear weapon proliferation, etc.

A major accident at Three Mile Island NPP (USA, 1978) and accidents at Chernobyl NPP (USSR, 1986) and Fukushima NPP (Japan, 2011) pointed out the unacceptable safety level of NPPs with the first generation thermal reactors.

Operational safety of the modern nuclear power is acceptable for the existing scale of its application subject to gradual replacement of the operating power units with the third generation reactors.

Resources of profitable mined natural uranium are limited. With the practice of uranium "burning" in TNRs prevailing at present these resources will be exhausted as early as this century, both in Russia and in the world. SNF processing in the course of plutonium recycling into new fuel for TNRs (MOX fuel, REMIX fuel) can extend these time limits just for a while increasing the expenditures and reducing the opportunity for subsequent development of fast reactors due to inefficient use of Pu [43].

No environmentally acceptable solution for the problem related to handling of SNF and long-lived RW of nuclear power was found.

The problem related to nonproliferation of fissile materials circulating within the NFC was not solved, thus aggravating the severe problem of nuclear weapon nonproliferation on top of everything else.

Construction of FNRs was limited to the first pilot power units because of their higher cost as compared with TNRs.

Safety of the present generation of TNRs is ensured mainly by increased quantity of various safety systems and radioactivity release limitation systems, as well as strengthening of requirements for the equipment and personnel. As a result, NPPs become more and more sophisticated and consequently more expensive. It may be said that nuclear power is close to its economic "limit" with the currently prevailing safety philosophy: further increase of safety systems will result in inevitable loss of the nuclear power competitiveness.

So, the first nuclear power development strategy - the strategy with FNR-based rapid growth - has not been implemented in any country.

Development of the safe growth strategy for nuclear power

In 1990s, specialists of the Russian industry-specific and academic institutes tried to propose the long-term development strategy for nuclear power based on the physical and engineering preconditions of NPP safety assurance, the commitment to cheapen their construction, maintenance of the natural radioactivity balance for RW disposal, prevention of proliferation of any weapon-grade materials, and complete involvement of the mined natural uranium into power production.

These works resulted in approval of the Strategy for Nuclear Power Development in Russia in the First Half of the 21st Century (hereinafter - Strategy-2000) [9] defining the main objectives and ways for implementation of nuclear power development within the period of up to 2050 by the Russian Government in May 2000.

Strategy-2000 became the first national-grade document declarative of commenced review of the attitude toward the use of NPPs supported by a number of actions of the leading European countries in the following years. It was possible to speak about renaissance of nuclear power. For Russia Strategy-2000 defined not only the possibility to return to new construction of improved NPPs of the previous type, though their completion on the previously developed sites, of course, was provided, and the period when development and replacement of power units would take place on the basis of improved NPPs with VVER reactors was foreseen. Deep insight into the previous nuclear power development period, analysis of any causes of the accompanying accidents, accumulating problems with SNF, inefficient use of natural uranium, as well as care for nonproliferation of any technologies resulting in development of nuclear weapons led to understanding of the technology upgrade priority justifying statement of the question with regard to large-scale development of nuclear power.

Solution of the safety problems eliminating any possibility for accidents resulting in evacuation of the public and alienation of significant land territories from practical use, involvement of all mined natural raw materials (instead of 0.7% of natural uranium) into power production with simultaneous prevention of continuous increase of accumulated SNF stock, the possibility for technology enhancement of the nonproliferation regime on the basis of scientific and technical approaches developed within the previous decade and called "inherent safety" became the basis of Strategy-2000. These provisions also formed the backbone of the Initiative of the RF President at the Millennium Summit of the UN (New York, September, 6, 2000) for commitment to nuclear power in the sustainable development within the new millennium the first decade of which strengthened the renaissance trends and gave rise to the series of long-term energy programs for development of nuclear power in some countries. Germany which had previously planned premature closure of several power units settled for extension of their service life, Great Britain reviewed the role of nuclear power in its long-term plans, Italy which had previously disregarded NPPs completely commenced preparation for arrangement of a new plebiscite. Priorities of energy development on the basis of nuclear power gained a new momentum in the programs of Japan and China.

Coping with crisis phenomena in the Russian economy, mainly due to sharp rise in the prices of exported raw material resources within the period of 2001–2008 enabled to return to completion of the previously commenced NPPs and to adopt the program for accelerated development of nuclear power in Russia in the second half of the decade with higher rates and scale than it was supposed in 2000. It was reflected in the Action Program of Rosatom defining commissioning of capacities up to 2020. In February 2010, the Federal Special Program "New Generation Nuclear Power Technologies for the Period of 2010–2015 with the Extension up to 2020" was adopted.

In March 2011, probably the largest accident in the history of the global nuclear power took place at Fukushima NPP (Japan). Three power units and SNF storage facilities were destroyed, radioactivity in the amount requiring massive evacuation of the public, as well as restriction of access for people and economic activities within the significant territory was released into the environment. The new revision of energy strategies in the majority of countries begins from this moment.

The key provisions of Strategy-2000 have maintained their relevancy up to the present moment, but require clarification and adjustment with due regard for the current status of nuclear power and its development opportunities, the current and expected growth rates for power generation and consumption in Russia and in the world, as well as some other global aspects of the worldwide energy market.

Basic principles of Strategy-2000

1. Nuclear fuel breeding

Breeding of fissile materials is one of the basic preconditions for developing nuclear power. This function can be implemented in FNRs. High breeding ratios, high power density and short plutonium doubling times that can be sought after in some other countries are not necessary under the existing conditions of the RF power engineering and economy. The breeding ratio of a little more than one and moderate power density are acceptable, so in the course of FNR development it is possible to focus mainly on solution of the cost-effectiveness and safety problems.

2. Inherent safety

The inherent safety principle is the consolidation of the inherent safety principle for reactor plants through extension to the entire nuclear fuel cycle with due regard for the RW issues and the nonproliferation regime. This principle includes:

- Prevention of severe reactor accidents and accidents at the nuclear fuel cycle facilities requiring evacuation of the public;
- Low-waste processing of SNF with radiation-equivalent RW disposal;
- Technology support of the nonproliferation regime.

Rejection of any dangerous engineering solutions is the best protection against hazards

Deterministic elimination of severe accidents should not be identified with unachievable absolute safety. Only catastrophic development of the most severe accidents shall be prevented in the deterministic manner, while the common probabilistic approach requiring credible probabilities of 10-3-10-4 known from the experience shall be applied to "ordinary" accidents.

According to their physical and technical principles FNRs with liquid-metal cooling have the maximum potential of inherent safety, and consequently cost-effectiveness, far from being completely implemented in their first generation.

Transition to dense thermally-conductive fuel and chemically-passive high-boiling heavy coolant with low neutron activation enables deterministic prevention of prompt neutron runaway accidents (due to arrangement of equilibrium cores with balance of burnt and reproduced fissile materials), loss-of-coolant accidents (LOCA), fires and hydrogen explosions in case of any human errors and equipment failures, as well as radioactivity releases requiring evacuation of the public even in case of any damage of the main external barriers (the containment, the reactor pressure vessel).

From "clean fuel - dirty waste" concept to "dirty fuel - clean waste" concept

Safe disposal of radioactive waste for decades and (according to the requirements of the USA regulatory authorities) millions of years causes quite reasonable doubts associated with reliability of such long-term estimates. Significant changes in the natural level of radiation and biological hazard can be avoided, and proofs of safe RW handling can be made persuasive subject to maintenance of the natural balance between radiological hazard of disposed RW and mined uranium raw material (radiation-equivalent RW disposal with their regional temporary concentration).

Excess of neutrons and their energy spectrum where all actinides are fissionable enable to arrange efficient burning of the most dangerous and long-lived radionuclides from the NFC waste in FNRs providing for radiation balance between disposed radioactive waste and mined uranium raw material without any special-purpose burner reactors up to the final stage of nuclear power development in many millennia.

The nonproliferation regime includes technology barriers and political surveillance.

Prevention of uranium blanket, transition to the fuel of equilibrium composition for FNRs and gradual rejection of uranium enrichment technologies for nuclear power create the necessary prerequisites for technology support of the nonproliferation regime.

The SNF processing technology shall eliminate any possibility for pure plutonium extraction. Besides, the possibility for surveillance over the configuration of any buildings and structures intended for processing of nuclear fuel by national means (for example, from the artificial satellites of the Earth) shall be provided.

Elimination of any transportation of fissile materials in the near-plant fuel cycle reduces the risk of their theft and enables to decrease the external fuel cycle time as well as the amount of fissile materials in it. The near-plant fuel cycle provides for the above-mentioned advantages, but does not eliminate the possibility for closure in case of centralized production facilities in any way.

The nonproliferation issue cannot be solved by tèchnical arrangements alone, as the possibilities for illegal obtaining and use of well-developed technologies for uranium isotopic separation and extraction of plutonium from SNF (particularly the SNF of light-water reactors) for weapon-grade materials remain regardless of the new nuclear power technology development. Complete solution of the problem requires the combination of technology and political arrangements and international control (IAEA).

3. **Competitiveness.** From "the more expensive the safer" stereotype to "the safer the cheaper" standard

Establishment of the technologies providing for large-scale development of nuclear power in accordance with the inherent safety principles shall be accompanied with enhancement of nuclear power competitiveness.

The cost of NPPs with FNRs, especially all CNFC components in comparison with the open NFC, shall be considerably reduced that is possible only on the basis of the inherent safety principle. Its consistent implementation, starting from the initial engineering solutions, creates the preconditions for simplification of the designs, requirements for the main and auxiliary equipment, the NPP structures and personnel, rejection of any additional safety system and reduction of the NPP costs.

Under the conditions of Russia improvement of the regulatory framework for the NPP design as well as rejection of expensive and unfeasibly large and bulky buildings, structures and infrastructure components represent the significant NPP cheapening reserve.

In the economic analysis of various fuel cycles increasing attention is paid recently not only directly to the process-related costs of power generation, but also to the total reimbursement costs of all losses (externalities) accompanying power production and distribution. Thus the market balance of the developing energy technologies should be considered with due regard for the externalities characteristic for each technology. At present rigorously substantiated assessments of externalities are almost absent.

Nuclear power should be committed to achieve economic competitiveness in the majority of regions even with the moderate fuel and energy prices expected in the nearest decades and not only in the indeterminate future when cheap fuel resources are exhausted, and greenhouse gas emissions are limited by quotas or penalized.

Alternative approaches to the nuclear power strategy

Analysis of approaches to the nuclear power development prospects in Russia and in other countries demonstrates two basic trends [56]:

1. Commitment to development of nuclear power on the basis of the existing and improved TNR types with the open NFC where ^{235}U is mainly burnt [89]. The options with the use of limited additional fuel reserve in the form of mixed uranium-plutonium oxide fuel (MOX fuel) produced by single-stage processing of the SNF from these reactors, extraction of the accumulated plutonium and its mixing with waste (depleted) uranium can be also included here.
2. Commitment to arrangement of the closed NFC with implementation of reactors providing for ordinary of extended breeding of nuclear fuel (BR

$\geq$ 1) [43]. They may include traditionally considered FNRs or light-water reactors with hard neutron spectrum discussed as early as 1970s and again proposed in the USA and the RF now.

In case of the first approach increasing quantities of natural uranium energetically utilized for less than 1% are engaged in handling, and the amount of SNF in storage is constantly growing. Nevertheless, this approach is considered as feasible under the current energy market conditions. The concept of further development in this direction has got widespread in the USA having the leading nuclear power in the world and is promoted by the prominent NPP designing corporations in the developing countries adopting the path of nuclear power acquisition. From the viewpoint of the USA the counted global uranium resources enable to pursue this course for a sufficient period of time.

It is obvious that large-scale nuclear power can be implemented only through the use of the second approach. However, the development strategies within the framework of this approach are conceptually different in various countries depending on the expected role of FNRs in the nuclear power structure. Three strategies for development of large-scale nuclear power with the following conventional names may be distinguished: "American", "French and Japanese," and "Russian" [19].

American strategy. The conceptual possibility for transition to processing of SNF from TNRs (accumulated in the temporary storage facilities for 100 years) in order to reduce the amount of high-level waste (HLW) subject to ultimate disposal through "after-burn" of HLW, primarily minor actinides, in FNRs is provided in the USA counting on TNRs with the open NFC in the immediate future. In this case FNRs are considered as noncompetitive energy generators and viewed as potential "cleaners" with prevailing TNRs. The BR below 1 is assumed for such FNRs, and their NFC remains open, as it requires continuous external (not from their NFC) make-up with fissile nuclides. The possibility to use FNRs with BR $\sim$ 1 and BR $>$1 is also considered, but their mission remains fundamentally the same.

French and Japanese strategy. France and Japan having no uranium deposits of their own traditionally established their development strategies providing for transition from the open NFC with TNRs to the closed NFC and FNRs with sodium coolant and the BR of much more than one capable of supplying the fuel for TNRs. The similar strategy has been considered and is still proposed by a group of specialists in Russia [89].

Russian strategy. The position formulated in Strategy-2000 [8] and detailed in [43] is based on the concept of large-scale nuclear power which can be developed by solution of its crucial tasks through the use of FNRs with moderate power density, without any excessive production of plutonium and sacrificing of the previously built NPPs with TNRs. In this case full internal breeding of plutonium in the core (BRC$>$1) with dense nitride fuel of equilibrium composition is important.

The development scenarios for the CNFC with FNRs shall be based on the actual existing structure of nuclear power. In this context it is important to understand the difference between two-component nuclear power with prevailing FNRs and two-component nuclear power with the dominant role of TNRs and FNRs used only for fuel make-up and HLW after-burning.

As mentioned above, the concept of two-component nuclear power was established in the second half of the past century, mainly under the influence of the following factors:

- Development of TNRs based on enriched uranium for military purposes and their subsequent refurbishment for civil power engineering;
- Understanding of the FNR necessity for development of nuclear power;
- Economic noncompetitiveness of the FNRs built in different countries with their peculiarities determined by the demand for high BR: high power density, use of sodium coolant, production of weapon-grade plutonium in the blanket.

In the course of the uranium enrichment technology assimilation TNRs became the basis of the modern nuclear power, and the function of fuel (plutonium) suppliers for TNRs in case of depletion of cheap natural uranium reserves was assigned to more expensive FNRs with higher BR. Large-scale circulation of SNF and MOX fuel between fast and thermal reactors and the centralized plants for SNF processing and nuclear fuel production is necessary for such two-component nuclear power.

Strategy-2000 considers two-component nuclear power as the stage in gradual transition of the dominant role to FNRs forming the basis for the future large-scale nuclear power.

Safety and economic competitiveness in comparison with other types of power generation remain the primary key factors for large-scale development of nuclear power in Russia and worldwide [84]. The technology platform of the Russian nuclear power based on VVER reactors is sufficient for the scope of NPP construction and export supplies predicted in the Energy Strategy of Russia for the period of up to 2030. However, its potential for solution of long-term strategical problems of the country is limited due to noncompliance of the safety level with the requirements for large-scale nuclear power, limited resources of natural uranium, accumulation of SNF, and reducing competitiveness due to enhancement of safety assurance arrangements.

Progress of Strategy-2000 implementation

The following provisions of Strategy-2000 have been successfully implemented by now [43]:

1. Safe and efficient operation and extension of the service life for the existing power units, particularly bringing of the capacity utilization factor to the level exceeding the design one.

2. Completion of construction and commissioning of power units with high and medium availability.
3. Bringing of the third generation power unit designs to the level of 3+ and construction of the reference third generation power units at the existing sites.
4. Commissioning of the power unit with BN-800 reactor.
5. Deployment of R&D works on the closed NFC technology for large-scale nuclear power.
6. Development of the design materials for the commercial power unit with BN-1200 reactor plant and the demo power unit with BREST-OD-300 reactor plant.
7. Development of conceptual designs for small-scale nuclear power installations with enhanced safety for remote regions and their service infrastructure.

The program of nuclear power development in Russia for the nearest two decades is aimed, first of all, at construction of the advanced third generation power units with thermal neutron reactors replacing the outdated power units as provided in Strategy-2000 [28].

Strategy-2000 today

However, considerable increase of capital costs for construction of NPP power units exceeding 3000 USD/kW (in 2000 ~1000 USD/kW) took place within the same period of time, the operating expenses and the prime cost of power generated at NPPs increased. All these factors reduce competitiveness of nuclear power in comparison with other types of power production, especially in comparison with modern combined cycle gas turbine units. The situation with the normal-mode ratio (the personnel size per an installed capacity unit) at the domestic NPPs amounting to 1.5 pers./MW has not improved (the average value for foreign NPPs is 0.3–0.5 pers./MW). Simple partial removal of the personnel outside the NPP periphery is not always accompanied with positive economic results, and in some cases leads to deterioration of the service quality. The annual output per a worker is 4.8 GW·h, while in non-nuclear power generating companies it reaches 10–13 GW·h/person. NPP designs with VVER reactors are inferior to the foreign ones in many parameters (metal consumption, the length of cable routes, construction volumes, the development areas, duration of construction, absence of the computer-based state-of-the-art CPS for commercial power units) as revealed in the course of VVER-TOI design review.

Disproportion in utilization of fuel resources in power generation (more than 65% – due to thermal power plants) has not been eliminated, and the trend toward increase of natural gas consumption in thermal power engineering has not been overcome for the past 18 years. The provision for reorientation toward export of electric power (particularly generated at NPPs) from export of raw materials has not been implemented. Despite the increase in the uranium mineral reserve base in Russia and worldwide, it is not capable of providing sustainable long-term development of nuclear power on the basis of TNRs initially made for military purposes (production of weapon-grade plutonium, reactors of nuclear submarines). The provision of Strategy-2000 on the necessity of transition to FNRs with the closed NFC becomes more and more urgent.

Subsequent to the accident at Fukushima NPP the basic strategic principles for safe growth of nuclear power capacities are still waiting for their engineering solutions and implementation in new installations.

Inherent safety - the necessary condition for large-scale nuclear power - cannot be reached only through further enhancement of "defense-in-depth" and wide implementation of passive safety means in the existing TNR designs, as proposed by some specialists. Experience has shown that it results in increase of capital costs and reduction of the NPP competitiveness, and the risk of repeated severe accidents remains.

Handling of SNF and RW remains one of the most hot-button problems. Any more appropriate solution than implementation of the closed NFC with return of valuable energy feedstock, extraction of useful isotopes, transmutation of neptunium, americium and isolation (disposal) of waste without any disturbance of the natural radiation balance has not been found yet.

Appearance of nuclear weapon in some countries or reaching of the threshold status in its development is indicative of the fact that just political and inspection arrangements are insufficient for implementation of the nonproliferation regime aims, as it is impossible to prevent enrichment of uranium from the traditional NFC and to prohibit construction of NPPs in certain countries for the benefit of independent energy development. The technology approaches to solution of this problem proposed in Strategy-2000 remain effective and require implementation.

It is obvious that large-scale nuclear power can develop only subject to its competitiveness at the energy market and investments with gradual shifting away from the common practice of governmental support. This refers, first of all, to the budgetary funds, the government guarantees for obtaining of low-interest loans may be reasonable in the longer run with due regard for the duration of nuclear power development programs. The inherent safety principle serves this purpose. Its consistent implementation, starting from the engineering solutions aimed to simplify the designs and to relax the requirements for equipment, buildings, the site and personnel, will enable to implement "the safer the cheaper" approach stated in Strategy-2000.

Strategy-2000 was focused on power generation, although some developments aimed at production of high-temperature industrial heat were also mentioned. Meanwhile, a half of the country's energy balance is related to heat generation, first of all for utility purposes, as well as to power generation outside well-developed grids. Strategy-2000 did not place the required attention on these problems to be solved, in particular, by development of low- and medium-capacity nuclear power plants.

It is impossible to develop new cheap NPPs with FNRs without update of the NPP design regulatory framework with involvement of the supervisory bodies and performance of R&D works, including the experience of prototypes. This objective remains among the most important for long-term development of nuclear power.

The need to maintain and replenish the workforce capacity in the nuclear industry, to preserve the crucial knowledge essential for significant development and to pass it down to the new generations of specialists was not mentioned in Strategy-2000. This problem has become even more drastic within the recent 15 years and requires special attention in development of new strategic documents.

Considerable positive changes took place upon approval of the Federal Special Program "New Generation Nuclear Power Technologies for the Period of 2010-2015 with the Extension up to 2020" [5] within the framework of which the top-priority task for the industry was determined - to implement Proryv project consolidating the projects on development of nuclear power technologies with inherent safety based on FNRs with the closed NFC.

The key provisions of Strategy-2000 gained widespread international support within the past years:

1. The requirements for innovative nuclear power systems complying with the sustainable development principles were determined within the framework of the international IAEA project INPRO arranged at the beginning of the century on the initiative of Russia. Joint research of the countries demonstrated the possibility for nuclear power based on FNR and CNFC technologies to comply with these requirements.
2. Within the framework of GENERATION IV International Forum, the countries developing nuclear technologies compiled the requirements for the fourth generation reactors. Six technologies were selected for joint development; four of them are FNR technologies.
3. The safety and cost-effectiveness requirements determined in INPRO and GENERATION IV coincided with the requirements of Strategy-2000.
4. At present the majority of countries with large NPP fleets (France, Japan, the Republic of Korea, the USA) or planning large-scale development of nuclear power generation (India, China) have their national programs for development of FNR and CNFC technologies.

Significance and urgency of the key provisions of Strategy-2000 increased subsequent to the accident at Fukushima NPP.

Part I

Global power generation and the role of nuclear power engineering

Chapter 1

Power generation and sustainable development

A human being as a biological object can rely only on his/her muscular power (~100 W) and uses stronger animals by domesticating them (the unit of power measurement named "horsepower" still exists: 1 hp = 736 W). Humans as social beings increased their energy power manyfold [176].

Specific human-induced energy security in 2015, kW/person:

Global average............................3.0
Russia......................................7.0
USA.......................................14.0
Developing countries 1.0

1.1 Modern energy sources

At present the humankind produces about 90% of energy by consumption of exhaustible energy sources (oil, coal, gas, nuclear fuel).

According to the IEA data (IEA, 2019) [176] the shares of various sources in the global power generation in 2016 were the following:

- Coal/peat (38.3%);
- Natural gas (23.1%);
- Hydro (16.7%);
- Nuclear (10.4%);
- Oil (3.7%);
- Other (renewable) (7.8%).

According to the IEA data energy consumption has increased by 52% within the period of 1990–2016 [190], while the global population increased by 41%. In this case average power consumption per capita increased by 51%.

Regional energy consumption has also increased within the period of 1990 to 2016: in the Middle East – by 247%, in China – by 238%, in India – by 182%, in Africa – by 108%, in Latin America – by 89%, in the USA – by 13%, in the EU-28 – by 3%.

Closed Nuclear Fuel Cycle with Fast Reactors: Handbook of Russian Nuclear Power.
DOI: https://doi.org/10.1016/B978-0-323-99308-1.00040-4

TABLE 1.1 Power generation broken down by sources, TW·h.

Year	Organic fossil fuel	Nuclear fuel	Renewable sources	Total
2000	9960	2591	2900	15,451
2016	16,320	2606	6011	24,937
Change in 2000–2016	6360	15	3111	9486

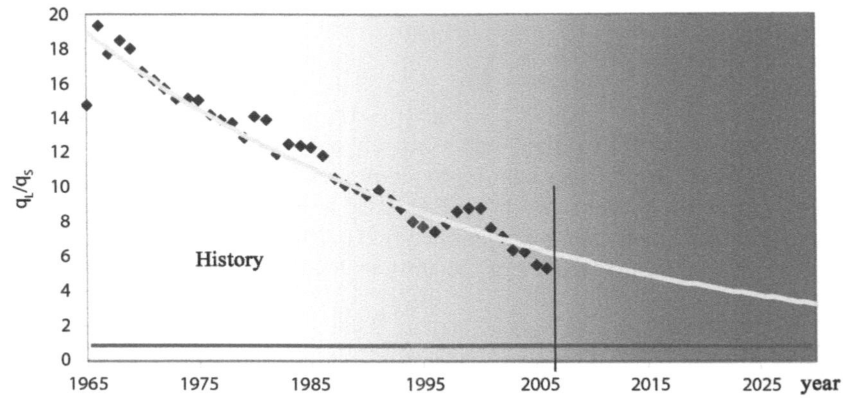

FIGURE 1.1 Ratio between the values of specific power consumption (q_L/q_s) in developed and developing countries.

The total global power generation in 2016 was 24,919 TW·h (Table 1.1).

1.2 Current peculiarities of energy consumption growth

According to the current evaluations, the population of the Earth will grow approximately to 10 bln., people by the middle of this century. Subject to implementation of strict measures for energy saving and maintenance of the current proportions in energy consumption between the developed and developing countries energy consumption in the world should increase 1.3–1.4 times approximately.

However, analysis shows [89] that the supposition on maintenance of the current proportions in energy consumption for extended periods of time is groundless (Fig. 1.1).

Considering the evolution of energy consumption in developed and developing countries it can be seen that the relative size of these groups scarcely changes over time and significant convergence in the specific power consumption is observed.

While in 1960s specific power consumption in these groups differed by 20 times, in 2006 this difference was reduced to 6 times. Taking into account higher population size and considerably higher economic development growth rates in the developing world it is obvious that this circumstance is the basic and constant cause of escalating tension in the global energy market. Balancing of specific power consumption in these groups implies increase in demand for energy resources by 2050 twice or thrice in comparison with the current level.

1.3 Fossil organic fuel

Combustion of coal, oil and natural gas provides for ~80% of global power consumption. Oil is in the first place in the total global annual power consumption equal to ~9.5 bln. tons of crude oil equivalent in 2016, followed by coal and gas. Global power generation in 2016 was also provided mainly through the use of fossil organic fuel (60–65%). In this case coal takes the leading position in the fuel balance of thermal power plants generating half as much power than petrochemical products.

Oil. Oil has remained the essential energy resource with the share of 33% [191] despite the fact that its share is constantly reducing. According to the data of British Petroleum (BP) the proven global reserves of oil in 2015 were approximately 232 bln. tons. In 2014, average daily consumption of oil in the world was about 13 mln. tons.

The established oil resources in Russia (2013) amount to approximately 13 bln. tons. Subject to stabilization of its annual extraction at the level of 0.5 bln. tons (2013) these resources will last for 25 years.

Natural gas. In 2015, the global reserves of natural gas were estimated by BP at the amount of 187 tln. m^3; subject to maintenance of the extraction level as of 2015 these resources should last for approximately 50 years.

The established recoverable resources of natural gas in Russia (2013) are estimated at the amount of 40–50 tln. m^3; subject to stabilization of its annual extraction at the level of 0.6–0.7 tln.m^3 these resources will be exhausted in 60–70 years.

Coal. Coal is taking on greater and greater importance in the global power consumption: in 2016 its share was 27% of the total power consumption. The global coal reserves are estimated at the level of 1 tln. tons (2009). Subject to stabilization of its extraction at the level of 8 bln. tons per a year (2013) these resources will last for 120 years.

The established coal resources in Russia are estimated at the amount of 150–170 bln. tons; subject to maintenance of its extraction at the level of 0.35 bln. tons per a year (2013) these resources will be exhausted in ~400 years.

1.4 Resource-related limitations of organic-based power engineering

Within the period from 1970s to the end of the second decade of the XXI century the global consumption of energy resources increased more than twice, and in 2015 it exceeded 10 bln. tons of crude oil equivalent.

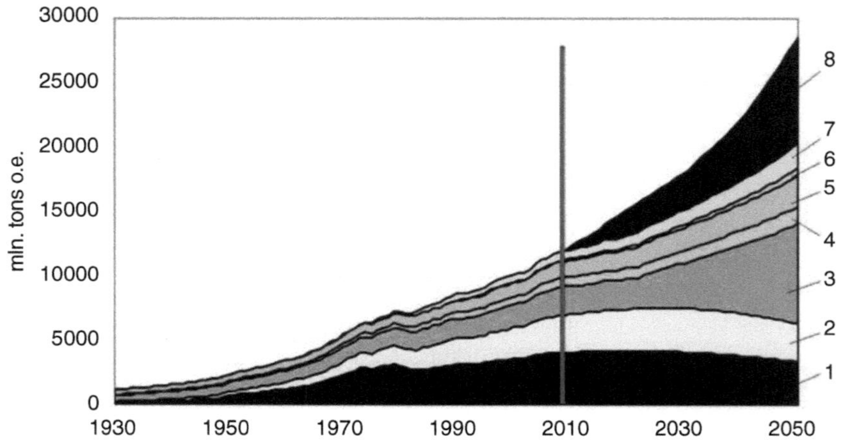

FIGURE 1.2 Energy demand and the possibilities to satisfy it through the use of various primary energy sources: 1 – oil; 2 – gas; 3 – coal; 4 – hydropower; 5 – biomass and waste; 6 – other renewable sources; 7 – atomic energy; 8 – unsatisfied demand.

The estimated demand for the primary energy sources in the first half of the XXI century [89] is shown in Fig. 1.2. It corresponds to the specific power consumption balancing rate represented with a yellow line in Fig. 1.1. It is the most optimistic scenario providing for considerable reduction of the economic growth rates in the developing countries and stabilization of power consumption in the developed countries.

This estimate assumes that the global coal extraction will increase fourfold by the mid-century as compared with 2009, use of biomass and waste will increase thrice, hydropower – twice, various renewable energy sources (without hydropower) – ninefold, and nuclear power – thrice [89]. Despite significant increase of energy resources deficiency of power will be keenly felt in the nearest future. The area of this unsatisfied demand is shown in dark color in Fig. 1.2. This estimate corresponds to the IEA forecast up to 2030. The estimated extraction of basic hydrocarbons (oil and natural gas) is shown in Fig. 1.3. Analysis of the global oil and gas extraction dynamics demonstrates that the maximum organic fuel extraction worldwide will be overpassed on the cusp of 2020–2030 [89].

The presented estimates are made on the basis of the assumption that the reserves of these energy sources not yet extracted exceed the proven modern resources approximately twice. The data up to 2015 correspond to the actual status [76].

Oil is the prevailing energy source for the global economic development in the beginning of the XXI century (Fig. 1.4). Firstly, it is the most significant in its energy capacity, and secondly, almost all transport (more than 95%) uses petroleum motor fuel as the energy source.

The energy consumption scale is constantly increasing, and the growth rate for consumption of organic fuel exceeds the rate of its resource base

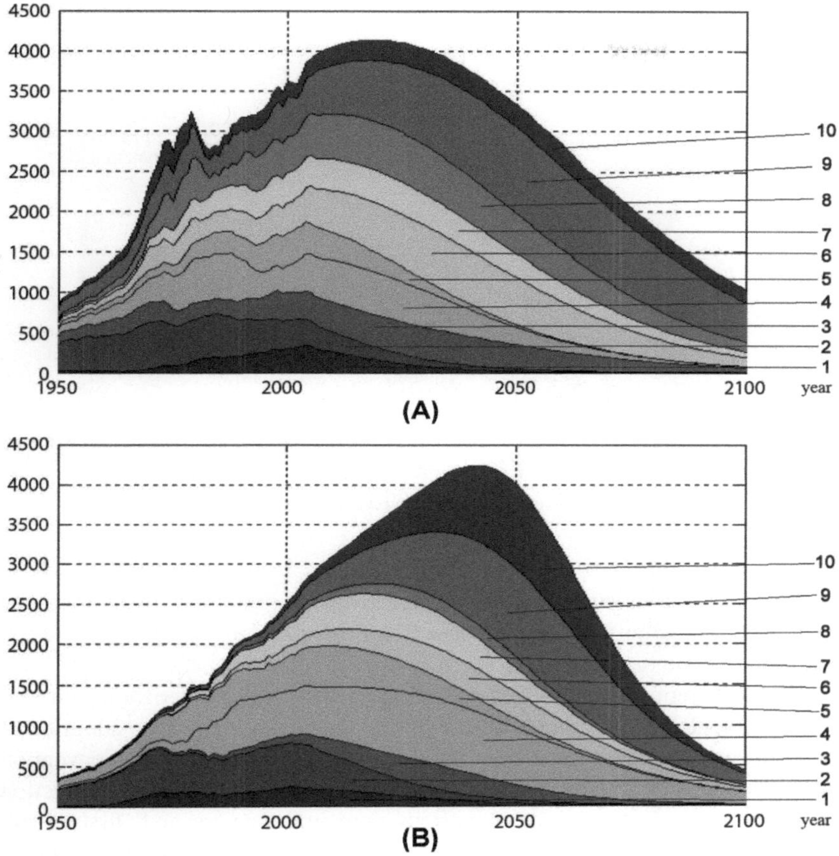

FIGURE 1.3 Estimated dynamics of oil (A) and natural gas (B) extraction in the current century: 1 – North America; 2 – USA; 3 – South America; 4 – Russia; 5 – Europe; 6 – Africa; 7 – Asia; 8 – Saudi Arabia; 9 – Middle East; – Iran

replenishment. It is quite probable that it would be impossible to satisfy the energy demand with the traditional technologies of organic fossil fuel utilization by the end of the current century.

1.5 Environmental restrictions of organic-based power engineering

Utilization of organic mineral fuel is arranged mainly through its combustion resulting in vast amounts of annual hazardous releases into the atmosphere.

In Russia, the fuel and energy industry facilities account for more than 30% of the total industrial releases of hazardous substances, more than 80% of industrial emissions of greenhouse gases, 70% of nitrogen oxide and 30% of sulfur dioxide.

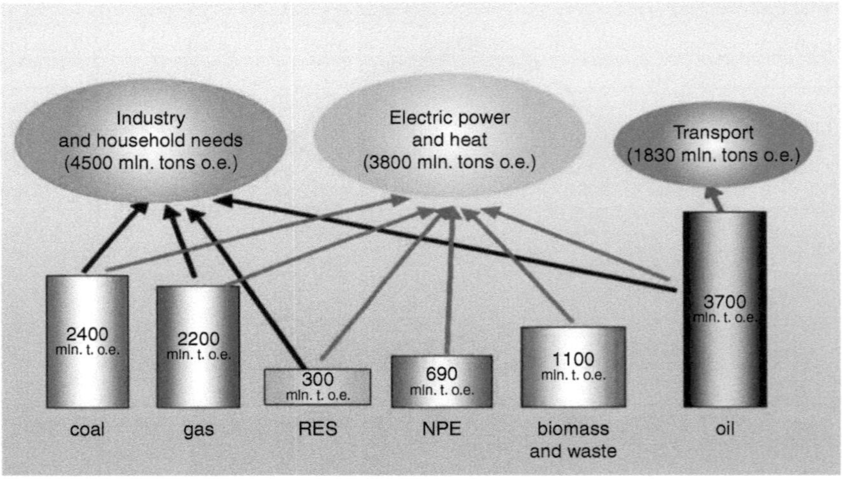

FIGURE 1.4 Consumption structure for energy sources (2010).

Development of such large-scale negative environmental phenomena as "rainfall acidification" and "greenhouse effect" is related to the power generation based on organic fuel.

Increased requirements for environmental safety are peculiar for the current stage of the global power engineering development. Subsequent to adoption of the Framework UN Convention and signing of Kyoto Protocol to the Convention on Climate Change implementation of the measures for limitation and reduction of greenhouse gas emissions in various economic sectors and, first of all, in the power generating industry became a significant trend in the economic development of countries. There were other conventions and protocols aimed to activate implementation of the Framework UN Convention (for example, Paris Agreement dated 2015). But the problem is yet to be solved. It is reasonable that the requirements for environmental safety of power engineering should become the limiters of its development as strict as availability of fuel resources and cost-effectiveness of production. Some states have already implemented penalties for emissions deteriorating the environment, but it has not yet become customary worldwide; transition to this practice is capable not only of practical solution of the environmental problems but also of fundamental change in the ratio between the utilized fuel resources. The power generating industry is the leader in the relative contribution to the total human-induced emissions of CO_2. The majority of these emissions are generated by coal combustion.

Despite the obvious environmental advantages of zero-carbon power generation sources at present they have not yet changed the ratio between power generation on the basis of clean technologies and technologies based on fossil fuel (Fig. 1.5).

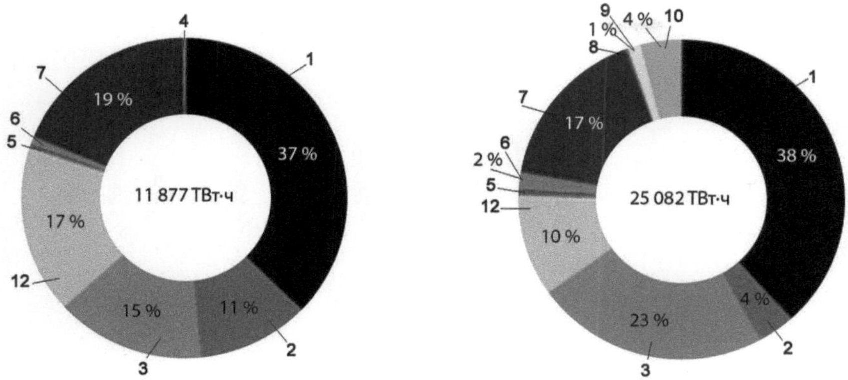

FIGURE 1.5 Global power generation structure in 1990 (A) and 2016 (B): 1 - coal; 2 - oil; 3 - gas; 4 - NPE; 5 - waste; 6 - biofuel; 7 - hydropower; 8 - geothermal power; 9 - solar photovoltaics; 10 - solar thermal power; 11 - RES; 12 - marine power engineering; 13 - other sources

Comparing the current established energy balance with the estimates of analytical agencies for the past years the prediction on reduction of the nuclear power share and retained leading positions of the coal and gas power engineering which has proved true in general comes under notice. At the same time CO_2 emissions from the power generation sector of the global economy increased by 1.7% in 2018 and reached their historical peak at the amount of $33–10^9$ tons in spite of increased installed capacity of RES-based generation (Fig. 1.6).

The IEA developed the sustainable development scenario aimed at drastic reduction of greenhouse gas emissions from power generation facilities providing for reduction of CO emissions mainly through further RES expansion with the annual commissioning of ~300 GW (e) from 2018. Practical implementation of this scenario with the current RES commissioning level (177 GW) seems to be impossible.

It is worth noting individually that only France among all major economically developed countries was able to achieve the specific CO_2 emissions per capita at the level of 4 tons. A large share of nuclear generation and a certain share of RES provide the opportunity for development of the economy with relatively high level of power consumption per capita where power generation leaves almost no so-called carbon footprint.

1.6 Mineral nuclear fuel

Raw material resources of nuclear fuel. The fuel made of the uranium isotope mixture (^{235}U and ^{238}U) produced in the course of natural uranium enrichment with ^{235}U isotope (c $x_5 = x_{nat} = 0.72\%$) to $x_5 = x_{en} = 3–5\%$ is used in thermal reactors. The main energy source in these reactors is the fission energy (with thermal neutrons) of ^{235}U nuclei loaded into the nuclear reactor within the fuel composition and ^{239}Pu nuclei produced in the course of the nuclear reactor

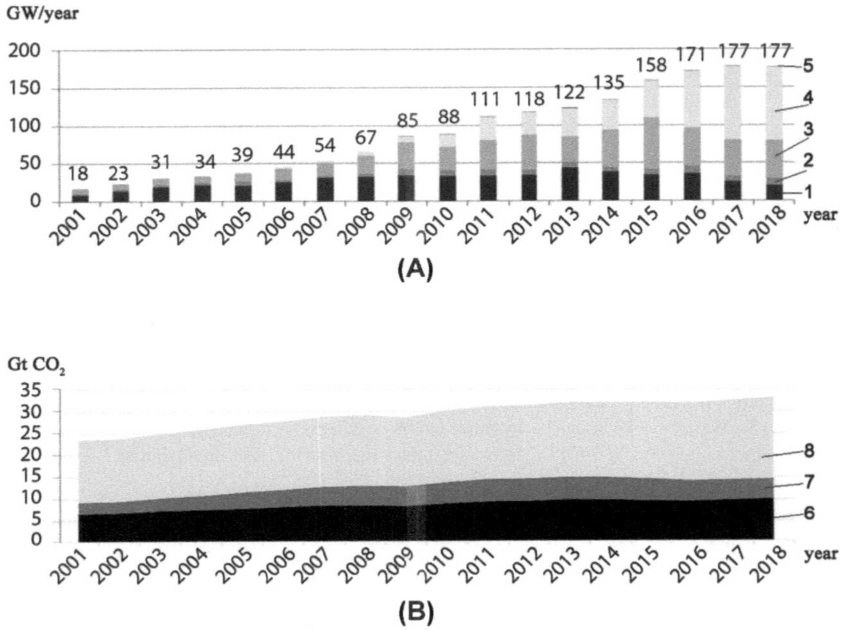

FIGURE 1.6 Commissioning of installed RES capacities (A) and the current level of CO_2 emissions (B) for the period of 2001–2018: 1 - hydropower; 2 - biofuel power engineering; 3 - wind power engineering; 4- solar photovoltaics; 5 - other RES; 6 - coal-based power engineering; 7 - other coal utilizations; 8 - other fossil fuel.

operation from ^{238}U nuclei. The additional insignificant energy source in them is fission of ^{238}U nuclei with fast neutrons.

In order to produce M_{nf} kg of nuclear fuel (c $x_5 = x_{en} = 3\text{–}5\%$) for a TNR from natural uranium (c $x_5 = x_{nat} = 0.72\%$)

$$M_{not} = M_{nf}(x_{en} - x_w)/(x_{nat} - x_w) \text{ kg of natural uranium are needed.}$$

where x_w is the content of ^{235}U in the enrichment production waste.

In the course of the TNR operation (BR < 1) energy is generated by "combustion" of

$$M_{fis} = M_{nf}(x_{en} - x_{snf})/(1 - BR) \text{ kg of issionable nuclei,}$$

where x_{snf} is the content of fissionable nuclides (^{235}U and ^{239}PH) in SNF.

Efficiency of natural uranium use in thermal reactors

$$M_{fis}/M_{nat} = [M_{nf}(x_{en} - x_{snf})/(1 - BR)]/[M_{nf}(x_{en} - x_w)/(x_{nat} - x_w)].$$

With $x_{en} = 5\%$, $x_w = 0.2\%$, $x_{snf} = 1.5\%$ and $BR = 0.5$ we obtain

$$M_{fis}/M_{nat} = (5\% - 1.5\%)(0.71\% - 0.20\%)/(1 - 0.5)(5\% - 0.2\%) = 0.7\%$$

In 2015, the established global reserves of natural uranium with the extraction cost of up to 80 USD/kg were estimated at the amount of 2.1 mln. tons, and the total reserves that can be utilized at the cost of 130 USD/kg – 5.7 mln. tons. If the modern nuclear power based on thermal reactors (with very low efficiency of natural uranium utilization ~0.7%) is preserved the uranium resources with the cost of up to 80 USD/kg will be exhausted in this century, provided that the rate of its consumption remains at the current level.

Nuclear fuel breeding. If only rare ^{235}U nuclei could be used for energy generation such nuclear power would be short-life and most likely useless, as there would be much more problems related to its arrangement than benefits from it.

Actually there is great amount of ^{238}U and thorium-232 (^{232}Th) in nature that can be used to produce fissionable nuclei: plutonium-239 and uranium-233 (^{239}Pu and ^{233}U respectively). As it was mentioned above, the mixture of fissionable nuclei (^{235}U, ^{239}Pu, ^{233}U) and fertile material nuclei (^{238}U, ^{232}Th) is called nuclear fuel.

At present thorium is not used as raw material for production of nuclear fuel due to the fact that nuclear power has inherited the uranium-plutonium cycle from nuclear weapon. Besides, the uranium reserves are large enough, and extraction of thorium is more complex and expensive due to absence of any established rich deposits.

^{239}Pu is generated in any thermal reactor where natural uranium enriched with ^{235}U is used as initial fuel. But generation of ^{239}Pu in the reactors where nuclei fission takes place in the fast neutron flux is the most efficient. Both natural uranium enriched with ^{235}U to 20% and the mixture of natural uranium and plutonium can be used as initial fuel in these reactors.

Only uranium-plutonium fuel has the chances for long-term use, because the reserves of ^{235}U are quite limited as it is mentioned above.

Fissionable ^{233}U nuclei can be produced in the reactors operating with uranium or uranium-plutonium fuel subject to arrangement of the thorium blanket at their core periphery.

The breeding ratio (BR) is the relation between the rate of fissionable nuclei number increase in the reactor due to radiation capture of neutrons with fertile nuclides (R^8) with their subsequent β-decay and the rate of fissionable nuclei number decrease in the reactor due to absorption of neutrons with fissionable nuclides ($R^5{}_a$ and $R^9{}_a$).

$$\mathrm{BR} = \mathrm{R^8}_c / \left(\mathrm{R^5}_a + \mathrm{R^9}_a\right)$$

So, by burning m_{fis} kg of initial fissionable material in the reactor we obtain $BR{\cdot}m_{fis}$ kg of new fissionable material which in its turn can produce $BR^2{\cdot}m_{fis}$ kg of new fissionable material through its burning, etc.

As a result, having m_{fis} kg of initial fissionable material and disregarding losses of fissionable material in the course of SNF processing the following can be burnt in the reactor:

$$m_{fis} + BR \cdot m_{fis} + BR^2 \cdot m_{fis} + \cdots \text{ kg of fissionable material.}$$

With BR < 1 not only initial fissionable materials m_{fis} loaded into the reactor, but also some (depending on the BR value) fertile materials in the reactor can be engaged in generation of nuclear fission energy

$$m_{fis} + BR \cdot m_{fis} + BR^2 \cdot m_{fis} + \cdots = m_{fis}/(1 - BR)$$

With BR $\geq$ 1 the breeding process enables to convert all fertile material into fissionable one and burn it (less the inevitable losses in the course of SNF processing). The value of $[(v_{ef})^{fis} - 1]$ with the superscript of the fissionable nuclide loaded into the reactor makes the main contribution to the BR.

For thermal (E = 0.025 eV) neutrons $(v_{ef})^5 = 2.071$; $(v_{ef})^9 = 2.106$.
For fast (E = 105 eV) neutrons $(v_{ef})^5 = 1.9$; $(v_{ef})^9 = 2.5$.

In thermal neutron reactors BR is < 1 both for uranium fuel and uranium-plutonium fuel. In pressurized water reactors BR is $\approx$ 0.5. BR is equal to $\approx$ 1.2 only in the thorium-uranium cycle in an ideal thermal reactor where neutron absorption takes place only in the fuel and no neutron leakage is present. In the real equipment with this cycle the BR is only a little higher than one.

In fast neutron reactors BR is > 1. The BR value is especially high in the reactors with uranium-plutonium fuel. In fast power reactors where the neutron spectrum is considerably softer than the fission spectrum the BR is within the range of 1.1 to 1.6 depending on the fuel composition.

At present fast reactors enabling to engage not only ^{235}U but almost all ^{238}U in the nuclear fuel cycle have the fuel reserves that will last for hundreds and even a thousand and more years (Table 1.2, the bottom line) [25].

1.7 Renewable energy sources

Serious changes have occurred in the growth rates of renewable energy sources (RES) within the recent decades. The share of RES in the total global power generation increased [43] (Fig. 1.7). According to the IEA estimates, a half of new capacities in the power generation sector commissioning worldwide by 2035 for satisfaction of the increasing demand will be attributed to RES, mainly hydropower and wind power engineering.

Contribution of hydropower engineering to the global power generation remains stable at the level of 15%. The share of RES, except for large-scale hydropower engineering, in power generation can increase from 3% in 2009 to 15% in 2035. China and the European Union will account for more than a half of this increase.

Within the period of 2000–2018 the aggregate capacities of new RES types (solar, wind, tide, geothermal, wave and other sources of energy) grew 21 times, from 56 GW in 2000 to 1179 GW in 2018. Their share (without hydropower engineering) in the final global primary energy consumption increased more than thrice and in power generation – from 3.4% in 2006 to 10.5% as of the end of

TABLE 1.2 "Magic skin" of power engineering, 2010.

Fuel	World			Russia		
	Reserves (established extractable), GW·year	Utilization intensity, GW	Period of exhaustion, years	Reserves (established extractable), GW·year	Utilization intensity, GW	Period of exhaustion, years
Oil	200,000	4600	40–50	10 000	400	20–30
Gas	150,000	2200	60–70	46,000	550	80–90
Coal	1,000,000	3000	100–200	110,000	170	400–500
TNR NF	90,000	750	120	9000	45	200
Σ	1,440,000	11,000	130	175,000	1200	150
FNR NF	15,000,000	11,000	1500	1,500,000	1200	300

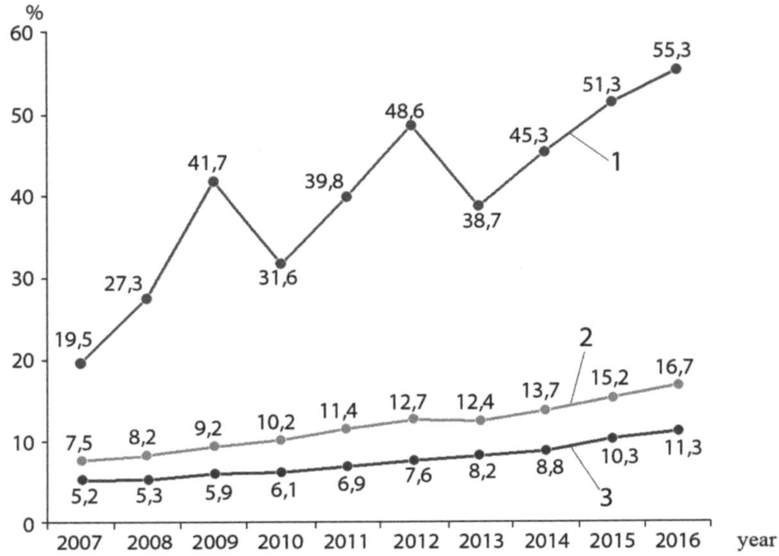

FIGURE 1.7 The contribution of RES to the global dynamics of installed capacities and power generation in 2007–2016 (according to the data of Bloomberg New Energy Finance, regardless of large-scale hydropower): *1* - RES contribution to the total growth of installed power generating capacities; *2* - share in the total installed power generating capacities; *3* - share in the total power generation.

2018. In this case the actual scope of the new RES commissioning regularly turned out to exceed the estimated one for the last 15 years.

By the beginning of 2019, the total installed capacity of all wind-powered generators exceeded 600 GW (with the average capacity utilization factor of 25%, 35% at the best turbines). The average increase in the total capacity of all wind-powered generators in the world from 2009 amounts to 38–40 GW per a year and is due to rapid development of wind power engineering in the USA, India, China and the EU.

According to the estimates of American Wind Energy Association the installed kilowatt cost of a wind turbine equal to 3000 USD and more in 1980 is reduced to ~1000 USD/kW by 2019 with due regard for the technical achievements in the area of wind turbine arrangement. In this case the average price of generated electric power will reduce to 3–15 cents/kW·h. The commercial potential of wind power engineering in the on-shore and off-shore zones of the Earth is estimated to be 20,000 GW.

The same pattern can be seen in solar power. In 2019, the aggregate installed capacity reached 500 GW. The growth of capacities based on photovoltaic converters was 100 GW just in 2018 with the potential for the electric power cost reduction to 3–13 cents/kWh within the nearest 5 years.

High development rates for new RES types can be attributed to the combination of rapid development of technologies supporting cost-cutting in production (particularly due to the constantly increasing economy of scale) and priority of this area in the energy policy of many countries worldwide. The low price of capital supported due to the national energy policy is an important success factor for the new RES types as it is crucial for such projects with high capital and low operating costs. However, development and cheapening of the technologies including widespread transfer of technologies became determinative for expansion in the RES utilization – particularly, the most drastic cost-cutting was achieved within the period when China localized and provided large-scale production of the main equipment components.

According to the existing estimates in Russia the resource of renewable energy sources (with the predominance of potential use of wind and solar power) amounts to at least 4.5 bln. tons of coal equivalent per a year that exceeds the national consumption of fuel and energy resources more than fourfold.

On the other side, RES have some drawbacks:

- Relatively short service life which means high frequency of capital maintenance under the conditions of the growing power grid;
- Upon increase of the res share in the power balance the expenditures for maintenance of the grid operability also grow (redundancy of capacities, load balancing, etc.);
- Gradual exhaustion of the feasible sites for res-based power generation;
- High material intensity;
- Low capacity utilization factor.

The global energy production as well as the final consumption is different from the global energy use due to various losses. For example, in 2008 the global energy production amounted to 143 PW·h, and the consumption – only to 98 PW·h. Energy losses depend on the energy source and the applied technologies. The efficiency factor of thermal power plants has fundamental limitations, for example NPPs lose about 70% for heating of the environment, and only approximately 30% are converted into electric power.

1.8 Thermonuclear fusion energy

Status of the problem. Heavy nuclei (uranium) fission reactions with generation of large quantities of radioactive fission fragments are used in the modern nuclear power. The nature provided for another energy generation opportunity – the fusion reaction for light element nuclei (first of all, hydrogen isotopes – deuterium and tritium). These reactions take place at extremely high temperatures when the substance is in the fully ionized (plasma) state and are called "thermonuclear fusion." Apart from emission of large amounts of energy inert gas helium is generated as the reaction product, no radioactive fragments are present. These particular reactions serve as the energy source for stars, including the Sun. Under the terrestrial conditions an intensive thermonuclear fusion reaction has been yet implemented only in the form of an explosion.

The main difficulty in arrangement of smooth thermonuclear fuel burning is to confine high-temperature plasma where a great number of plasma instabilities occur. In the course of long-term research several dozens of various systems were proposed; the systems with magnetic plasma confinement "tokamak" as well as the plants with inertial confinement (laser thermonuclear fusion) enabled to achieve the maximum success by the beginning of the XXI century.

The systems of "tokamak" type were proposed and implemented for the first time at Kurchatov Institute in the late 1950s. The success achieved by the Soviet scientists under the guidance of Academician L.A. Artsimovich was so significant that widespread construction of tokamaks also started in other countries. Wide-scale international cooperation was deployed for the most rapid achievement of the objective and saving of resources. The results enabling to start the designing of a fusion reactor were obtained on some large tokamaks in the USSR, the EU, the USA, and Japan.

ITER. In 1988–1990, the conceptual design of the international experimental thermonuclear reactor which was called ITER was developed by joint efforts, and its engineering design was developed in 1992–2001.

Within the period of 2001–2006 the seven-sided negotiations with involvement of Russia, the EU, India, Korea, China, the USA and Japan were held and finished in Paris by signing of the intergovernmental Agreement on Arrangement of the International ITER Organization on Thermonuclear Energy for joint implementation of the ITER project on November 21, 2006. The site located near Cadarache, the French nuclear center, was selected for the ITER construction. This Agreement was ratified by all project participants, particularly in Russia by Federal Law No. 143-FZ dated 19.07.2007.

In accordance with the schedule approved in 2016 completion of the construction and production of the first plasma are planned to take place in December 2025. Construction and installation works at the ITER construction site are well under way, and works for manufacturing and supply of the ITER equipment are in progress among all partners. As of 01.01.2019, the completed scope of works amounts to approximately 60% of the total scope of works for the first plasma production. Re-equipment of the ITER and trial of the operation modes in order to obtain the thermonuclear capacity of 500 MW will be carried out within the period of 2026 through 2035.

The responsibility for fulfillment of the obligations of the Russian Federation specified in the Agreement is assigned to Rosatom by Order of the Government of the Russian Federation No. 1508-p dated 15.10.2008.

The obligations for payment of the contribution of the Russian Federation in kind to the International ITER Organization are assigned to the Private Institution of Rosatom State Atomic Energy Corporation "ITER Project Center" (ITER Center Private Institution) by Order of the Government of the Russian

TABLE 1.3 ITER financing.

Year	Scope, mln. RUB	Year	Scope, mln. RUB
2018	4792.4	2022	5912.9
2019	5287.4	2023	5862.9
2020	5942.2	2024	5962.6
2021	5942.2	2025	5890.8

Federation No. 75-p dated 26.01.2011. According to the same Order scientific and methodological support of the research and development works performed by the Russian organizations within the framework of the ITER Project implementation is assigned to Kurchatov Institute.

ITER Center Private Institution together with the International ITER Organization compiles technical assignments for development and manufacturing of the particular ITER equipment and concludes Supply Agreements with the International ITER Organization. Besides, ITER Center Private Institution concludes the relevant contracts with the Russian enterprises, carries out the manufacturing quality control and acceptance of the works.

Financing of the works is arranged within the framework of the public contract between Rosatom and ITER Center Private Institution. The scope of the required financing in 2018–2025 is given in Table 1.3.

The Russian Federation is responsible for manufacturing and supply of 25 special-purpose equipment systems for the ITER (Table 1.4). Works for development of these systems are going on successfully from the very beginning of the ITER project. Performance of the works for the ITER project supports and forms scientific, engineering and technology teams and schools at the enterprises of the established cooperation and provides for training of highly skilled professionals in certain universities (MIPT, MEPhI, Novosibirsk State University, Leningrad State University, Moscow State University, etc.). All unique systems developed at the Russian enterprises are manufactured at the highest level of quality in accordance with the requirements of the International ITER Organization, the IAEA and the French nuclear regulatory authority as the ITER is a nuclear facility within the territory of France.

The Russian Federation fully completed manufacturing and supply of superconductors for the winding of the ITER magnetic coils – one of the most important equipment types – at the amount of 200 tons. The best gyrotrons in the world with the capacity of 1 MW for plasma heating to the thermonuclear temperature are developed, manufactured (2 of 8 pcs) and tested in the Russian Federation. The Russian Federation supplies a large quantity of various equipment for the ITER power supply system. Manufacturing of the internal energy-intensive components (the diverter and the blanket) is in progress.

TABLE 1.4 Main systems supplied by Russia.

System number	ITER system	Result	Cooperation
1–2	Nb3Sn and NbTi superconductors	Production of the best low-temperature superconductors in the world is arranged (225 tons are manufactured)	VNIINM, TVEL, CMP, VNIIKP, Institute of High Energy Physics, Kurchatov Institute
3	Poloidal field winding PF 1	A superconductive poloidal coil is developed, and its manufacturing commenced	NIIEFA, the Middle Neva Shipbuilding Plant
4	Upper nozzles	The design is developed, and manufacturing is started	NIIEFA, MAN Turbo Diesel
5	First wall (FW)	The design is developed, and manufacturing of the most energy-intensive 40% of the FW is started	NIIEFA, NIKIET, VNIINM, Bazalt
6	Blanket module connectors	Mechanical and electrical connectors are developed, and their production is arranged	NIKIET
7	Central diverter assembly (CDA)	CDA modules are developed, and their production is arranged	NIIEFA
8	Tests of the FW and diverter components	Benches are arranged, and tests of the FW and diverter components from Russia, Japan and the EU are started	NIIEFA, TRINITI JSC, VNIINM
9	Switching equipment	Developed, manufacturing commenced	NIIEFA, Siemens
10	Gyrotrons	The best gyrotrons in the world are manufactured and tested (170 GHz >1 MW, 1000 s)	RAS Institute of Applied Physics, Gikom, Kurchatov Institute
11	Reflectometry	A reflectometer/refractometer is developed	CTF-Center, Kurchatov Institute

(*continued on next page*)

TABLE 1.4 Main systems supplied by Russia—cont'd

System number	ITER system	Result	Cooperation
12	Overcharging atom analyzer	The best atom analyzers, neutron and gamma-quantum spectrometers in the world are developed	RAS Physical and Technical Institute, Tekhno-Eksan, ITER Center, NIIEFA
13	Neutron flux monitor	Neutron flux monitors with ^{235}U and ^{238}U for burning control are developed	ITER Center, TRINITI, NIITFA, Institute of Nuclear Physics of the Siberian RAS Division
14	Thomson scattering	Unique lasers, the primary mirror and polychromators are developed	RAS Physical and Technical Institute, Tekhno-Eksan
15	Spectroscopy of hydrogen lines	The unique primary mirror made of monocrystal molybdenum is developed	CTF-Center, Luch, Kurchatov Institute
16	Active spectroscopy	Unique spectrometers, the molybdenum primary mirror, fiber bundles are developed	ITER Center, Solar, TRINITI, GIPO, Luch
17	Vertical neutron chamber	The best diamond neutron spectrometers, U-238 fission chambers, MI cables are developed	ITER Center, TRINITI, Institute of Nuclear Physics of the Siberian RAS Division, NIITFA
18	γ-spectrometry	The best γ-quantum spectrometers arc developed	RAS Physical and Technical Institute, Tekhno-Eksan
19	Laser-induced fluorescence	Unique lasers are developed	RAS Physical and Technical Institute, Tekhno-Eksan
20	Port plug test benches	Port plug testing techniques are developed	Cryogenmash
21–25	Port plugs E-11; U- 2,7,8; L-8	Port plug manufacturing technologies are developed, integration of diagnostics is started	Institute of Nuclear Physics of the Siberian RAS Division, Tekhno-Eksan

As of June 2018, 37 Russian experts (21 skilled specialists and 16 persons from the support personnel) are engaged in the ITER project, including one director of the department, one head of the section, 15 engineers as well as specialists in scientific research, financing and IT.

1.9 Role of radiation risks in nuclear power and human-induced risks for the public

Nowadays it is needless to prove that any state-of-the-art technology introduces additional health risks into our lives. Radiologists were probably the first to raise this issue and the first to develop the practical tools for comparative risk assessment with regard to application of various technologies (41, 95, 114, 172–174). Assessment of damage for the environment and human health due to power generation in the nuclear and alternative power engineering remains the subject of research and widespread public discussions. One of the main difficulties resides in the fact that the effects caused by the emissions from different types of power engineering in the human organism are often complex by nature and thus hardly comparable, like, for example, leukemia and bronchitis. Another problem is rather of psychological nature and related to the fact that our consciousness does not accept risks in the course of ordinary life. As a result, humans make every effort to reduce these risks through the use of additional protection means, and nevertheless the risks remain.

The scientific approach to assessment of the human-induced impact consequences and environmental safety management based on analysis of risks for the public health was proposed through the efforts of scientists from many countries in the last quarter of the XX century. This risk-oriented approach provided the possibility to compare the consequences of factors different in their nature. The methodology based on evidentiary medicine including such disciplines as evidentiary toxicology and evidence-based risk assessment became one of the areas developing the commitment to socially acceptable risk levels in the XXI century [41,90]. These areas are still thinly represented in the national science.

Publication of the assessment data on the potential damage from the technology in the USA sets an example of evaluation of the impact of thermal power plant emissions on the public [131], Table 1.5 and Fig. 1.8. So, almost 2800 people annually die in the USA from lung cancer induced by this hazardous factor. In comparison, the number of deaths caused by all accidents at nuclear facilities is almost 50-fold less.

At present the results of the public health risk assessment are the mandatory components of environmental impact audit for new construction projects, refurbishment of the existing production facilities and rehabilitation of the contaminated territories in the USA and Western Europe countries. The foreign methodology described in detail in the study [65] is getting more and more widespread in Russia as well. Transfer to regulation of the human-induced impact on the basis of the public health risks in our country is determined by the

TABLE 1.5 Impact of thermal power plants on the public health in the USA [131].

Parameter	Number of cases per year
Death	23,600
Death from lung cancer	2826
Hospital confinement	21,850
Seeking urgent medical assistance due to an asthmatic attack	26,000
Heart attacks	38,200
Chronic bronchitis	16,200
Asthmatic attacks	554,000
Lost working days	3,186,000

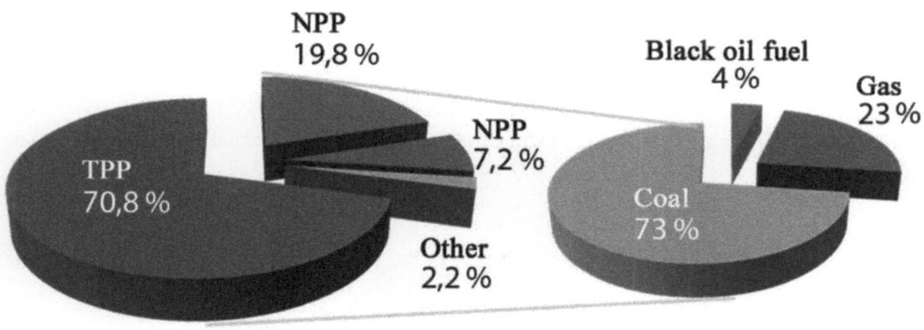

FIGURE 1.8 General power generation structure in the USA [131].

Fundamental Principles of the National Policy for Environmental Development of the Russian Federation within the Period of up to 2030 [2].

Meanwhile, analysis of the current hygienic standards demonstrates that the approaches to regulation, in particular for chemical impacts in the Russian Federation are based on absolutely different principles than for radiation impacts which has been rightly mentioned by the national patriarchs of radiology [69]. Difference in regulation of carcinogenic substances in the atmospheric air can reach up to 6 orders depending on the approach (Fig. 1.9).

The fact that the public opinion still perceives radiation as an extremely hazardous factor accompanying the nuclear industry since the first atomic bombing and the cold war period should be also reckoned with. The accidents at Three Mile Island NPP (the USA) in 1979, Chernobyl NPP in 1986 and then at Fukushima NPP (Japan) in 2011 have certainly made the significant contribution to enhancement of this opinion.

Analysis of risks in the area of atomic energy use [116] and comparative assessment of the impact of power generation systems on the environment and human health [192] give evidence of the fact that the contribution of nuclear

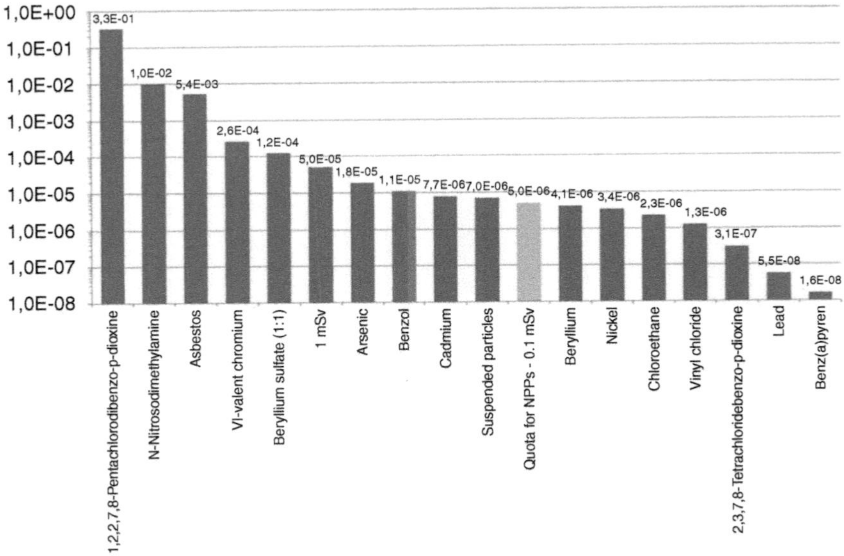

FIGURE 1.9 Life-long carcinogenic risk due to inhalation with the annual exposure at the level of MAC_{AD}.

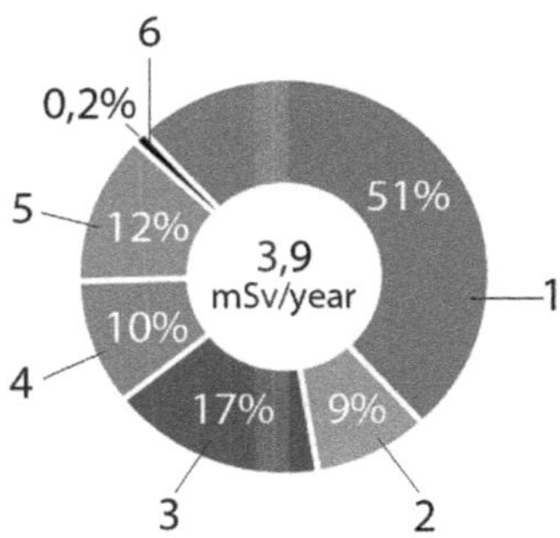

FIGURE 1.10 Structure of the average annual exposure dose for the population of Russia in 2015 [30]: 1 – radon; 2 – natural isotopes (internal exposure, except for radon); 3 – natural isotopes (external exposure); 4 – cosmic radiation; 5 – medical sources; 6 – other human-induced sources.

facilities under normal operation conditions is negligibly small as compared with the risks from other radiological factors such as background radiation exposure and radiological diagnostic procedures.

According to the data of Ramzaev Radiation Hygiene Research Institute the structure of the average annual exposure dose for the population of Russia, for example in 2015, was due to radon radiation by more than 50%, see Fig. 1.10 [30].

The public exposure doses due to operation of nuclear industry facilities are below 0.01 mSv/year [30,79,87] while the acceptable dose is 1 mSv/year [15], and the average variability of background radiation exposure is 2 to 10 mSv/year.

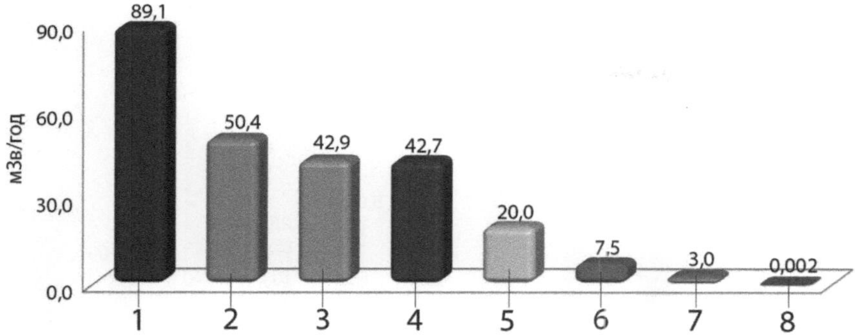

FIGURE 1.11 Average exposure doses due to equilibrium equivalent concentration of radon isotopes for individual groups of the most exposed inhabitants in various constituent entities of the Russian Federation and Finland as compared with the exposure doses due to nuclear industry facilities [79]: 1 – Chelyabinsk region; 2 – St. Petersburg; 3 – the Altai Republic; 4 – the Republic of Adygeya; 5 – Annual dose limit for professionals (NRB-99/2009); 6 – Finland; 7 – Average annual doses for professionals; 8 – Doses for the public from the facilities with ionizing radiation sources.

The share of medical exposure which had remained rather stable till 2015 then began to increase on the model of the advanced countries and reached 13.6% in 2016. It is worth reminding that in the USA the average annual dose due to medical diagnostics increased from 3.6 mSv/year in 1980 to 6.2 mSv/year in 2006 [133]. Such increasing trend is certainly expected in our country as well.

Individual exposure doses from human-induced radiation sources can be considerably different. For example, the aircraft crews spending the significant part of their work life at the heights where the cosmic radiation level is 20 times higher than on the Earth surface get the exposure dose of several mSv within a year. This dose is comparable with the average individual exposure dose of any NPP personnel per a year. Spacemen get the dose of approximately 200 mSv per a year of flight.

Radon radiation still remains the determinative radiological factor in the Russian Federation as well as in many other countries of the world (Fig. 1.11).

This obvious disproportion in formation of dose burdens for humans and other inhabitants of the planet should have stimulated the national efforts aimed to reduce the determinative factors, but here economic, bureaucratic and other reasons that make "temporarily" turn a blind eye to these facts break in.

In this case significant efforts are directed to reduce the negligible share of the human-induced factor of radiation nature. It is beyond argument that application of the best technologies providing for the minimum environmental contamination is a progressive trend. But in this case the efforts shall be primarily aimed at the priority factors defining the environmental situation to a great extent. It is not a big secret that chemical pollution is the determinative factor affecting the quality of the environment in the residential territories of the Russian Federation.

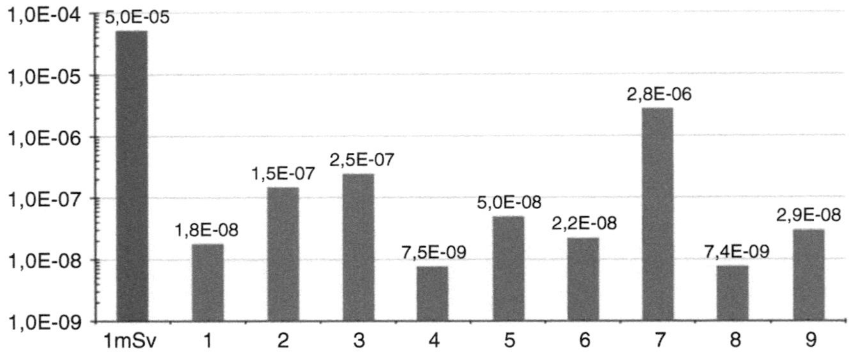

FIGURE 1.12 Assessment of life-long carcinogenic risk as compared with the standard value due to radiation impact on the critical groups of the public from the annual releases and discharges of some organizations of Rosatom: 1 – Beloyarsk NPP, Zarechny; 2 – Novovoronezh NPP, Novovoronezh; 3 – Leningrad NPP and North-West section of RosRAO FSUE, Sosnovy Bor; 4 – Angarsk, 2015–2016; 5 – Kirovo-Chepetsk Chemical Plant, Kirovo-Chepetsk, 2008; 6 – Siberian Chemical Plant, village of Chernilshchikovo; 7 – MCP, critical group, village of Balchug; 8 – Obninsk, 2014–2017; 9 – Dimitrovgrad, 2015–2018.

Nevertheless, the society in the location areas of nuclear facilities often perceives radiation exposure and the associated risks as the priority ones. Comparison of radiation risks with other environmental risks enables to determine the place of the radiation risks in the structure of environmental risks within any territory.

The works for comparative analysis of the risks due to impact of radiological and chemical factors on the public health in the location areas of nuclear industry facilities have been carried out since 2000s in the Nuclear Safety Institute of the Russian Academy of Sciences together with the Human Ecology and Environmental Hygiene Research Institute n.a. A.N. Sysin supported by the Russian FMBA directorates. Performance of such integrated studies also forms the scientific base for more adequate public perception of the information on the environmental aspects of industrial activities. Besides, scientifically founded determination of the priority risk factors and development of the arrangements affecting thereof can provide for improvement of the environment quality and reduction of actual risks for the public health.

Quantitative assessment of the impact of toxic substances as compared with radiation risks was performed in the location areas of nuclear industry facilities: Angarsk, Novouralsk and Electrostal (nuclear fuel production), Novovoronezh, Sosnovy Bor, Zarechny (Novovoronezh NPP, Leningrad NPP and Beloyarsk NPP), Zheleznogorsk (MCP), and Dimitrovgrad (Research Institute of Nuclear Reactors) [21,62,91,93,97,111].

Basically, the radiation risk assessment results presented in Fig. 1.12 show that the risks due to annual releases and discharges of radioactive substances into the environment are negligibly low (below 1×10^6 almost everywhere).

TABLE 1.6 Ranking of factors affecting the public health in Angarsk according to the risk level [93].

Parameter	Human-induced risk	Contribution to human-induced risk, %
Suspended particulate	3.0E-03	85.7
Benzol	2.5E-04	7.1
Sulfur dioxide	2.0E-04	5.7
Formaldehyde	2.7E-05	0.8
Carbon oxide	2.0E-05	5.7E-01
Nickel*	2.9E-06	8.2E-02
Benz(a)pyren	1.3E-06	3.8E-02
Lead	1.6E-07	4.5E-03
Uranium	1.5E-07	4.3E-O3
Total for human-induced ingredients	3.50E-03	100.0

**Risks due to nickel are specified only to demonstrate the position of this substance on the scale of risks. Human-induced nature of nickel present in the air of Angarsk remains unproved, and its physiological necessity for humans makes classification of the amounts capable of ingress into the human organism as harmful for health unjustified.*

Comparative analysis of the risks due to chemical and radioactive substances is based on the common methodological framework related to the dispersion of releases in the environment thus enabling to define the health risks from various release sources of different nature located in different parts of the populated location or even outside it for individual points of the residential area (Fig. 1.13).

Although risks due to various pollutants may be added thus forming the evaluation for an individual area or the entire populated location, it is more useful to rank the risks for individual substances in order to make administrative decisions at the municipal level. An example of such ranking is given in Table 1.6.

In general, Fig. 1.14 and 1.15 illustrate the role of individual factors affecting the public health, for example, in Angarsk, Zheleznogorsk and Krasnoyarsk, quite demonstrably.

The levels of noncarcinogenic risk due to pollution of the atmospheric air in Angarsk with suspended particulate and sulfur dioxide and carcinogenic risk related to air pollution with formaldehyde, benz(a)pyren and benzol exceed the target risk level. The aggregate value of human-induced risk due to impact of the controlled chemical factors is at the edge of unacceptable risk.

Assessment of life-long risks from stationary sources of chemical releases into the atmosphere for the population of some cities with nuclear industry facilities and for large cities situated in the location areas of such facilities was

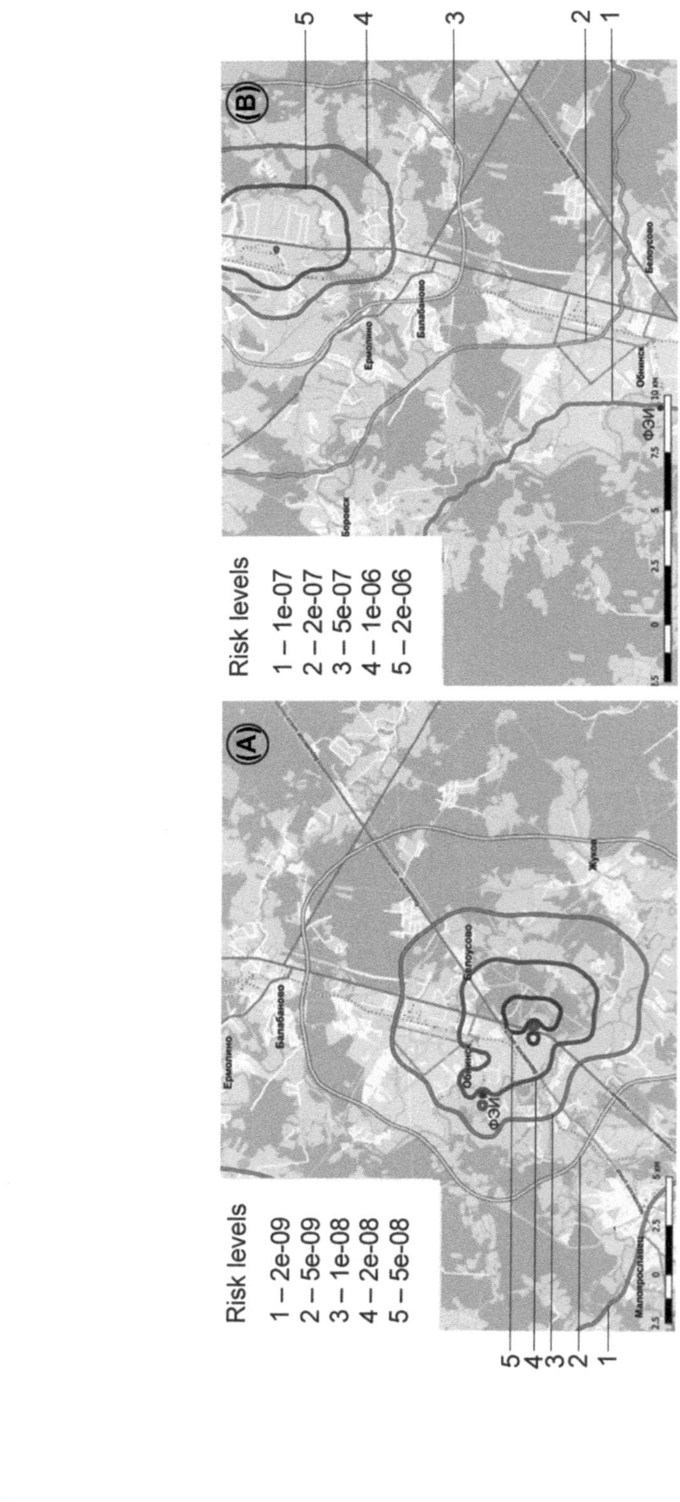

FIGURE 1.13 Assessment of the radiation risk levels due to radioactive releases of nuclear facilities in Obninsk (A) and the risks due to releases of suspended particulate from the facilities in Obninsk and Balabanovo (B).

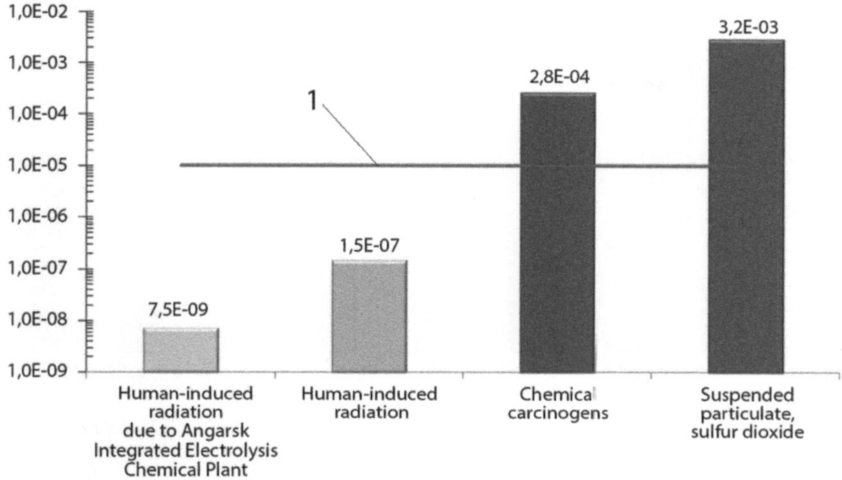

FIGURE 1.14 Risks from the regular affecting factors in Angarsk [93]: 1- socially acceptable risk.

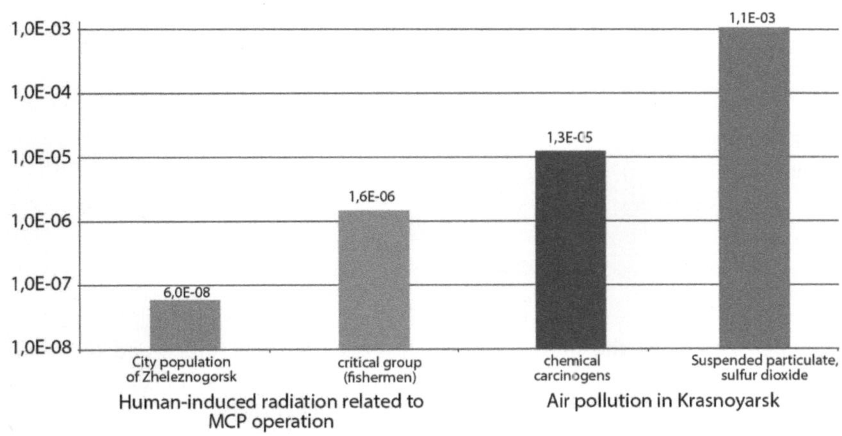

FIGURE 1.15 Comparative assessment of the impact of human-induced radiation and chemical contamination of the atmospheric air on the public health in Zheleznogorsk and Krasnoyarsk [91].

performed in the Institute for Safe Development of Nuclear power within the period of 2008–2018. Comparative analysis demonstrates the different degree of well-being with the atmospheric air condition in these populated localities. Assessment of risks due to releases of chemical substances for the population of some cities is shown in Fig. 1.16.

Except for Obninsk, life-long risks due to releases of chemical substances are quite high, and for many cities these risks exceed the upper limit values of socially acceptable risks (1×10^4) that are consequently indicative of the

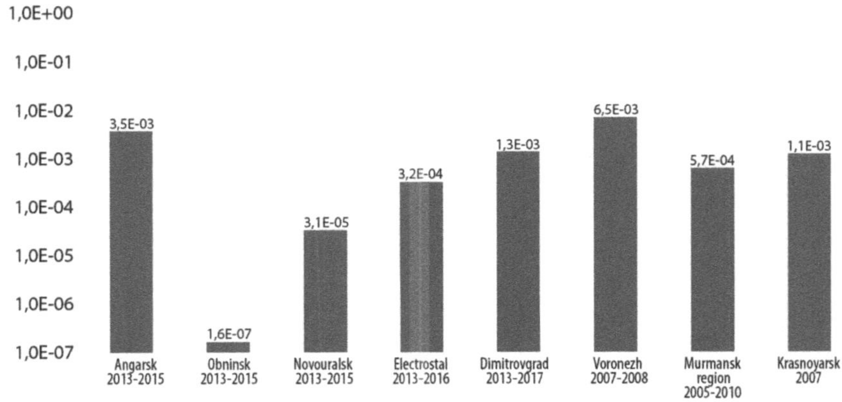

FIGURE 1.16 Assessment of chemical impact risks for the population of some Russian cities.

TABLE 1.7 Death risk among the population of the cities with large coal-operated CHPPs [155].

City	Population size, thous. people	Individual death risk
Ulan-Ude	371.4	5.1×10^{4}
Cheremkhovo	50.0	1.9×10^{3}
Chita	316.7	8.8×10^{4}
Novocherkassk (Rostov SDPP)	188.7	3.2×10^{4}
Ussuriysk	158.4	1.0×10^{3}

necessity to change the production technologies in these cities and to install more efficient devices for collection of hazardous substances. It is clear that this cannot be done within a short period of time, but the performed analysis enables to define the priority actions.

It should be noted that the existing risk levels for the public health in the considered large urban agglomerations are not of exclusive nature. The high air pollution level in the cities also poses a significant threat for the public health in other large industrial centers of the country. According to the data of the national observation system of the Russian Meteorological Service the size of the Russian population exposed to the chronic impact of controlled pollutants with the concentrations exceeding the acceptable values reaches 65 mln. people, including about 20 mln. people exposed to one-time concentrations exceeding 10 MACs. Risks exceeding the upper limit of socially acceptable risk are registered in the cities with large-scale coal-operated traditional thermal power engineering facilities (CHPPs) (Table 1.7).

TABLE 1.8 Death rate of the RF population in 2016 due to various causes and estimated values of the population risks due to chemical contamination of the environment and radiation exposure [59].

Causes	Number of people	Percentage,%
RF population size as of 01.01.2016	146,544,700	
Total deaths in 2016	1,891,015	100
Malignant tumors	295,729	15.6
External causes of death,	167,543	8.9
Including suicides	23,119	1.2
Causes of death due to alcohol	56,283	3.0
Population risk of death		
From the atmospheric air pollution	4,800	0.25
Drinking water pollution	18,900	0.99
Soil pollution (death rate from tumors)	3,200	0.17
Natural and medical radiation exposure	11,300	0.59
Human-induced radiation exposure	27	0.001

So, under normal operation conditions of the facilities radiation risks for the public are several orders lower that the risk of chemical impact. Summing up the above-mentioned aspects it is useful to cast a glance over the general pattern of death risks in the Russian Federation (Table 1.8).

As it can be seen from the above-mentioned data, the relative contribution of human-induced radiation exposure for the population of the country is very low.

The risks associated with emergency situations both at any facilities engaged in the activities accompanied with releases of toxic substances and at radiation-hazardous facilities should be particularly mentioned.

The objective data on the radiological consequences of peaceful atomic energy use related to radiological accidents and incidents from 1945 to 2007 are provided in the essential appendix to the report of UNSCEAR for 2008 (Table 1.9).

The consolidated data on major accidents at power generation facilities only from 1969 to 2000 according to the analysis of the Paul Scherrer Institute (PSI), Switzerland, are provided in Table 1.10.

The death toll due to accidents at nuclear installations since 1945 is 500-fold lower than the death toll due to accidents (with the number of victims exceeding 5 persons, and only instant deaths) in the traditional power engineering for the period twice as short.

In spite of the available data convincingly demonstrating the high environmental safety level of nuclear technologies the problem of negative attitude towards everything related to the nuclear technologies still exists in the public

TABLE 1.9 Number of early effects and deaths in radiological accidents*.

Accident type and consequences	1945–1965	1966–1986	1987–2007	Total	Conclusion of the Committee with regard to the report completeness
Accidents at NFs					
Early effects Deaths	46 16	227** 40	2 3	275 59	It is likely that the majority of deaths and injuries are reported
Industrial accidents					
Early effects Deaths	8 0	109 20	49 5	166 25	It is likely that some deaths and injuries were not reported
Incidents with unattended IRSs					
Early effects Deaths	5 7	60 10	204 16	269 33	It is likely that some deaths and injuries were not reported
Accidents in the course of R&DW					
Early effects Deaths	1 0	21 0	5 0	27 0	It is likely that some deaths and injuries were not reported
Accidents in the course of medical application					
Early effects Deaths	N N	470 3	143 42	613 45	It is obvious that many deaths and significant quantity of injuries were not reported
Total					
Early effects	60	887	403	1 350	
Deaths	23	73	66	162	

**Based on the published information (except for malicious actions and nuclear weapon tests).*
***Table 10 on p. 52 of Appendix R.671 to the UNSCEAR report for 2008.Notes: NF – Nuclear Facilities; IRS – Ionizing Radiation Sources;N – no data available.*

TABLE 1.10 Consolidated data on major (>5 victims) accidents in power engineering.

	OECD countries			Non-OECD countries		
Power engineering type	**Accidents**	**Victims**	**Victims/GW**	**Accidents**	**Victims**	**Victims/GW**
Coal*, including the data for 1994–1999:	75	2259	0.157	1044	18,017	0.597
				819	11,334	6.169
China excluding China				102	4831	0.597
Oil	165	3713	0.132	232	16,505	0.897
Natural gas	90	1043	0.085	45	1000	0.111
LPG	59	1905	1.957	46	2016	14.896
Hydropower industry	1	14	0.003	10	29,924	10.285
Nuclear	0	0	-	1	31**	0.048
TOTAL	390	8934		1378	67,493	

**In 1969–2000.*
***Only instant deaths.*

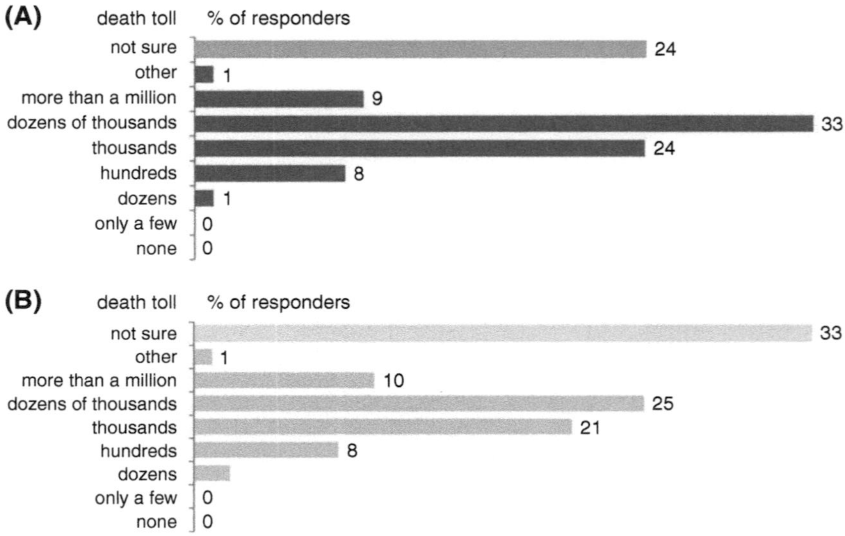

FIGURE 1.17 Distribution of the answers of the responders as percentage of the total number of people polled for the following questions: (A) – the accident at Chernobyl NPP took place in 1986. How many people in total died from radiation exposure as a result of this accident, in your opinion? [107] (B) – the accident at Fukushima NPP in Japan took place in 2011. How many people in total died from radiation exposure as a result of this accident, in your opinion? [58].

opinion. This is due, inter alia, to the past radiological incidents in the USSR (the accidents at Mayak Production Association FSUE and Chernobyl NPP), the USA (the accident at Three Mile Island NPP) and Japan (the accident at Fukushima-1 NPP) and results from exaggerated nonprofessional evaluation of their consequences and politicization of the problem in general together with the low level of education.

Consequently, twisted perception of the radiation factor impact on the human health both during normal operation of the facilities and in case of severe accidents is formed within the general public. This is expressly demonstrated by the results of the all-Russian opinion poll carried out by the Nuclear Safety Institute of the Russian Academy of Sciences in 2012 [58,107]. In the course of the poll more than a half of Russians answered that radiation in Chernobyl became the cause of death for thousands, dozens of thousands and even more people. Distribution of the answers for the accident at Fukushima-1 NPP was similar to the distribution for Chernobyl (Fig. 1.17). In both cases distribution of the answers scarcely depended on the age, education, financial position, the place of residence and occupation of the responders [58].

Almost complete coincidence in distribution of the answers of the responders for the consequences of accidents at Chernobyl and Fukushima is even more surprising, as the radiological consequences of the accident at Fukushima-1 NPP are practically negligible. This fact vividly characterizes the established consistently twisted public perception of the radiological consequences of radiation accidents.

So, the public perception of the radiation damage scale after the accident at Chernobyl NPP is overstated by 2–3 orders as compared with the actual data. No deaths due to the radiation factor among the personnel and the public are registered in Japan [167,171]. The professional community itself often provides quite ambiguous data to the public. For example, the joint press release of the IAEA, the WHO and the UNDP compiled subsequent to Chernobyl Forum in 2005 under the title: "Chernobyl: True Extent of the Accident" informed the global mass media that "a total up to four thousand people could have died in the long run...subsequent to the accident at Chernobyl nuclear power plant. However, as of the mid-2005 less than 50 death cases can be directly attributed to the radiation exposure during the catastrophe". In the similar manner, only "hypothetic" estimated long-term effects of the accident (7 to 33 thousand of oncological diseases followed by death) and not the actual data were included into the database of the Paul Scherrer Institute mentioned above.

In the more general context, in spite of the fact that radiological consequences of the past severe accidents at nuclear power plants were limited or negligibly small they were perceived as national tragedies at the level of the entire society.

In this case all people affected by any long-term protective arrangements were classified as "suffered from radiation" regardless of the actual small exposure doses obtained by them. As a result, the number of "sufferers" after the Japanese accident in 2011 was evaluated at the amount of 200 th., and in case of the accident at Chernobyl NPP – 7–8 mln.

In-depth analysis carried out in [107] showed that the extent of the long-term protective arrangements implemented in both cases was due to keen perception of radiation hazard typical for the society in general, including the persons authorized to make any decisions for protection of the public and any specialists engaged by them. It should be recognized that regulatory authorities and governments of the nuclear countries have not yet learnt this lesson – the radiation protection system for the public and territories around the world considerably lags behind other four components of the NPP defense-in-depth according to its development level. Lack of attention to the scientific, technical, organizational, administrative, regulatory, social and psychological problems faced by the managers at the federal and regional levels in making of any decisions for protection of the public and territories, particularly for selection of the dose criteria and the derivative intervention levels, results in multi-fold scale-up of the radiation accident consequences. The objective to increase the safety standard level up to the fifth one becomes the priority under the conditions of large-scale nuclear power development.

The estimates for the scale of nuclear power development are quite uncertain: from the pessimistic ones predicting the insignificant share of nuclear power in the total energy supply balance for the humankind, to the optimistic ones defining the position of nuclear power as the main energy supply base by the middle of the XXII century. This considerable uncertainty of the estimates is related not only to the general difficulties in prediction of long-term scenarios of social, economic and industrial development of the humankind, uncertain estimates for the reserves of hydrocarbon energy sources, but also to ambiguous interpretation of the comparative analysis results for the safety parameters of nuclear and traditional power engineering. Based on the systemic analysis results with the use of vast data array and evaluation of the impact of power generation facilities on the public health and the environment, with due regard for the normal operation conditions and accidents in thermal, hydropower and nuclear power, including the accident in Chernobyl and the accident at Fukushima-1 NPP the firm conclusion can be made that the role of radiation risks in the negative impact on the public health and the environment is negligibly small on the local, regional and global scale.

In summary it should be re-emphasized that unbiased assessment of the achieved safety level for nuclear technologies as compared with other energy production technologies, plus strict compliance with the safety priority, adequate development of the protection system for the public and territories in case of any beyond design basis accidents, implementation of the nuclear legacy programs, development of the RW and SNF handling systems as well as decommissioning of nuclear- and radiation-hazardous facilities are the reasonably required preconditions for sustainable development of nuclear power in the future.

Chapter 2

Role of nuclear power in the Russian fuel and energy industry

2.1 State of nuclear power in Russia

Nowadays the Russian nuclear power is based on NPPs with water-cooled thermal neutron reactors of VVER and RBMK types operated in the open nuclear fuel cycle. As of 2019 ten NPPs (36 power units) with the aggregate installed power capacity of 30.2 GW (11.98% of the total installed capacity of power plants in the national grid of Russia) are operated in Russia and generate approximately 19% of all electricity produced in the country (TPPs – 63.7%, HPPs – 17.1%, wind, and solar power plants – 0.1%). Three power units with VVER reactors and the aggregate capacity of 3.6 GW (e) are under construction at the RF territory. In December 2019, the first floating nuclear cogeneration plant "Academician Lomonosov" was connected to the grid and by January, 23, 2020 delivered the first 10 mln. kW·h of power to the network of Chaun-Bilibino load center (the Chukotka Autonomous District).

Russia is the world leader in the number of nuclear power units constructed abroad (16% of the global market of NPP construction services). As of the end of 2019 the portfolio of foreign orders of Rosatom for construction of NPPs amounted to 36 power units at different implementation stages.

In 2018, uranium mining by Rosatom reached 7304 tons (2904 tons from mining in Russia and 4400 tons from foreign assets), or 13% of the global uranium mining enabling the country to be included into the top five uranium producers in the world. Russia takes the second place in the world in the amount of natural uranium reserves (about 740 thous. tons of balance reserves as of the end of 2018, including 216 thous. tons in the mineral raw material base of Uranium One Holding uniting the foreign uranium mining assets of Rosatom). Russia accounts for approximately 36% of the global uranium enrichment market and 17% of the global nuclear fuel production for NPPs.

The power engineering development benchmarks at the first stage of the long-term strategy are set in the draft Energy Strategy of Russia for the period of up

Closed Nuclear Fuel Cycle with Fast Reactors: Handbook of Russian Nuclear Power.
DOI: https://doi.org/10.1016/B978-0-323-99308-1.00020-9

to 2035 (hereinafter – ES-2035) [4]. The following basic parameters of the fuel and energy industry are specified in the draft[1]:

- Reduction of the following by 2035 (as compared with 2015):
  - The share of electric power industry in the total amount of capital investments in the industry from 22 to 13–14%;
  - The GDP energy intensity – by 1.3–1.5 times;
  - the GDP electric intensity – by at least 1.1–1.3 times;
- The greenhouse gas emission level – not more than 70% of the level in 1990;
- Increase in the installed capacity of NPPs by 1.3 times with the relevant decommissioning of the power units built in the Soviet period;
- Increase in gas production of up to 40% subject to favorable conditions at the global and internal market;
- Maintenance of the gas share in the structure of internal consumption of fuel and energy resources at the level of 51–52%;
- The average growth rate of energy consumption of 1.4–1.6% with the average GDP growth rate of 2–3% per a year;
- Increase in the crude energy production by 25%.

The target benchmarks for the electrical power industry of Russia (including nuclear power) specified in ES-2035 are characterized by the parameters presented in Table 2.1.

Increase in the installed NPP capacity to 31 GW by 2030 and ~34–35 GW by 2035 was provided in the draft of ES-2035 as well as the Land-Use Planning Scheme for the Russian Federation Power Industry [7]. The technology basis of nuclear power in Russia is sufficient to provide the estimated NPP construction amounts and performance of the export commitments, but significant expansion of the machine-building capacities is required, first of all, as well as qualitative improvement of the building complex parameters. However, safety, raw material resources and economic competitiveness first and foremost remain the key factors for large-scale development in the wider prospect.

2.2 Forecast of the Energy Research Institute of the Russian Academy of Sciences-2016

All considered scenarios for further extension up to 2040 [76] provide for the growth of atomic energy use worldwide with the rates higher than the energy consumption growth rates thus enabling nuclear power to increase its share in the crude energy consumption structure from 5 to 7%.

1. The draft Energy Strategy of Russia for the period of up to 2035 was developed in accordance with the Federal Law "On strategic planning in the Russian Federation" (as amended on 01.02.2017).

TABLE 2.1 Target parameters of the Russian Energy Strategy for the period up to 2035.

Parameter	2015 (act.)	2035
Aggregate electric power production, bln. kW·h, including:	1064	1380–1470
Hydropower plants	170	201–230
Nuclear power plants	195	246–269
Thermal power plants	697	904–925
Renewable energy sources	2	29–46
Crude energy export volume, mln. tons o.e.	961	1009–1201
Share of gas in the total export of fuel and energy, %	24	30–33
Share of Asia Pacific in the total export of fuel and energy, %	15	30–40
Specific fuel consumption for electric power supply, g o.e./kW·h	323	290
Power losses in the grids from the total electric power output, %	11.1	8
Minimum share of expenses for technology innovations in the total production expenses, %	1	3

The share of nuclear power in electricity generation will increase slightly from 10.5% in 2015 to 11.0–11.4% in 2040. In this case the trends in nuclear power development will be different for developed and developing countries. Reduction of the NPP share in electricity generation from 17.9% in 2015 to 17.5–17.7% in 2040 is expected in the OECD countries. On the contrary, the share of nuclear power in developing countries will increase from 4.5% in 2015 to 7.9–8.1% in 2040.

The necessity for decommissioning of the significant number of NPP units with expiring design service life becomes urgent for many countries within the nearest decades. Thus the service life of many operating nuclear power units is extended, generally from 40 to 60 years. About 20% of 450 power units operating worldwide (as of the end of 2018) have been operated for 40 years and longer (Fig. 2.1), and about 30% of the existing nuclear capacities are to be decommissioned by 2040.

Considerable decrease of nuclear capacities (by 20%) will take place in the European countries (Belgium, Germany, Spain, France, Sweden, Switzerland) by 2040 due to the adopted political decisions on refusal or reduction in atomic energy use. In North America the nuclear capacities will decrease within the entire period under consideration according to the probable and critical scenarios

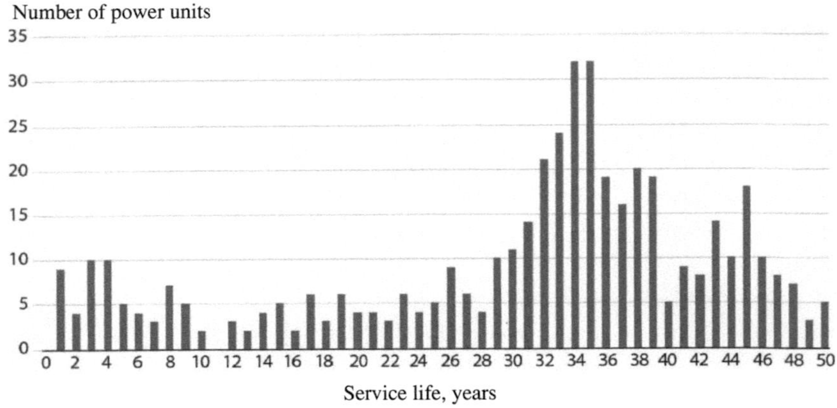

FIGURE 2.1 Number of operating nuclear power units worldwide as of the middle of 2019 depending on the service life (Source: PRIS database of the IAEA, March 2019).

FIGURE 2.2 Application of nuclear capacities from 2016 to 2040 with breakdown by the countries of the world, probable scenario [76]: – absent and not expected; – absent but planned; – present but reduction is expected; – present and expansion is expected.

by 0.2 and 1.8% respectively, and according to the favorable scenario they will increase by 3% reflecting the difference in the energy demand in various scenarios. In general, the nuclear power prospects around the world are diverse, but the number of countries operating NPPs will increase (Fig. 2.2).

The developing countries of Asia, mainly China and India, will account for significant increase in nuclear capacities. Apart from them, some countries in

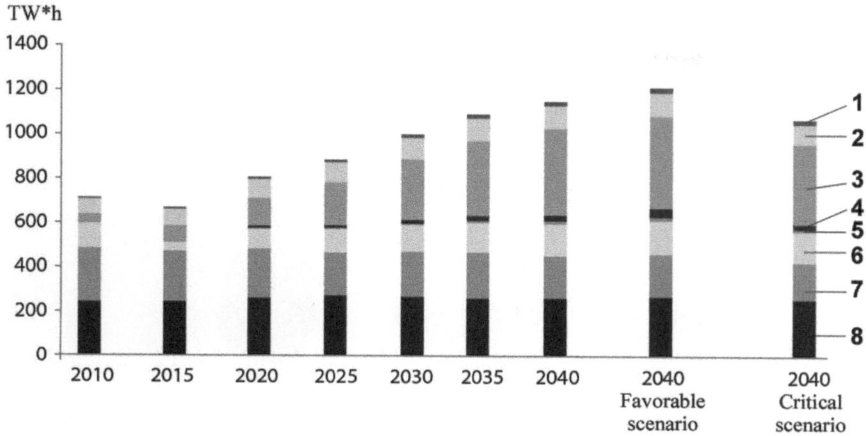

FIGURE 2.3 Electricity generation at NPPs with breakdown by the regions of the world [76]: 1 – South and Central America; 2 – CIS; 3 – developing Asian countries; 4 – Middle East; 5 – Africa; 6 – developed Asian countries; 7 – OECD, Europe; 8 – North America.

this region also plan to commence development of their own nuclear industry (Indonesia, Malaysia, etc.).

Lithuania, Poland and Turkey are planning to join the list of European countries using nuclear power generation. Despite the appearance of new players, power generation at NPPs in Europe within the period under consideration will reduce by 19% according to the probable scenario, 15% – according to the favorable scenario and 27% – according to the critical scenario.

Commissioning of NPPs for the first time is expected in some CIS countries (Belarus, Kazakhstan); the production in this region will increase by 49% according to the probable scenario, 53% – according to the favorable scenario and 32% – according to the critical scenario.

Increase in generation is expected in the Middle East within the estimated period from 0 to 102 TW·h according to the probable scenario (comparable with the current generation in Canada) and to 160 TW·h according to the favorable scenario (comparable with the current generation in the Republic of Korea), see Fig. 2.3.

Construction of the first nuclear power units is expected in Saudi Arabia, Israel, and Jordan; the first four-unit nuclear power plant in under construction in the UAE.

Increase of nuclear capacities in the developing countries will enable them to approximate to the parameters of the OECD countries with regard to nuclear power generation amounts by 2040. According to the probable scenario the share of developing countries in the global nuclear power generation will increase thrice – from 17% in 2010 to 51% in 2040.

Among the developed Asian countries significant increase in nuclear power generation by 2040 is expected only in Japan as compared with 2015 when

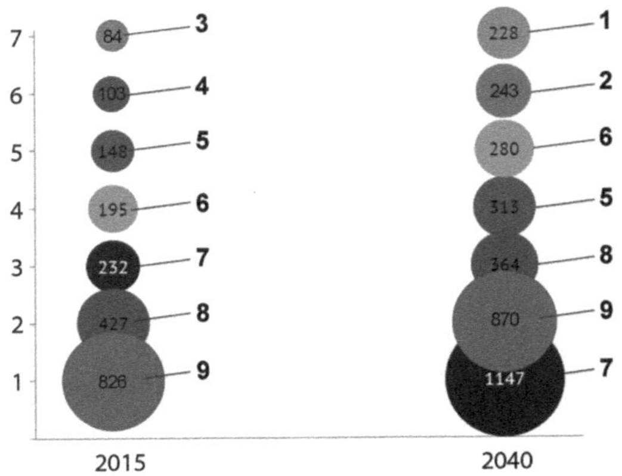

FIGURE 2.4 Places of countries in the world according to power generation at NPPs (the left scale) and the output amounts in TW·h (the circle size), probable scenario [76]: 1 – India; 2 – Japan; 3 – Ukraine; 4 – Canada; 5 – Republic of Korea; 6 – Russia; 7 – China; 8 – France; 9 – USA.

nuclear power generation was basically suspended and carried out only at two restarted power units. Japan will reach the level of 2003 with regard to power generation at NPPs only by 2040 because of the accident at Fukushima NPP in 2011 and the related consequences, and will not be able to achieve the level of 2010 within the estimated period.

The USA, France and China were the leaders in electricity generation at NPPs in 2015. By 2040 China is expected to take the first place through fivefold increase in nuclear generation. The USA will rank the second. France will shift from the second place to the third one, reduction of nuclear generation in the country will amount to 15%. Russia will shift down to the fifth place from the fourth one even in spite of generation increase by 43% up to 2040. Power generation at NPPs in the Republic of Korea will increase twice by 2040, and the country will take the 4th place (Fig. 2.4)[2]. Japan is expected to occupy the 6th place through restarting of its nuclear reactors subsequent to completion of the certification procedure for compliance with the new safety rules. India will take the 7th place due to fivefold increase in nuclear power generation by 2040.

Emergence of new generation technologies in the industry is probable within the period under consideration. Commissioning of the reactors with enhanced safety characteristics facilitating solution of the nuclear weapon nonproliferation problem and operated with spent uranium may open new opportunities for the nuclear industry subject to prevention of energy generation cost increase. The Forecast assumes the possibility for emergence of new generation reactors and

2. The possible paradigm shift in energy development of the Republic of Korea is not taken into account in the forecast. Moon Jae-In, President of the Republic of Korea, who won the elections in May 2017 declared the focus on abandonment of nuclear and coal-powered generation and expansion of solar and wind energy use.

gradual transfer to this type in construction of new NPPs. But these technologies will not be able to transform the world of power engineering within the nearest 25 years taking into account long-term investment cycles in the nuclear industry, although significant expansion in the use of these power units may be expected in the future subject to successful research and practical application of the first prototypes.

Small-scale power units may be one of the promising trends in nuclear power application. They are the most advantageous in the areas with limited access to power generation in large amounts through the use of any other energy technologies with adequate cost. But higher expenditures for power generation in comparison with large power units and the necessity for significant efforts in order to ensure safety (including the threats of terrorism) will constrain widespread application of such solutions. All three scenarios of the Forecast assume expansion in the use of small-scale NPPs in individual countries but in a limited way.

At present peaceful atomic energy is mainly used for electricity generation and at the special-purpose water transport (icebreakers, lighter carrier ships, maintenance ships), but expansion of its use in heat generation, for the hydrogen production process and water desalinization is expected in the long run. The latter is especially relevant for the Middle East and African countries.

2.3 Estimates of nuclear power development in the world

The IAEA is the most respected agency in the studies of nuclear power industry development worldwide; it regularly publishes the estimates of nuclear power development in the mid-term. The "low" estimate [154] assumes reduction of the global nuclear power capacities from 392 GW (e) in 2017 (with the share of 5.7% in the installed capacity of all power plants) to 356 GW (e) in 2050 (with the share of 2.8%) taking into account the planned decommissioning of the power units with expired service life. The estimate data are characterized by considerable reduction of the installed capacities in the USA, Europe and the OECD countries from the Pacific region. The most significant increase of capacities is expected in the Middle East and non-OECD Asian countries. Expansion in the shale gas mining in various regions of the world and RES financing are the main factors restricting potential development of nuclear power for the purpose of this estimate.

The "high" IAEA estimate assumes for rapid overcoming of the financial and economic crisis and recovery of the previous power demand parameters. Together with many other predictions implying application of the political mechanisms in order to mitigate the consequences of the environmental impact the "high" estimate also assumes increase in nuclear capacities to 748 GW (e) in 2050 with the NPP share of 5.8% in the installed power generating capacities of the global power engineering. Power generation at NPPs worldwide amounting to 2500 TW·h or 10.3% of the aggregate output in 2017 will not exceed

2870 TW·h (5.6%) according to the "low" estimate and will reach 6030 TW·h with the share of 11.7% in the total global power generation according to the "high" one.

The previous estimates of the global power industry development (including nuclear power) are considerable revised downwards in the report of the International Energy Agency (IEA) World Energy Outlook 2018 [222]. Three development scenarios are considered: Current Policies Scenario, New Policies Scenario and Sustainable Development Scenario.

In 2017, the global carbon dioxide (CO_2) emissions related to power generation increased by 1.6% subsequent to three years of stability. According to the preliminary data this growth will continue in the nearest future which is a significant deviation from the path corresponding to the climatic targets. Implementation of the current strategy scenario will result in growing tension almost in all aspects of energy security.

The global energy consumption growth by a quarter predicted by the IEA in 2040 [222] is fully accounted for by developing countries headed by China and India. The in-depth shift of energy consumption towards Asia is observable for all fuel types and technologies and also for the investments in power engineering. Asia accounts for a half of the global natural gas consumption increase, 60% of wind and solar energy increase, more than 80% of oil consumption increase and more than 100% of coal and atomic energy consumption increase (with due regard for reduction in other regions). Fifteen years ago European companies dominated among the major ones in electrical power sector with regard to their installed capacity; nowadays six of the ten largest power supply companies are Chinese ones.

Power industry is going through the deepest transformations from the moment of its emergence more than a century ago. Electricity is getting more and more preferable as "fuel" for the countries relying mainly on the consumer goods industries, services and digital technologies. Its share in the global final consumption is close to 20% and promises further growth. Political support and reduction in the cost of technologies result in rapid growth of various renewable electric energy sources bringing the power sector to the forefront of the efforts for emission reduction but requiring to change the functioning of the entire system in order to ensure reliable power supply.

Electricity generation growth will reach 14–50% within the period up to 2040 depending on the scenario under consideration. According to New Policies Scenario nuclear generation will grow to 3726 TW·h by 2040 (1120 TWh or 43% more in comparison with 2016) thus requiring increase of the installed NPP capacities by 100 GW (Fig. 2.5A). According to Sustainable Development Scenario power generation at NPPs will reach 4960 TW·h in 2040 (2355 TW·h or 90% more in comparison with 2016), and 265 GW of new NPP capacities will be commissioned (Fig. 2.5B). The electricity demand growth in the developed countries is expected to be moderate, but the investment needs are still vast due to modification of the generation structure and the infrastructure refurbishment.

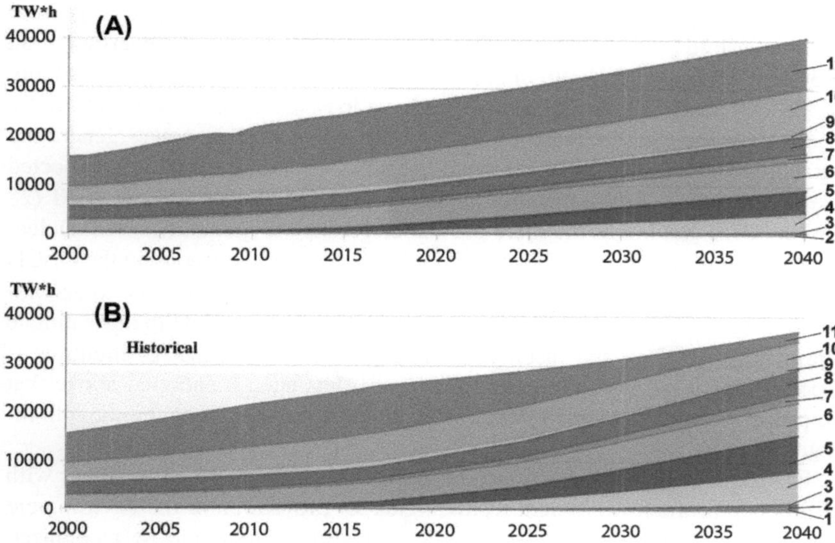

FIGURE 2.5 Electricity generation growth dynamics according to New Policies Scenario (a) and Sustainable Development Scenario (b). Energy types: 1 – concentrated solar; 2 – marine; 3 – geothermal; 4 – photovoltaic solar; 5 – wind; 6 – hydro; 7 – bio-energy; 8 – atomic; 9 – oil; 10 – natural gas; 11 – coal.

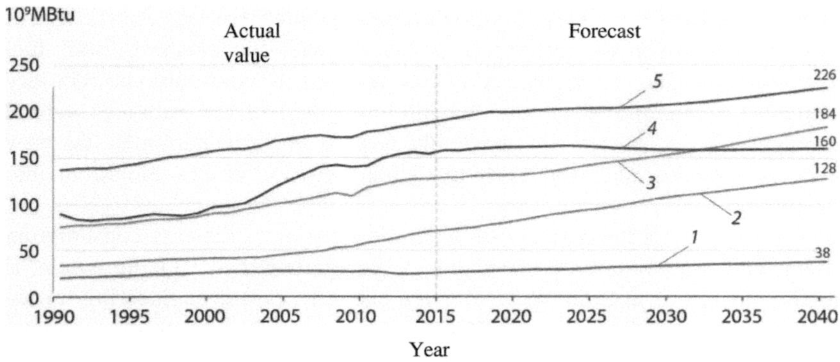

FIGURE 2.6 Global energy consumption forecast with breakdown by fuel types: *1* – nuclear power; *2* – RES; 3 – natural gas; *4* – coal; *5* – oil and other liquid fuel (basic scenario of EIA 2017). 1 MBtu = 947.8 TJ = 3412 GW·h.

It is noted in the estimate of the US Energy Information Administration dated 2017 [178] that according to the basic scenario (Reference Case) energy consumption in the world will grow for all types of primary energy sources, except for coal. In this case the maximum growth dynamics is expected for natural gas and RES (Fig. 2.6).

The aggregate installed power generating capacities worldwide in general will increase to 8819 GW by 2040 (6383 GW in 2015) with the energy output of 34,050 TWh (23,430 TWh in 2015). The NPP capacities in 2040 will amount to 508 GW (343 GW in 2015) and produce 3657 TWh (2510 TWh in 2015). In this case the total increase of energy output and nuclear power generating capacities is expected to take place in the non-OECD countries. It is expected that the installed NPP capacities in China will reach 139 GW in 2040 (28 GW in 2015) serving as the main growth driver for the global installed nuclear power generating capacity. The cumulative installed NPP capacity in the OECD countries will reduce by 14 GW within the same period of time due to decommissioning of NPPs in the USA, Canada and Europe. According to this estimate significant growth of the installed NPP capacities in the OECD countries is expected only in the Republic of Korea but it has been mentioned above that the power engineering development concept in the country is changing with re-targeting to RES.

For some countries large-scale development of nuclear power together with RES is one of the few efficient technologies for prevention of the environment pollution and reduction of CO_2 emissions. However, it should be recognized that development of alternative generating technologies, unfavorable political situation and negative public opinion are the factors considerably restricting increase of nuclear capacities in the world. Slowdown in commissioning of new NPPs in the developed OECD countries, and in some cases the decisions on complete shutdown or restriction in development of nuclear programs shift the nuclear power development center to developing countries.

Considerable opportunities are associated with the cleanest fossil fuel type – natural gas, especially taking into account its nontraditional recovery sources (for example, shale gas) constituting a half of the estimated natural resources of gas, but it will inevitably require implementation of the new recovery standards and is unlikely to enable reduction of CO_2 emissions into the atmosphere.

The world is on the critical path from the viewpoint of energy consumption extent, and the window of opportunities for limitation of CO_2 emissions is rapidly closing. If no resolute actions are taken the power engineering infrastructure will reach the ceiling of CO_2 emissions prescribed in the development scenarios up to 2040.

It is clear that zero-carbon power generation comes to the fore in this situation, and nuclear power is the most developed section thereof, particularly as the growing demand in the transport sector and increasing costs of field development and extraction of raw materials confirm the end of the cheap oil era.

But speaking about the role of nuclear power drastic changes in the growth rates for renewable energy sources and their increasing contribution to the total global energy generation should not go unmentioned (Fig. 2.7).

Although the aggregate investments (Fig. 2.8) to the renewable energy sector (without large-scale hydropower) have reduced by 8% in 2018 (332.1 bln.

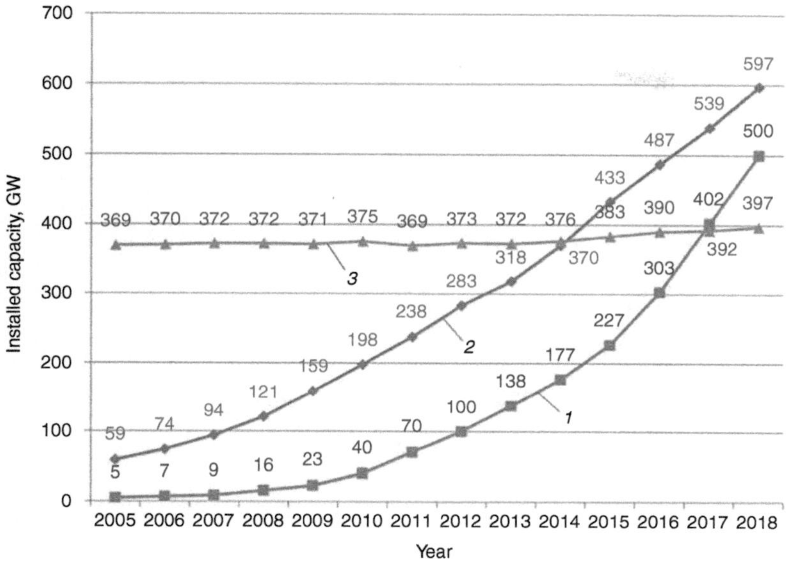

FIGURE 2.7 Dynamic pattern for NPP and RES installed capacities (without large-scale hydropower) in 2005–2018: 1 – solar power plants; 2 – wind power plants; 3 – NPPs.

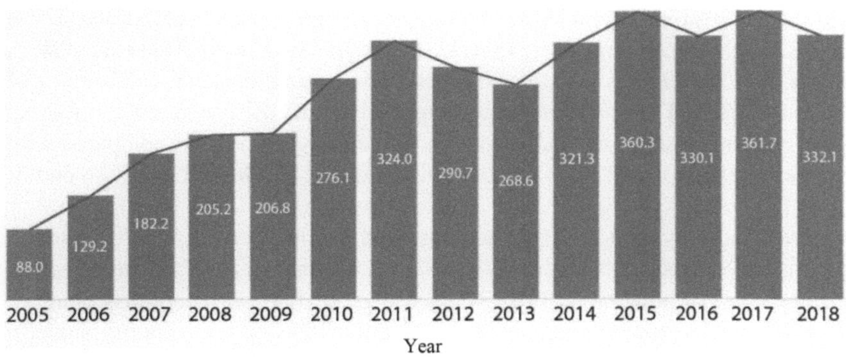

FIGURE 2.8 Investments to renewable energy sources worldwide, bln. USD [132].

USD as compared with 361.7 bln. USD in 2017) they still consistently exceed the annual level of 300 bln. USD in 2014–2018. According to the opinion of Bloomberg New Energy Finance analysts [187], the aggregate investments to the new power generating capacities in the world will amount to 11.5 tln. USD within the period up to 2050, including 8.4 tln. USD for wind and solar power with 50% contribution of these energy sources into the global power generation by 2050.

According to the data of the World Wind Energy Association [223] the aggregate installed capacity of wind power installations in the world amounted

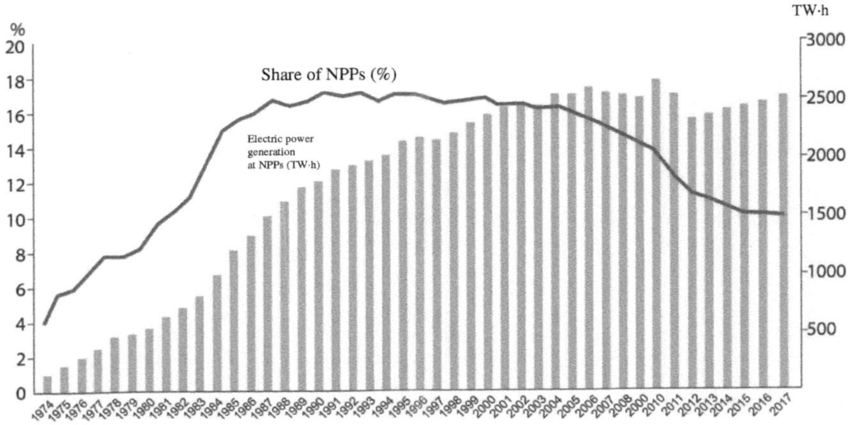

FIGURE 2.9 Share of NPPs in the global electricity generation.

to approximately 597 GW as of the end of 2018. Only in 2018, more than 210 GW were commissioned making wind power one of the most fast-growing segments of the energy market. Reduction of specific capital costs played a big role in the global expansion of wind power installations. According to the IEA data [199] overnight costs for on-shore wind power installations in the OECD countries amount from 1570 USD/kW in the USA to 3000 USD/kW in Japan. In China it was possible to reduce this parameter down to the level of 1200 USD/kW. The levelized cost of electricity (LCOE) generated at on-shore wind power installations worldwide has reduced by 37–56% as compared with the peak values of 2007–2010 and amounts to 60 USD/MW·h as of the end of 2017 [201].

Even more dynamic changes are observed in solar power. According to the preliminary assessment of the IEA [177] by the end of 2018 the installed capacities of solar power plants (SPPs) exceeded 500 GW with commissioning of 98 GW in 2018 (according to another evaluation 109 GW were commissioned). The levelized cost of electricity (LCOE) generated at the advanced large-scale SPPs around the world reduced by 73% in comparison with 2010 and amounted to 100 USD/MW·h as of the end of 2017 thus enabling them to compete with thermal power plants in many geographical territories with regard to the LCOE parameter [201]. The value of overnight costs for large-scale industrial SPPs (>1 MW) may be different for various regions of the world: from 940 USD/kW in China and 1200 USD/kW in Germany to 2560 USD/kW in Japan [199].

The global nuclear power is in the state of crisis. By the end of 2017 it reduced to 10.5% from the maximum 18% of the aggregate global power generation in the beginning of 1990s (Fig. 2.9). The estimates of the reputable power organizations indicate the probability of further reduction of this share to the level of single digits in case of certain development scenarios.

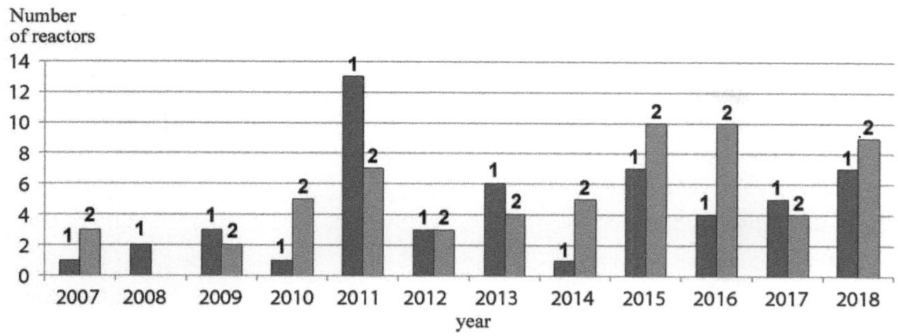

FIGURE 2.10 Commissioning of new NPP power units worldwide in comparison with decommissioning of old power units (PRIS database of the IAEA): 1 – final shutdown; 2 – connected to the grid (new power units).

Commissioning of new NPP units within the recent 10 years is approximately equal to decommissioning and lags considerably behind the commissioning of alternative power facilities (Fig. 2.10).

Competitiveness resting on the safety problem is the main barrier in the path of the modern nuclear power development. Any attempts to solve the safety problem by arrangement of additional active protection means resulted in reduction of the nuclear power competitiveness in comparison with the power generation based on organic fuel and brought competitiveness against RES to the brink.

2.4 Competitiveness of nuclear power in Russia

Assessment of competitiveness for the existing power generation technologies presented in this section is carried out for the Russian conditions in accordance with the LCOE parameter with due regard for the estimated dynamics of deflator indices as per the data of the RF Ministry of Economic Development.

Technical and economic indices (TEI) of RES are assumed on the basis of the global data with due regard for the typical for Russia capacity utilization factor of wind and solar power plants in the Russian national grid (Table 2.2). The results of competitive selection of the investment projects aimed at RES arrangement in the RF territory demonstrate considerable and increasing scattering in the declared capital intensity of individual projects. For the purpose of competitiveness research it is assumed that the lower limit of relative capital investments into construction of wind and solar power plants is close to the average value subsequent to the results of competitive selection of projects in 2018, and the upper limit – to the same parameters in 2017.

The initial TEI of combined cycle turbine gas units (CCGTs) are specified in Table 2.2 and assumed for the Russian conditions with due regard for the best practices applied at the existing natural gas fueled power plants.

TABLE 2.2 Consolidated TEI of alternative generation in Russia.

Parameter	WPPs	SPPs	CCGTs
Installed unit capacity, MW (e)	100	100	410
Relative capital investments, thous. RUB/kW (e) in 2018 prices, VAT excl.	75–100	80–105	45
Semi-fixed costs, % of line 2	1.5	2.1	3.0
Duration of construction, years	2	2	2
Type of fuel	–	–	Natural gas
Specific fuel consumption, kg o.e./kW·h	–	–	0.23
Auxiliary power consumption, %	0.3	0.3	3.3
Service life, years	25	25	30
Capacity utilization factor, %	23	17	68–76

Dependence on the gas prices in the course of operation is the key peculiarity of CCGTs significantly affecting their competitiveness. The annual gas price growth rates are assumed in accordance with the valid estimates of the Energy Research Institute of the Russian Academy of Sciences and amount to the average of ~2% in the minimal option providing for regulation of the internal prices, and ~3% in the maximal option reflecting long-term change of the global prices and providing for approximation of the internal prices to the equipotency conditions for the internal and export supplies (Table 2.3).

The TEI of NPPs with thermal neutron reactors (TNRs) constructed or planned to be constructed in the RF in the immediate future are specified in Table 2.4. In the calculations presented in this section the parameters of thermal neutron NPPs are considered for NPP-2006 and VVER-TOI projects in order to evaluate their competitiveness.

Regional variations in the construction costs for CCGT and NPP power units under the conditions in three zones of the Russian national grid – the Center, Ural and Siberia – were taken into account in calculations of the LCOE value.

The LCOE parameter calculation results for various generation types under the current conditions of the Central European part of Russia, Ural and Siberia are presented in Table 2.5. The calculations are carried out according to the industry's methodology of "Rosatom" Corporation with the nominal discount rate of 10, 7 and 5%. The calculations for regions are different in the LCOE parameters for CCGTs (the capital investments and natural gas prices vary) and NPPs (the capital investments vary) (Fig. 2.11).

The following basic conclusions may be made subsequent to the calculation results for the Russian conditions:

TABLE 2.3 Estimated dynamics of natural gas prices for industrial consumers according to the data of the Energy Research Institute of the Russian Academy of Sciences, RUB/thous. m^3 (VAT excl.).

	Center		Ural		Siberia	
Year	Min.	Max.	Min.	Max.	Min.	Max.
2018	5308		4501		3347	
2020	5412	5516	4570	4639	3554	3762
2025	5516	5828	4639	4651	3762	4916
2030	5672	7028	4645	5885	4339	6053
2035	5828	8320	4651	6739	4916	6970
2045	7028	9082	5885	8505	6053	7611
2050	8320	9924	6739	8932	6970	8309

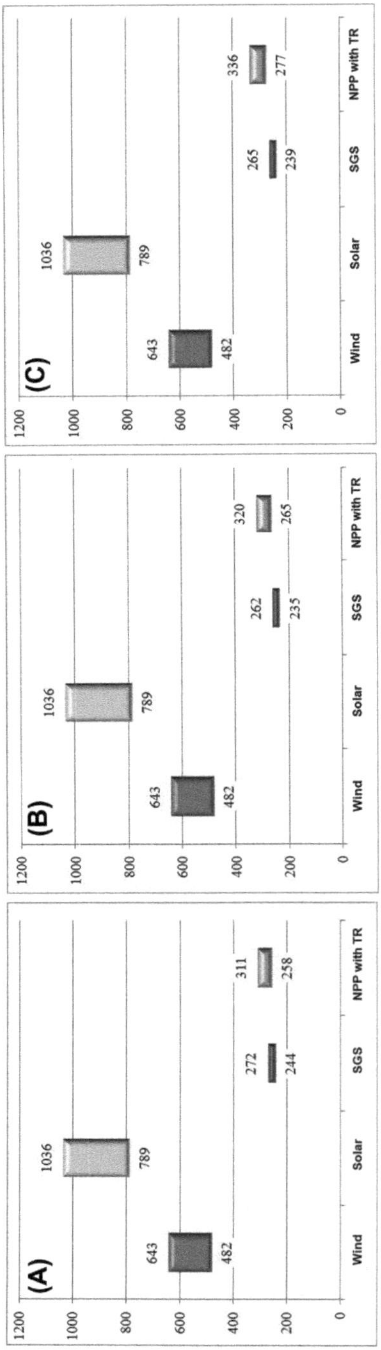

FIGURE 2.11 – LCOE of power plants in Central-European part of Russia, the Urals and the Siberia for technical and economical parameters of 2018.

TABLE 2.4 TEI of NPPs with TNRs (cost parameters at 2018 values, VAT excl.).

Parameter	NPP-2006	VVER TOI
Installed capacity, MW (e)	2 × 1197	2 × 1255
Relative capital investments (for the Center conditions), thous. RUB/kW (e)	133	114
Semi-fixed costs, % of line 2	2.3	2.2
Duration of the first power unit construction, years	6	6
Type of fuel	UO_2	UO_2
Fuel factor, kop./kW·h	23	21
Auxiliary power consumption, %	5.5	6.4
Service life, years	50	60
Capacity utilization factor, %	85	~90

TABLE 2.5 LCOE of power plants in the Central European part of the RF, Ural and Siberia with different discount rates for TEI of 2018, kop./kW·h.

	Center		Ural		Siberia	
Generation type	Min.	Max.	Min.	Max.	Min.	Max.
Nominal discount of 10%						
WPPs	482	643	482	643	482	643
SPPs	789	1036	789	1036	789	1036
CCGTs	244	272	235	262	239	265
NPPs with TNRs	258	311	265	320	277	336
Nominal discount of 7%						
WPPs	392	522	392	522	392	522
SPPs	639	839	639	839	639	839
CCGTs	221	247	210	234	211	237
NPPs with TNRs	200	241	205	247	213	258
Nominal discount of 5%						
WPPs	325	433	325	433	325	433
SPPs	531	697	531	697	531	697
CCGTs	203	228	191	214	192	216
NPPs with TNRs	162	195	165	199	170	207

1. At the current development stage RES in Russia are inferior to NPPs with TNRs and power units with CCGTs.
2. At present power units with CCGTs are the most cost-effective power generation technology for high discount rates.

3. NPPs with TNRs operated in the open nuclear fuel cycle cannot guarantee further efficient competitive development of nuclear power in Russia with the achieved TEI.

Development of nuclear power in Russia should be predicted with due regard both for the current cost-efficiency in comparison with other generation types and the expected energy demand growth, and for technology competitiveness of the new closed cycle nuclear technologies providing for environmental efficiency and safety of energy development at the global market.

The future transfer from export of raw energy resources and even export of electric power to export of new technologies necessitates the most rapid acquisition of the closed cycle nuclear technologies where Russia has the required head start at the global level.

Part II

Basic components of a new technology platform for nuclear power engineering

Chapter 3

Fuel cycles of nuclear power

3.1 Classification of nuclear fuel cycles

A nuclear fuel cycle (NFC) is the set of technology processes required for:

- Production of nuclear fuel;
- Burning of nuclear fuel in the reactor plant;
- SNF unloading from the core (for removal of fission products and minor actinides (mas));
- Specific-purpose SNF handling (fractioning into fuel nuclides [Pu, U] and rw [fission products, mas] or storage for final disposal);
- Make-up of the core with fissionable and fertile nuclides in the amount compensating for their burnup in the course of NF combustion;
- Storage and final disposal of the SNF RW.

Nuclear fuel is the mixture of fissionable and fertile nuclides. Depending on the nuclide composition of this mixture, nuclear fuel cycles are classified into uranium-plutonium and thorium-uranium fuel cycles.

The uranium-plutonium fuel cycle is the cycle where ^{238}U is a fertile nuclide, and ^{239}Pu is a fissionable one, that is, the cycle where ^{239}Pu nuclei are generated in the reactor through irradiation of ^{238}U nuclei with neutrons. The mixture of fissionable ^{235}U nuclides and fertile ^{238}U nuclides is taken as the initial nuclear fuel in order to implement the uranium-plutonium fuel cycle in practice. In the countries with considerable amount of accumulated TNR SNF and its processing facilities the initial NFC closing stage may be passed by usage of plutonium from processing products in TNRs or FNRs and make-up with natural or depleted (by ^{235}U isotope) uranium.

The thorium-uranium fuel cycle is the cycle where ^{232}Th is a fertile nuclide, and ^{233}U is a fissionable one, that is, the cycle where ^{233}U nuclei are generated in the reactor through irradiation of ^{232}Th nuclei with neutrons. The mixture of fissionable ^{235}U (or ^{239}Pu) nuclides and fertile ^{232}Th nuclides may be taken as the initial nuclear fuel in order to implement the thorium-uranium fuel cycle in practice.

At present only the uranium-plutonium cycle is used due to economic reasons.

Closed Nuclear Fuel Cycle with Fast Reactors: Handbook of Russian Nuclear Power.
DOI: https://doi.org/10.1016/B978-0-323-99308-1.00018-0

Depending on the nuclear reactor core feed with fissionable nuclides in the amount compensating for their burnup in the course of NF combustion nuclear fuel cycles are classified into the closed NFC and the open NFC.

The closed nuclear fuel cycle is the NFC without any need for usage of external fissionable nuclides.

The open nuclear fuel cycle is the NFC with the need for external fissionable nuclides.

3.2 Open nuclear fuel cycle

The open NFC is classified into nonrecycling, partially recycling and recycling one according to the SNF handling nature.

The open NFC is the NFC where SNF is considered as radioactive waste subject to final disposal. The nonrecycling open NFC is characterized by the following cycle:

- Natural uranium extraction;
- Natural uranium enrichment;
- Production of nuclear fuel from enriched uranium;
- NF burning in the nuclear reactor;
- SNF storage;
- RW (SNF) disposal;
- Natural uranium extraction, etc. in a new circle (cycle).

The partially recycling NFC is the NFC where fissionable and fertile nuclides are extracted from its SNF for nuclear fuel production, and fission products and minor actinides are subject to final disposal. In other words, the partially recycling NFC is the NFC where only fissionable and fertile nuclides are recycled. The partially recycling open NFC is characterized by the following cycle:

- Natural uranium extraction;
- Natural uranium enrichment;
- Production of nuclear fuel from enriched uranium and SNF;
- NF burning in the nuclear reactor;
- SNF storage;
- SNF processing;
- Disposal of RW (fission products and MAs);
- Natural uranium extraction, etc.

The recycling open NFC is the open NFC where fissionable and fertile nuclides are extracted from SNF for production of nuclear fuel, and minor actinides are extracted for after-burning (transmutation) in nuclear reactors (within nuclear fuel or in neutral matrices). The recycling open NFC is characterized by the following cycle:

- Natural uranium extraction;
- Natural uranium enrichment;
- Production of nuclear fuel from enriched uranium and SNF;
- Burning of NF (with MAs) in the nuclear reactor;
- SNF storage;
- SNF processing;
- Disposal of RW (fission products);
- Natural uranium extraction, etc.

3.3 Closed nuclear fuel cycle

The closed NFC is classified into the classic one and the transmutation one according to the MA handling nature.

The classic closed NFC is the closed NFC where minor actinides are subject to final disposal. The classic closed NFC is characterized by the following cycle:

- Natural uranium extraction;
- Production of nuclear fuel from SNF and natural uranium;
- NF burning in the nuclear reactor;
- SNF storage;
- SNF processing;
- Disposal of RW (fission products and MAs);
- Production of nuclear fuel from SNF and natural uranium, etc.

The transmutation closed NFC is the closed NFC where minor actinides are subject to burn-out (transmutation) in nuclear reactors (within nuclear fuel or in neutral matrices). The transmutation closed NFC is characterized by the following cycle:

- Natural uranium extraction;
- Production of nuclear fuel from SNF and natural uranium;
- Burning of NF (with MAs) in the nuclear reactor;
- SNF storage;
- SNF processing;
- Disposal of RW (fission products);
- Production of nuclear fuel from SNF and natural (or waste) uranium, etc.

The main classification of nuclear fuel cycles [25] is presented in Table 3.1. The NFC subclassification (according to the SNF fractioning depth) is presented in Table 3.2. The NFC evolution is presented in Table 3.3.

TABLE 3.1 NFC classification.

NFC	NFC supply	Reactor plant (BR)	NFC make-up	Recycle	NFC waste
Open nonrecycling	$U_{nat.en.}O_2$	TNR (BR < 1)	$U_{nat.en.}$	–	SNF
Open recycling	$(U_{nat.en.}Pu)O_2$ $U_{r.en.}\ O_2$	TNR BN-600 (BR < 1)	$U_{nat.en.}\ U_{r.en.}$	U, Pu U	Fission products, MAs
Closed classic	$(U_wPu)O_2$ $(U_wPu)N$ $(U_wPu)_{met}$	FNR (BR > 1)	$U_{nat.}(U_w)$	U, Pu	Fission products, MAs
Closed transmutation	$(U_wPu)O_2$ $(U_wPu)N$ $(U_wPu)_{met}$	FNR (BR > 1)	$U_{nat.}(U_w)$	U, Pu, MAs	Fission products

TABLE 3.2 NFC subclassification (according to the SNF fractioning level).

NFC	SNF	U 95% (wt)	HLW 5% (wt)	Pu	U–Pu	Cs–Sr	Np-Am-Cm-Tc-I	Np-Pu	T_c–I	Am-Cm
					98% (activity)					
Open nonrecycling	Disposal	–	–	–	–	–	–	–	–	–
		Storage	Disposal	–	–	–	–	–	–	–
		Storage	Fraction. →	Storage	Storage	Storage	Disposal	–	–	–
Open recycling	–	Recycle 1 MOX TNR	Fraction. →	Recycle 1 MOX TNR	Recycle 1 MOX TNR	Storage	Disposal	–	–	–
Closed classic	–	Recycle 1 NF FNR	Fraction. →	Recycle 1 NF FNR	Recycle 2 NF FNR	Storage	Disposal	–	–	–
Closed transmutation	–	Recycle 1 NF FNR	Fraction. →	Recycle 1 NF FNR	Recycle 2 NF FNR	Storage	Fraction.	Recycle 3 NF FNR	Storage	Recycle: after-burn. in FNR

TABLE 3.3 NFC evolution.

Open nonrecycling NFC	Open recycling NFC	Closed classic NFC	Closed transmutation NFC	Ideal NFC
$U_{nat.}$	$U_{nat.}$	U_w, $U_{nat.}$	U_w, $U_{nat.}$	$U_{nat.}$
NFC: Enrich. prod. TNR (BR < 1) – – –	NFC: Enrich. prod. TNR (BR < 1) Proc. SNF Recycle U, Pu -	NFC: – FNR (BR > 1) Proc. SNF Recycle U, Pu -	NFC: – FNR (BR > 1) Proc. SNF Recycle U, Pu Recycle MA	NFC: – NR (BR > 1) Proc. SNF Recycle U, Pu Recycle MA
Enrichment production waste	Enrichment production waste	–	–	–
$(SNF)_{disp.}$	–	–	–	–
–	Losses in processing SNF	Losses in processing SNF	Losses in processing SNF	–
–	Fission products $(Cs, Sr)_{st.}$	Fission products $(Cs, Sr)_{st.}$	Fission products $(Cs, Sr)_{st.}$	Disposal of fission products
–	$(MA, Tc, I)_{disp.}$	$(MA, Tc, I)_{disp.}$	Fission products $(Tc, I)_{st.}$	
–	Fission products $(rest)_{disp.}$	Fission products $(rest)_{disp.}$	Fission products $(rest)_{disp.}$	

3.4 Tasks solved in the closed NFC

The closed nuclear fuel cycle (CNFC) using fast reactors with inherent safety is a new integrated product in the area of nuclear power technologies with the following unique properties:

- Deterministic prevention of severe accidents;
- Consistent reduction of accumulated SNF and transfer to generation of radioactive waste equivalent in radiation and toxicity to the relevant characteristics of raw uranium (radiation-equivalent RW disposal);
- Use of the natural resources without any other industrial application as fuel;
- Absence of any real fuel base limitations (on the scale of several thousand years);
- Technology support of the nonproliferation regime;
- Competitiveness in comparison with other large-scale power generation technologies.

Implementation of these advantages is possible on the basis of FNRs with inherent safety preventing accidents that result in the necessity for evacuation of the population.

In comparison with non-nuclear energy generation techniques nuclear power under the CNFC conditions claims to become the first environmentally preferential energy generation technique:

- Without any waste disturbing the ecological balance (including CO_2);
- Without any negative impact on the environment (including the greenhouse effect);
- Without any need for combustion of vast amounts of hydrocarbons more beneficial to be used for other purposes.

The CNFC on the basis of fast reactors solves the basic problems of nuclear power generation revealed at the first stage of its acquisition and hampering the large-scale nuclear power development in the fundamental and systemic manner [45].

Objectives with regard to fuel resources. The CNFC on the basis of fast reactors fundamentally removes the problem of fuel resources for nuclear power through the use of ^{238}U increasing their useful base by approximately 150 times and bringing nuclear power to the top position in the total resource energy balance (Fig. 3.1).

In comparison with the TNR-based open NFC the CNFC with FNRs has undeniable advantages with regard to lesser amount of the consumed resources, accumulation of SNF and RW returned to the environment that may be characterized with the following data (Table 3.4) [45].

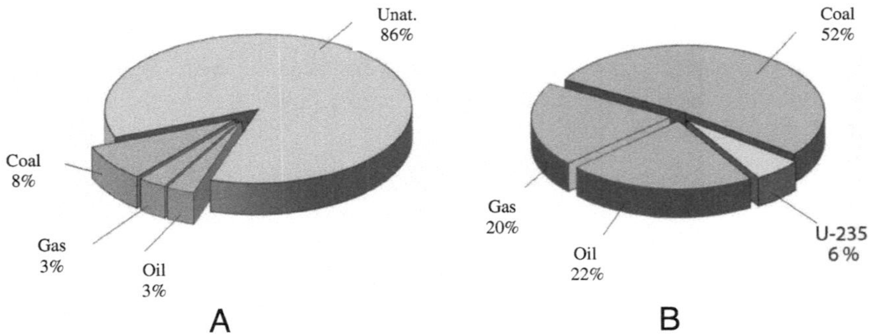

FIGURE 3.1 Comparison of energy intensity for different natural energy resources: (A) – with due regard for ^{238}U (closed NFC with FNRs); (B) – regardless of ^{238}U (open NFC with TNRs).

TABLE 3.4 Conceptual (according to the order of values) comparisons of the consumption of natural resources and accumulation of SNF and RW in the open and closed NFC.

Parameter	ONFC	CNFC	Note
Annual consumption of U per 1 GW (e) · year	150 t	1 t	
Consumption of U for 60 years per 60 GW (e) · year	9,000 t	60 t[1]	
Maximum capacity of NPE with 600–700 th.t of natural U	68–80 GW (e) for 60 years	600–700 GW (e) for 1000 years[2]	
Annual U extraction for 25 GW	3750 t	Not required	26.7 GW – the NPE capacity in the RF; 3100 t – current extraction in the RF
Amount of generated fuel resources (Pu) per 1 GW (e) · year	0.15 t	1 t	
SNF accumulation per 1 GW (e) · year	15 t	None	
RW (fission fragments) per 1 GW (e) · year	1 t	1 t	
Potential biological hazard of HLW, Sv/kg SNF			
Pu	$\sim 10^5$	~ 100	
Am	$\sim 10^4$	~10	

[1] *In the steady-state mode.*
[2] *With due regard for start of FNRs with UN-fuel or with BR> 1.*

It follows from these data that the closed NFC:

- Requires 150-fold less natural uranium;
- Subject to FNR start with uranium produces 6-7-fold more secondary fuel resources (plutonium) than in case of uranium use in TNRs;
- Expands the nuclear power existence duration from hundreds to thousands years;
- Enables to give up uranium extraction at all (subject to the steady level of the aggregate NPE capacity and operation with waste uranium) or to increase the NPE capacity with the same level of extraction;
- Reduces potential biological hazard of HLW.

The following fundamental principles of the NFC arrangement may be compiled in the context of optimal utilization of fuel resources:

a) All actinides (both plutonium and uranium) should be used, as a rule, in FNRs;
b) Utilization of uranium in TNRs is the relic of nuclear power establishment;
c) Utilization of plutonium in VVERs (MOX) is not feasible due to the following reasons:
  - The base for development of the CNFC with FNRs is reduced;
  - The properties of plutonium are deteriorated (energy-grade plutonium from uranium SNF of VVERs without any recycling is optimal for FNRs);
  - Utilization of plutonium is not efficient;
  - The amount of fuel factor for NPPs with TNRs increases (taking into account the cost of plutonium).

Objectives with regard for selection of the fuel composition. Advantages of dense fuel in comparison with oxide one are obvious. Taking into account the risk factors MOX fuel may be considered as back-up one subject to development of the improved core for BN-1200 reactor plant. At the same time the NFC closure tasks may be solved through the use of BN-800 and MOX fuel (due to similarity of the MNIT and MOX fuel technologies).

Objectives with regard to irradiated nuclear fuel. Tasks to be solved through the use of the CNFC with regard to SNF have three aspects:

1) The CNFC eliminates accumulation of FNR SNF;
2) Step-by-step transition to the CNFC enables to stop accumulation of TNR SNF at the first stage and to reduce the SNF handling costs as replacement of a TNR with a FNR:
  - Prevents generation of 900 t (15 t/year–60 years) of VVER SNF and expenditures for its storage prior to processing evaluated at the amount of 384 thous. RUB/t/year by Rosatom in 2015;
  - Reduces the annual amount of processed SNF;
  - Reduces the storage volume of irradiated uranium.

3) Utilization of the SNF processing products in FNRs is an efficient way to solve the problem of accumulated VVER SNF (at the second stage);
   - A new FNR can utilize trans-uranium elements from all SNF generated during the service life of a VVER reactor;
   - Replacement of 10 GW of thermal reactors with fast ones will almost entirely solve the problem of accumulated VVER SNF in Russia.

Objectives with regard to high-level waste. The radiation-equivalent approach in the CNFC is the main technique to solve any potential environmental problems in the course of RW handling and also the basic argument in the struggle against radiophobia. It actually means that radiation safety of the environment is guaranteed not through the use of any hardware and techniques but by absence of any activity exceeding the existing natural levels.

Objectives with regard to reactor plants. The CNFC can be arranged only on the basis of FNRs with complete fuel breeding. BR $>$ 1 which is unachievable in TNRs is necessary and sufficient to use ^{238}U and obtain high parameters with regard to fuel resources.

No FNRs with mixed fuel have been yet developed in Russia although the scientific and technical basis for this is present. It may be demonstrated for the first time in BN-800 reactor plant with full loading of MOX fuel.

Transmutation of minor actinides in FNRs is necessary to solve the problems with HLW requiring generation of excessive (not engaged in thermal energy production) neutrons which can be arranged only in FNRs.

Transmutation of minor actinides still remains an unsolved technical and technology problem.

Chapter 4

Fuel supply

4.1 Effect of burnup depth

The nuclear fuel burnup depth in NPP reactors is one of the key factors affecting technical and economic characteristics of the NFC. In traditional nuclear power with the open NFC burnup depth directly determines the consumption of uranium resources and SNF accumulation amounts accounting for the trend of its increase. So, for example, in the early XXI century with the burnup depth of ~40 GW·day/t typical for light water reactors in the global nuclear power the opportunity to increase this level to 100 GW·day/t was considered in the systematic study of MIT [214]. Reduction on the fuel recycling volume with increase of burnup depth is of vital importance for the NFC closure. At the same time the set of restricting factors to be clarified in the course of design and R&D works should be also taken into account. Accumulation of even uranium isotopes deteriorating the characteristics of regenerated fuel is one of these restrictions for light water reactors using upgraded regenerated uranium from their SNF, and so at present the permissible average burnup depth for them is assumed to be about 50 GW·day/t.

The optimal burnup depth for the fuel of fast reactors with inherent safety is determined in view of their wide-scale implementation in order to obtain the relevant target for transition of nuclear power to the new technology platform (NTP), including potential export of VVERs and FNRs depending on the adopted development scenarios for the Russian nuclear power. Economic advantages achieved due to increase of fuel burnup depth are related to reduction in the fuel consumption amounts which in its turn affects the expenditures for both initial and final part of the nuclear fuel cycle. The restricting factors such as the temperature of fuel and fuel rod cladding, the damaging dose for the fuel assembly structural materials and corrosive impact of the sodium or lead coolant as well as the required allowance for fuel swelling between the fuel rod cladding and the fuel kernel for FNR reactors make it more difficult to achieve the high parameters (~12%) of average fuel burnup for innovative facilities within the shortest period of time. According to the opinion of designers, increase of burnup in excess of 7.3% will require development of new structural materials which will eventually cause fuel fabrication cost increase. In this context assessment

Closed Nuclear Fuel Cycle with Fast Reactors: Handbook of Russian Nuclear Power.
DOI: https://doi.org/10.1016/B978-0-323-99308-1.00013-1

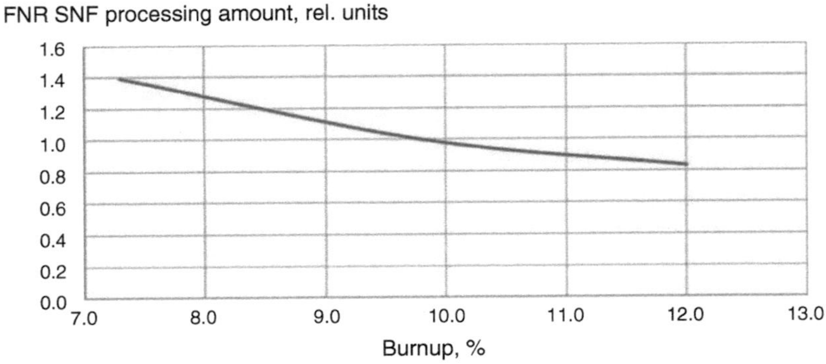

FIGURE 4.1 Effect of FNR fuel burnup depth on the SNF processing volume.

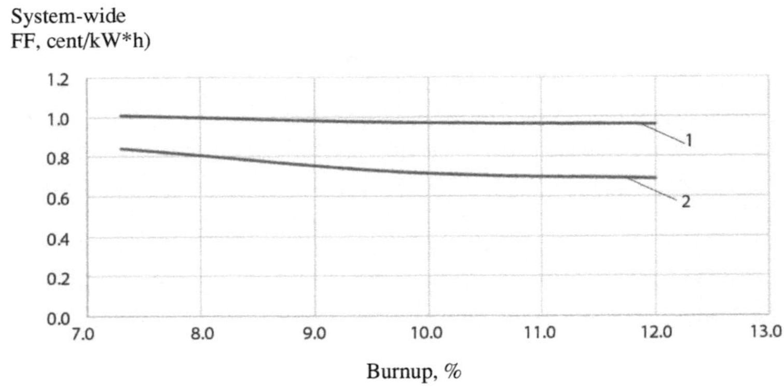

FIGURE 4.2 Effect of FNR fuel burnup depth on the specific system-wide energy cost fuel factor: 1 – FF (d = 0 %); 2 – FF (d = 5 %).

of the economic NFC parameters both with due regard for the latter factor and in the absence of any transition to higher fuel burnup depth parameters is relevant. The systemic study carried out in the Innovation and Technical Center of Proryv Project on the basis of the design results obtained by the beginning of 2017 confirmed feasibility of achievement of the previously recommended average FNR fuel burnup depth at the level of 12% for all nuclear power development options. Reduction in the integral FNR SNF processing volume with increase of burnup depth as exemplified by the scenario with the Russian nuclear power transition to the NTP by the end of the XXI century and achievement of the installed capacity level of 70 GW (e) is shown in Fig. 4.1.

The relevant economic parameter of the NFC is shown in Fig. 4.2 with the specific system-wide energy cost fuel factor values (FF) determined in the similar way as the specific system-wide energy cost but with due regard only for the expenditures directly associated with nuclear fuel production and consumption. The specific FF gives the average (with due regard for discount) value of the current parameters for the nuclear power system in general within the time period under consideration. The obvious advantage may be recognized in case of the FNR fuel burnup depth increase from 7 to 10%; the tendency towards the FF decrease is also observed in the course of further increase in the achieved burnup depth to 12%. The effect of reduced FNR fuel consumption with increase of burnup depth is manifested to the fullest extent in the second half of the XXI century when their share in the installed NPP capacities becomes dominant, and thus the economic effect in the form of the FF reduction in Fig. 4.2 is demonstrated more prominently in case of the discount rate of less than 5% as recommended for long-term projects.

4.2 Role of uranium-plutonium fuel for thermal reactors

It is worth reminding that MOX fuel is produced when all plutonium from SNF is mixed with depleted uranium: 6–8% of plutonium and 92–94% of depleted uranium. REMIX fuel is produced when regenerated uranium and plutonium (plutonium extracted from SNF together with approximately the same amount of uranium) is mixed with upgraded remaining regenerated uranium (4–6% of ^{235}U) and an additional portion of fresh enriched uranium in order to make the mixture consisting of approximately 4% of ^{235}U, 1–2% of plutonium, and ~94–95% of other uranium isotopes, mainly ^{238}U.

There is certain saving of natural uranium – approximately 12.5% for MOX fuel and about 22% for REMIX fuel.

The abstract technical question as to which nuclear fuel is more preferable for FNRs – uranium oxide (UOX fuel) or mixed oxide (MOX fuel) – has the absolute answer: uranium oxide. This answer is based on the need for saving of energy-grade plutonium accumulated in TNR SNF for start of the nuclear power with the closed NFC and FNRs where plutonium is utilized with the maximum efficiency. Besides, the cost of MOX fuel is considerably higher than the cost of uranium oxide fuel.

At present MOX fuel is used for TNRs in several countries. It is related to the policy for disposal of large amounts of plutonium accumulated in TNR SNF and not demanded for FNRs as their commercial start-up was planned significantly earlier and then indefinitely postponed. The countries that arranged processing facilities and MOX production had to look for the opportunities to use them, and for this purpose ignored even the capital investments for their construction in determination of the MOX price actually subsidizing production of MOX fuel.

4.3 Systemic evaluation of the Russian nuclear power development scenarios with and without the use of REMIX fuel for VVERs

Numerical evaluation of the VVER REMIX fuel implementation effects was performed in the systemic study of the Innovation and Technical Center of Proryv

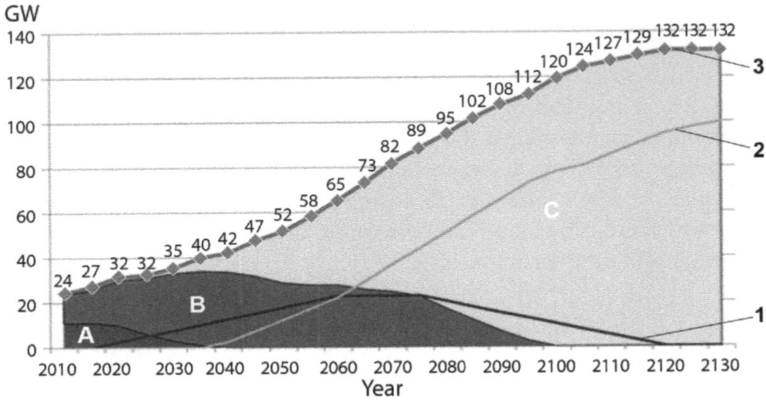

FIGURE 4.3 Installed NPP capacities: A – RBMK; B – BN+BR; C – VVER; 1 - export VVER; 2 -export BR; 3 - total in Russia.

Project in 2018 through comparison of the two scenarios of the Russian nuclear power transfer to the NTP by the end of the XXI century with predominating FNRs with inherent safety and the closed NFC and passing through the stage of two-component structure TNR+FNR in order to reach the installed NPP capacity level of 120 GW (e) by 2100. The assumed capacity growth dynamics with gradual decommissioning of TNRs replaced with FNRs is shown in Fig. 4.3, including also the curves of the expected changes in the installed capacities of export TNRs and FNRs.

Two scenarios are compared:

1) Basic (without the use of REMIX fuel): The VVER fuel resource is enriched uranium, particularly with the maximum possible utilization of upgraded regenerated uranium produced in the course of VVER SNF processing;
2) REMIX scenario: At the initial stage REMIX fuel for VVER is produced from stock plutonium and regenerated uranium from VVER SNF and then - from regenerated U-Pu produced in the course of VVER SNF processing mixed with upgraded (up to 5.7% by ^{235}U) regenerated uranium and fresh enriched uranium (4.4% by ^{235}U). REMIX fuel is mainly supplied for export.

The FNR fuel resource is provided through the use of stock energy-grade and ex-weapon-grade plutonium (50 + 50 tons), new supplies of Pu and regenerated U-Pu from VVER SNF processing (including the one returned from export NPPs), regenerated U-Pu from processing of domestic FNR SNF and also (upon completion of VVER SNF reserve utilization) the marginal fuel on the basis of enriched uranium.

Changes in the fuel consumption volume and structure in two development scenarios for two-component nuclear power are compared in Fig. 4.4.

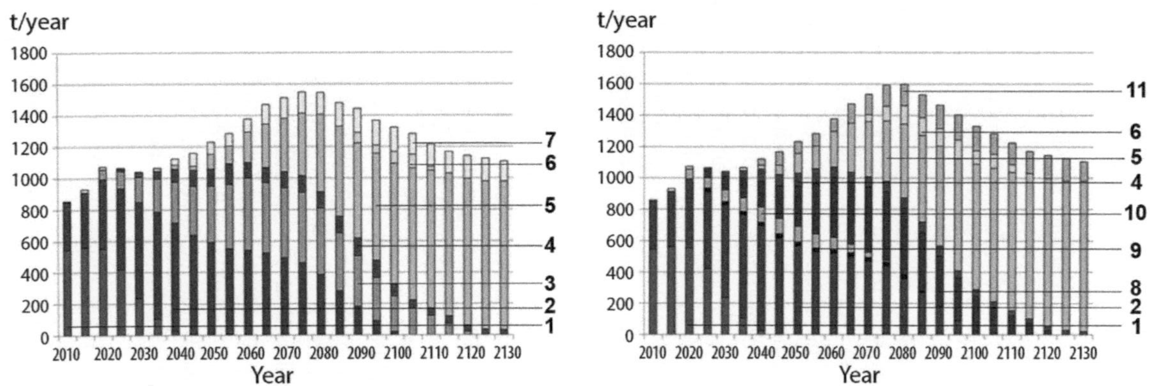

FIGURE 4.4 NPP fuel consumption: basic (a) and REMIX (b) scenarios: 1 – RBMK UOX; 2 – VVER UOX; 3,10 – export VVER UOX; 4 – fresh MNIT; 5 – regenerated MNIT; 6 – UN/UOX; 7 – export BR UN; 8 – VVER REMIX; 9 – export VVER REMIX; 11 – export FNR UN.

The main systemic effect of VVER REMIX fuel introduction is to reduce the plutonium resource base for increase of FNR capacities (the same is also true for VVER MOX fuel utilization). As a result, additional amount of marginal uranium fuel is required to continue increase of FNR capacities at the stage of the VVER fleet decommissioning completion and utilization of their accumulated SNF.

It increases the integral system demand for natural uranium. The alternative solution with increase of the plutonium resource base by its enhanced breeding in FNRs through the use of the breeding blanket may be economically unfeasible due to significant increase of FNR SNF processing amounts (this issue requires further investigation).

The thesis on the possibility for return of the generated plutonium and regenerated uranium to the VVER fuel cycle is one of the main arguments for REMIX fuel introduction. But this thesis is perfectly true only for single-component nuclear power with TNRs. Utilization of plutonium from VVER SNF (as well as uranium) in two-component power engineering is much more efficient in FNRs. Use of upgraded regenerated uranium (both in VVERs and FNRs) provides for significant saving of natural uranium in any case (with or without REMIX fuel). But then this saving is greater if the plutonium resource is used in FNRs and not in REMIX fuel for VVERs. Fig. 4.5 where integral consumption of natural uranium in two scenarios is compared illustrates the above-mentioned aspects.

Fig. 4.5 shows that eventually integral consumption of natural uranium for the internal needs of Russia in the REMIX scenario will increase by 8% (~40 kt) by 2100, and for the needs of fuel export (including the main amount of REMIX) it will reduce by 3% (~12 kt), but in general the natural uranium resource consumption in the REMIX scenario are 3% (~26 kt) more. The demand for separation works also increases.

Comparison of the structure and dynamic pattern of the cost of ordinary UOX and REMIX fuel for VVERs in the Russian nuclear power development scenario with transition to the NTP with FNRs and the closed NFC is given in Fig. 4.6. With the assumed technical and economic indices of the NFC process stages the cost of REMIX fuel at the initial stage (with the use of "free-of-charge" stock plutonium) remains close to the cost of UOX fuel in spite of much higher production (fabrication) cost. But in the future REMIX fuel becomes much more expensive due to addition of upgraded regenerated uranium and fresh enriched uranium.

Both fuel types rise in price more intensively upon commencement of utilization of FNR uranium fuel for which the resource of upgraded regenerated uranium is allocated in priority (as the maximum saving is achieved by production of uranium with the enrichment of ~13% for FNR fuel which is higher than 4.5–5% for VVER fuel).

Introduction of REMIX fuel will demand additional investments into the NFC production facilities. Assessment in accordance with the TEI data shows

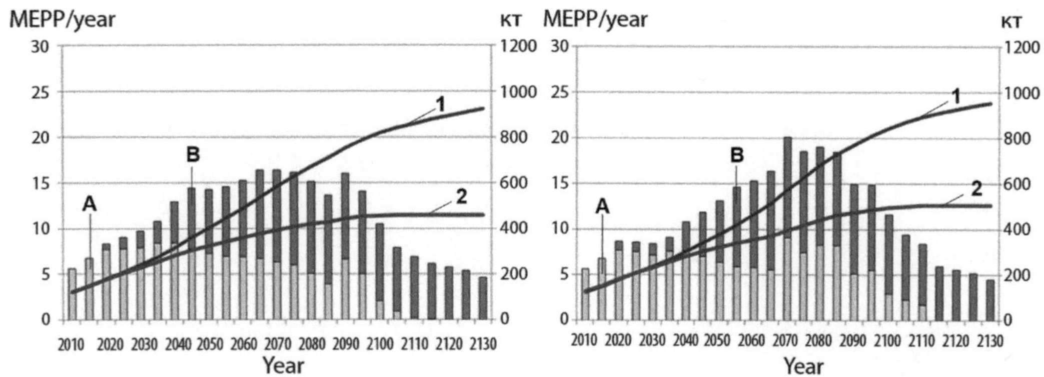

FIGURE 4.5 Integral uranium consumption (kt) and separation works (MEPP/year): *a* - basic scenario; *b* - REMIX scenario: A – separation works for Russia; B – separation works for export NPPs; 1 - total integral; 2 - integral for Russia.

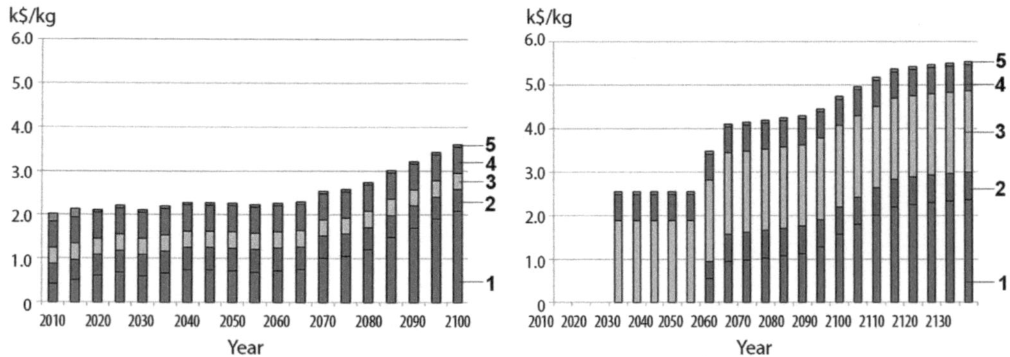

FIGURE 4.6 Cost of UOX fuel (a) and REMIX fuel (b) for VVERs:1 – contribution of Unat.; 2 – contribution of enrichment; 3 – contribution of production; 4 – allowances for SNF processing; 5 – allowances for RW management.

that the specific capital investments into the ordinary UOX VVER fuel production module with the capacity of 250 t/year can reach ~3.6 mln. USD/(t/year), and for production of REMIX fuel with the same capacity – 4.75 mln. USD/(t/year). Another assessment for REMIX fuel by conversion calculation on the basis of design data for the IPC FRM gives ~12.5 mln. USD/(t/year).

From the viewpoint of system-wide fuel balance utilization of VVER REMIX fuel reduces the external plutonium resources for FNRs; in this case more marginal uranium nitride fuel is required, and the need for it arises 20 years earlier as can be seen in Fig. 4.4.

Involvement of upgraded regenerated uranium into the closed NFC as prescribed by the concept of REMIX fuel is an important factor for efficient utilization of fuel resources. This provides for considerable saving of natural uranium in any case, with or without REMIX fuel. But in two-component nuclear power the needs of the system in general for natural uranium increase with due regard for development of the FNR fleet upon introduction of VVER REMIX fuel.

Chapter 5

Prevention of severe reactivity-related accidents

5.1 Chernobyl catastrophe

Destruction of the fourth power unit took place on April 26, 1986 at Chernobyl NPP (Ukraine, 110 km from Kyiv). Destruction had the explosive nature, the reactor was completely destroyed, and a large amount of radioactive substances was released into the environment. Thirty-one persons died within the first three months after the accident. More than 115 thous. people were evacuated from the 30 km zone. Long-term effects of exposure revealed within the subsequent 15 years caused death of 60–80 persons. One hundred and thirty four persons suffered from radiation disease of certain severity. Significant resources were mobilized for mitigation of the accident consequences, more than 600 thous. people were engaged in the postaccident clean-up. The cloud generated by the burning reactor spread radioactive materials (primarily iodine and cesium radionuclides) across the significant part of the European territory. The maximum fallout was observed in the regions of the Soviet Union situated in the vicinity of the reactor and at present belonging to the territories of Belarus, Russia, and Ukraine.

5.2 Dense fuel as a nuclear safety factor

Reactivity-related accidents are accidents initiated by unauthorized transition of the reactor into the uncontrolled above-critical state ($\rho > \beta_{ef}$) with subsequent instantaneous increase of the reactor thermal power to the values unsuitable for its integrity.

The possibility for a reactivity-related accident in a FNR is related to:

1) In-process insertion of positive reactivity equal to the operational positive reactivity margin with the value exceeding the effective delayed neutron fraction β_{ef} into the reactor;

Closed Nuclear Fuel Cycle with Fast Reactors: White Book of Russian Nuclear Power.
DOI: https://doi.org/10.1016/B978-0-323-99308-1.00016-7

2) Appearing of the positive reactivity exceeding the effective delayed neutron fraction β_{ef} in the reactor due to:
- The positive void reactivity effect of the liquid-metal coolant;
- The negative temperature reactivity effect of the coolant.

First of all, let us consider the problem associated with the operational positive reactivity margin. In general, the operational positive reactivity margin is necessary to compensate for reactivity decrease due to:

- NF burnup (i.e., reduction of the fissionable nuclide concentration in the fuel);
- Fouling of nuclear fuel (i.e., increase of the fission product nuclei [moderate neutron absorbers] concentration in the fuel);
- Poisoning of nuclear fuel (i.e., increase of the fission product nuclei (intensive neutron absorbers) concentration in the fuel);
- Increase of the NF temperature with the negative temperature reactivity effect.

No poisoning of nuclear fuel takes place in FNRs. Such fission products as xenon and samarium have ordinary moderate cross-sections in them, that is, belong to slags, and thus no operational positive reactivity margin is necessary to compensate for fuel poisoning.

In the course of the reactor operation with the rated power the operational positive reactivity margin for compensation of the "fast" negative temperature reactivity effect of nuclear fuel shall be completely utilized and absent on the absorbing control rods.

There is no need to have any excessive fuel loading in FNRs above the critical one, as nuclear fuel burnup is compensated through appearance of new fissionable nuclides due to the NF breeding process. With the core breeding ratio of rather more than one (up to 1.05) the new fissionable nuclides are sufficient for both compensation of reactivity reduction due to NF burnup and compensation of reactivity reduction due to NF fouling. Thus, the operational reactivity margin for NF burnup and fouling may be below β_{ef} eliminating any possibility for instantaneous uncontrolled runaway of the reactor because of in-process insertion of the positive reactivity exceeding β_{ef}.

Three fundamental properties of dense nuclear fuel play a significant role from the viewpoint of their effect on the basic core characteristics and safety: density, thermal conductivity and specific quantity of dispersive light elements (oxygen, carbon and nitrogen).

Higher density of fuel and lesser content of light elements facilitate increase of the breeding ratio in the reactor core (BRC) and the integral breeding ratio of the reactor plant. The cores with so-called equilibrium fuel where the fissionable substance burnup is fully compensated by its breeding gain peculiar unique properties. Cores with the BRC $\sim$1 provide:

- Complete breeding of fissionable nuclides being the sufficient precondition for practical utilization of the energy resource of ^{238}U;

- Operation without any uranium blanket in order to eliminate production of low-background plutonium with the quality similar to weapon-grade one;
- Absence of any need for extraction of plutonium from SNF and the possibility to apply the technologies without uranium and plutonium separation thus ensuring technology support of the nonproliferation regime together with the above-mentioned advantage;
- Minimization of the reactivity margin for burnup reducing the maximum reactivity margin of reactor plants and enhancing their nuclear safety;
- Unique stability of heat emission of fuel rods and fuel assemblies in the course of their operation.

The terms "equilibrium fuel" or "fuel of equilibrium composition" mean uranium and plutonium fuel where the composition and mass of plutonium remain almost unchanged in the course of the fast reactor operation, or rather change so slightly that the reactivity changes associated with fuel burnup are much less than β_{ef}. These concepts are inextricably linked to the wider term "equilibrium reactor operation mode" which means operation of the fast reactor within the closed NFC with BR $\sim$1 when burning of fissionable nuclides is fully compensated by their breeding, mainly due to conversion of ^{238}U to ^{239}Pu, and the possibility to operate with the low reactivity margin appears thus preventing the reactor runaway due to prompt neutrons. Processing of the fuel unloaded from the reactor in the closed NFC involves only extraction of fission products and their replacement with depleted (waste) uranium without any adjustment of the isotope composition of plutonium and minor actinides, density and mass of the loaded fuel.

The concept of equilibrium operation mode also applies to operation of FNRs started with enriched uranium subject to gradual transition to uranium and plutonium fuel, although the composition of unloaded fuel changes in this case. In case of such start-up enrichment of uranium in the starting fuel loading shall be selected with due regard for the possibility to operate the FNR with low (less than β_{ef}) reactivity change due to full compensation of reactivity reduction in the course of ^{235}U burnup by its increase because of ^{238}U conversion into ^{239}Pu. Stabilization of the low reactivity margin shall be provided within the entire transition period (4–5 campaigns with the duration of 5–6 years) up to ^{235}U burnup and transition to uranium and plutonium fuel without any adjustment of the isotope composition of plutonium and minor actinides as well as fuel density and mass in the course of its processing within the closed NFC.

Of course, the condition of the low reactivity margin steady within the campaign in the course of the reactor operation requires high accuracy in production of the start-up loading and the first batches of additional fuel, especially with regard to the isotope composition of supplied plutonium. These loadings shall comply not only with the condition for the reactor critical state assurance within the period of operation between the regular refuelings, but also with the condition for achievement of the average BR value of $\sim$1.05 within the campaign.

Higher thermal conductivity of fuel can be used either to reduce the fuel temperature, or to increase the thermal load on the fuel depending on its utilization concept.

Fuel temperature reduction:

- Enhances efficiency of feedback due to the Doppler effect;
- Reduces the energy stored in the fuel thus enhancing safety of reactor plants in case of any potential emergencies associated with the core cooling loss;
- Increases efficiency of the first safety barrier - the fuel matrix - with regard to reduction of any gaseous radioactive fission product releases under the fuel rod cladding.

5.3 Heavy coolant as a nuclear safety factor

There is a problem related to the void reactivity effect of the liquid-metal coolant. So, for example, BN-600 reactor operated at Beloyarsk NPP has the positive void reactivity effect of the coolant.

The fact of the matter is that sodium moderates neutrons although it is not qualified as moderator. So in case of sodium "discharge" moderation of neutrons is reduced, and the neutron spectrum in the reactor is shifted towards higher energies resulting in increase of ^{238}U fission with fast super-threshold neutrons and, consequently, an increase of the fast fission factor and therefore - in reactivity increase. Besides, sodium absorbs neutrons (although poorly), and so sodium "discharge" from the core results in reduction of "parasitic" absorption of fast neutrons thus causing increase of the fast fission factor and reactivity of the reactor. It is a weak point of sodium reactors, as in case of any severe accidents with loss of heat removal the sodium "discharge" effect and the relevant positive void reactivity effect can occur due to its evaporation because sodium boils at the temperature of just ~900°C.

In case of light ("greatly" moderating) neutron-absorbing sodium replacement with heavy ("poorly" moderating) nonabsorbing lead (lead and bismuth) the problem of the positive void reactivity effect will not occur, as in this case enhancement of neutron leakage from the core and not increase of the fast fission factor becomes the dominant "void" effect.

So, the problem related to the positive void reactivity effect of the liquid-metal coolant may be solved by transition from the sodium coolant to the lead (lead and bismuth) one.

Now let us discuss the problem of the large negative temperature reactivity effect. The large negative temperature reactivity effect creates the possibility for so-called cold reactivity-related accidents when rapid cooldown of the core takes place. Cooldown can occur, for example, in case of any steam pipeline rupture. In this case pressure in the steam generator falls to the atmospheric one at once, the water boiling point becomes 100°C instead of 320°C, heat removal from the primary circuit drastically increases - thus cooldown of the primary circuit and insertion of positive reactivity take place. That is why in the ideal case the absolute value of the negative temperature reactivity effect shall be sufficient to reduce the power upon any decrease in heat removal (prevention of "hot" accidents) but shall not be so high as to cause "cold" accidents when any excessive positive reactivity is released in case of sharp cooldown of the core.

It is worth to be noted that the coolant temperature coefficient of reactivity in the fast reactors with the liquid-metal coolant is by an order lower than in thermal reactors where water density is changed considerably depending on the temperature.

Chapter 6

Prevention of severe heat removal accidents

6.1 Accident at EBR-1

On November 29, 1955, human factor caused an accident at the American experimental reactor EBR-1 (Idaho, USA). In the course of an experiment with plutonium the reactor was self-destroyed because of erroneous actions of the operator, and 40% of its core burnt out.

Within the service life the reactor was operated with the cores of four different configurations, and metal fuel was used in each of them. Highly enriched uranium consisting mainly of ^{235}U isotope was used in the first three cores. The fuel in the form of uranium-zirconium alloy containing 2% of zirconium was used in the second core. The diameter of fuel rods was 1.25 cm, and 217 such rods arranged in the form of a triangular grid formed the hexagonal core located in the center with the cross-sectional size of 19 cm. The small core size is indicative of considerable compactness of fast reactors with liquid metal cooling. The blanket containing rods with natural uranium is located around this central core with high enrichment of ^{235}U. The mixture of sodium and potassium remaining liquid at the room temperature was used as the coolant in the reactor.

Power oscillations were observed in this core with very low coolant flow rates. The experiment for investigation of this effect was started on November 29, 1955. In the course of the experiment which was carried out with complete termination of the coolant circulation through the core and disabling of the certain safety system locking devices abrupt increase of power was performed in order to determine the rate of reactivity increase together with the temperature. It was planned to stop the experiment upon reaching of the fuel temperature equal to 500°C, but due to coincidence of the investigated effect and the operator's error the temperature increased above 720°C. Interaction between the metal uranium fuel and the stainless steel cladding began at this temperature, and consequently approximately 40% of the core melted; however, it did not cause an explosion, damage of the plant or occurrence of any radiation hazard.

Approximation of fuel rods in a fast reactor results in increase of reactivity or neutron density. This is directly related to the incident which has caused the

Closed Nuclear Fuel Cycle with Fast Reactors: Handbook of Russian Nuclear Power.
DOI: https://doi.org/10.1016/B978-0-323-99308-1.00033-7

core melting in EBR-1. The fuel rods were able to bend, and it caused increase of reactivity; in this case the process turned out to be spontaneously developing as the bend increased in the course of temperature growth. This was the cause of the temperature effect investigated in the course of the experiment which was later explained from the theoretical viewpoint.

Later the core of EBR-1 reactor was dismantled and replaced with another one with spacing ribs used in its design for prevention of the temperature effect. Expansion of these ribs upon increase of the temperature results in expansion of the core thus creating the negative temperature reactivity coefficient instead of the positive one previously observed. Operation of EBR-1 reactor which was finally shut down in December 1963 provided useful information on the peculiarities of the fast reactor designs. At present the cores of all reactors are designed with significant number of imposed restrictions, and thus they always have the negative temperature reactivity coefficient.

6.2 Accident at Three Mile Island NPP (USA)

Prior to the events in Chernobyl the accident at Three Mile Island NPP was considered as the largest one in the history of global nuclear power. It is still considered as the most severe nuclear accident in the USA. A major accident took place on March 28, 1979 early in the morning at Power Unit 2 of Three Mile Island NPP with the capacity of 880 MW (e) situated in 20 km from Harrisburg (Pennsylvania) and owned by Metropolitan Edison Company. As it turned out, Power Unit 2 was not equipped with an additional safety system, although such systems were installed on some power units of this NPP. In spite of partial nuclear fuel melting it did not burn through the reactor vessel, and radioactive substances mainly remained inside. According to various evaluations, radioactivity of noble gases released into the atmosphere was 2.5–13 mln. Ci, but releases of hazardous nuclides such as Iodine-131 were insignificant. The power plant territory was also contaminated with radioactive water discharged from the primary circuit. It was decided that evacuation of the people living near the power plant was not necessary, but the authorities recommended pregnant women and pre-school children to leave the 8-kilometer zone. Officially the works for the postaccident clean-up were completed in December 1993. Decontamination of the power plant territory was carried out, and fuel was unloaded from the reactor. However, radioactive water had partially soaked into the containment concrete, and this radioactivity was almost impossible to remove. Operation of another reactor at the power plant (TMI-1) was resumed in 1985.

6.3 Accident at Mayak PA (South Ural, Russia)

The Kyshtym accident is a major radiological human-induced accident which took place on September 29, 1957 at Mayak Chemical Plant situated in a restricted-access town of Chelyabinsk-40. Nowadays this town is named Ozersk.

The accident was called "Kyshtym" because Ozersk was security-restricted and absent on the maps till 1990. Kyshtym is the nearest town. On September 29, 1957 at 04:22 p.m., explosion of a vessel with the volume of 300 m^3 containing about 80 m^3 of high-level radioactive waste occurred because of the cooling system malfunction. The vessel was destroyed by the explosion evaluated as equivalent to dozens of tons of TNT, the concrete slab with the thickness of 1 meter and the weight of 160 tons was thrown away, and about 20 mln. Ci of radioactive substances was released into the atmosphere. Some radioactive substances were raised by the explosion to the height of 1–2 km and formed a cloud consisting of liquid and solid aerosols. Within the period of 10–11 hours radioactive substances fell out at the length of 300–350 km north-eastward from the explosion area (according to the wind direction). The territory of several facilities of Mayak plant, a permanent military camp, a fire station, a prison colony and then the territory with the area of 23 thous. km^2 and the population of 270 thous. people in 217 populated localities of three regions: Chelyabinsk, Sverdlovsk, and Tyumen fell into the radioactive contamination zone. Chelyabinsk-40 itself was unaffected. A total of 90% of radioactive contaminations fell out at the territory of the restricted administrative unit of Mayak Chemical Plant, and the remaining part was dispersed further. In the course of the post-accident clean-up 23 villages from the most contaminated areas with the population of 10–12 thous. people were resettled, and buildings, property, and cattle were destroyed. The sanitary protective area was arranged within the most contaminated section of the radioactive trail in 1959 according to the resolution of the government in order to prevent radiation spreading, and all economic activities were prohibited within this area. In 1968, the Eastern Ural national park was arranged within this territory. As of the moment of this book publication the contamination zone is called the Eastern Ural radioactive trail. The total length is approximately 300 km with the width of 5–10 km.

6.4 Fukushima catastrophe (Japan)

The earthquake which happened on March, 11, 2011 caused seismic ground vibrations affecting the structures, systems and components of the power plant. Then a series of tsunami waves followed, and one of them flooded the site. The amplitude of the recorded ground vibrations and the height of tsunami waves considerably exceeded the expected magnitude of danger used as the basis for the initial power plant design. Several power units of Fukushima Daiichi NPP were affected by the earthquake and the accompanying waves.

The earthquake damaged the power supply lines at the site of Fukushima Daiichi NPP operated by Tokyo Power Company, and tsunami caused significant destruction of the operation infrastructure as well as the safety assurance infrastructure at the site. In the aggregate it resulted in loss of external power supply and failure of the power supply sources at the site. This, in its turn, caused loss of the cooling function at three operating power units and also in the spent fuel pools. Four other nuclear power plants situated along the coast were also affected by the earthquake and tsunami to the different extent. But all operating power units of these power plants were shut down in a safe manner.

Despite the efforts of Fukushima Daiichi NPP operators aimed to maintain control over the power plant overheating of the core occurred in the reactors of power units 1–3, nuclear fuel melted, and integrity of the three containments was compromised. Hydrogen releases from the high-pressure reactor vessels took place and caused explosions inside the reactor buildings of power units 1, 3 and 4 resulting in destruction of the structures and equipment as well as injury of the power plant personnel. Radionuclides from the power plant were released into the atmosphere and then fell out on the land and ocean surface. Besides, direct discharges to the sea took place.

More than 150 thous. people left the contaminated territories within the radius of 50 km from Fukushima Daiichi NPP. According to the calculations of the Japanese Nuclear and Industrial Safety Agency (NISA) the amount of radioactive Cesium-137 released into the atmosphere during the accident is comparable with 168 bombs dropped on Hiroshima in 1945. In April 2011, the level of Cesium-137 in the ocean water samples taken near the Fukushima coast exceeded the pre-accident value by 50 mln. times.

6.5 Heavy coolant as a factor for prevention of severe heat removal accidents and explosions at NPPs

In the majority of cases, heat removal accidents are related to residual heat removal in the reactor subsequent to its shutdown (planned or emergency), that is, after termination of the chain nuclear fission reaction. Prevention of such accidents is associated, on the one side, with arrangement of natural circulation in the primary circuit with the intensity sufficient for residual heat removal, and on the other side – with the use of the sustainable ultimate heat sink. The atmospheric air may serve as such sustainable heat sink.

Reactors with liquid-metal coolants (sodium, lead, eutectic alloy of lead and bismuth, etc.) are planned to be used as FNRs at present and in the long run. Neutron physics and thermal hydraulics of the reactor core are traditionally considered as the main components of safety analysis for NPPs with the water coolant. Maintenance of the required water chemistry for the water coolant at any NPP is almost no different from the set of measures implemented at TPPs for this purpose.

Upon emergence of liquid metal coolants necessary for nuclear power installations with FNRs the problems related to the coolant quality maintenance were also classified as the key components of safety analysis. Maintenance of the required physical and chemical regime of the coolant (or the coolant technology) evolved from an important but additional operation into one of the main components in substantiation of the design and operating parameters of the reactor.

As far as the heavy liquid-metal coolant (lead, lead and bismuth) is concerned we may discuss not only the coolant quality maintenance technology but also its corrosion and slag safety.

Sodium coolant. Alkali metals: sodium-potassium (EBR-1) and sodium are historically the first liquid-metal coolants for fast reactors. Selection of sodium as the coolant for a power unit can seem strange, particularly taking into account the fact that sodium reacts with water in a very violent way: the reaction is accompanied with release of hydrogen which can ignite and explode. Nevertheless, sodium claims to be a good coolant for FNRs due to many reasons.

Firstly, the sodium nuclei absorb and moderate neutrons very poorly (as against the hydrogen nuclei included into water) that is crucial for FNRs.

Secondly, sodium has high thermal conductivity and low vapor pressure (0.12 MPa at 900°C), thus enabling to arrange the core with high power density of 1000 kW/L with much lower pressure in the primary (reactor) circuit than in water-cooled reactors. The pressure in the sodium circuit is determined only by pressure drop in the circuit which is usually less than 0.7 MPa. Low pressure considerably simplifies the design and operation of both the reactor and auxiliary equipment of the power plant.

Thirdly, there are no chemical reactions with hydrogen generation in the sodium circuits at the contact points between sodium and hot fuel rod claddings, as against the steam-zirconium reaction in water-cooled reactors.

Fourthly, relatively high boiling temperature of liquid sodium (883°C) provides a large time margin for the operators in order to eliminate any potential emergencies in the course of transient processes. For example, in case of any significant increase of the coolant temperature at the reactor outlet no fuel rod melting due to formation of the steam film will take place as against water cooling. The permissible thermal fluxes are almost unlimited with any critical thermal loads at the operating thermal flux values of up to 2.3×10^6 kcal/m^2·h.

All above-mentioned properties of the sodium coolant make its usage more beneficial in comparison with the water one.

Other liquid metal coolants have the same properties, but low corrosive and erosive impact on the structural materials is the advantage of sodium. This unique property together with high thermal conductivity (ninefold higher than for lead) and thermal capacity (fourfold higher than for lead) have determined selection of sodium as one of the first liquid metal coolants.

High chemical reactivity, of course, takes the first place among the drawbacks of sodium as coolant. Reaction with water poses the maximum hazard. That is why the systems with steam-water cycles shall be equipped with the devices ensuring explosion safety. Contact of sodium with steam or oxygen in the air is not so dangerous because no self-sustained combustion reaction occurs. Nevertheless, metal oxidation upon contact with air shall be prevented as sodium oxide is not soluble in liquid sodium, and formation of insoluble oxides can result in clogging of individual channels. Besides, sodium oxides deteriorate the corrosive properties of the liquid coolant.

Considerable radiation activation is the second significant drawback of sodium as coolant because it necessitates arrangement of biological protection for the external part of the heat-transfer system circuit. Sodium isotopes have short decay periods, but in case of metal contamination with any impurities isotopes with longer decay periods are generated. The task of radiation protection gets more complicated, and arrangement of the design enabling to remove all liquid metal from the system in case of its repair is required. The above-mentioned circumstances encourage imposing of strict requirements for chemical purity of liquid metals.

Slight but physically significant moderation of neutrons with sodium nuclei necessitates usage of the cores with the compact grid of fuel rods in the reactor thus considerably reducing the natural coolant circulation ratio required for cooldown of the reactor plant from any initial state. Besides, the compact grid of fuel rods results in significant power consumption for the coolant pumping.

The drawbacks and peculiarities of the sodium coolant listed above encourage to search for liquid metal having the above-mentioned advantages but free of the sodium drawbacks.

Lead and bismuth coolant. Eutectic alloy of lead and bismuth with the melting temperature of ~ 125°C and the operating temperature range of 200–650°C [33] was the first candidate for this role. Interaction of this coolant with air goes on with low intensity and with formation of the oxide film on the liquid metal surface which blocks further development of the reaction. The coolant is almost inert to water thus completely eliminating any possibility for generation of explosive hydrogen amounts in the reactor circuit. Long-term experience in application of this coolant conclusively confirmed these important physical and chemical peculiarities and advantages. High boiling temperature (~ 1670°C) and high thermal capacity of the lead and bismuth coolant prevent any accidents related to departure from nucleate boiling and also solve the problem of the positive void reactivity effect associated with its boiling. Lower moderating capacity of heavy nuclei of the lead and bismuth coolant as compared with light sodium nuclei not only solves the problem of the positive void reactivity effect but also enables to use the wide grid of fuel rods in the core thus eliminating the restrictions for the natural coolant circulation ratio which plays an essential role in prevention of heat removal accidents, and at the same time enables to reduce power consumption for the coolant pumping considerably. One more important property of the lead and bismuth coolant, namely its capability to catch fission products (iodine, cesium, etc. except for inert gases) thus reducing the possibility and severity of radioactive material leakages into the environment is also worth to be noted. But the problems related to corrosion resistance of the structural steels and maintenance of the coolant purity in the course of long-term operation should be solved for transition from chemically reactive sodium to the noncombustible and explosion-safe lead and bismuth coolant, that is, the coolant technology problem shall be solved.

The initial stage in application of the lead and bismuth coolant when no attention was paid to the coolant technology issues was accompanied with tragic events. An accident on the nuclear-powered submarine with death of the crew members took place in 1968. Fuel rods melted because of the reactor core slag clogging, fuel entered the circulating coolant with the relevant consequences [33].

Causes and development mechanism of the emergency with partial slag clogging of the reactor core are well-studied and published. The main cause involved generation of significant amounts of lead and bismuth oxides in the course of repair and refueling works and their subsequent ingress to the reactor core with the coolant flow after repair and refueling. In this situation the reactor core acted as a filter for the coolant with all severe consequences. Nowadays it is obvious that the required scope of knowledge with regard to physical and chemical properties of the coolant and the relevant experience were absent, thus it was impossible to predict and prevent this negative page in the history of the lead and bismuth coolant application.

Within the subsequent period from 1968 to the present day fundamental changes have taken place in the heavy coolant handling culture and understanding of the mechanism of physical and chemical processes going on in the circulating liquid metal circuit in the course of long-term operation. The causes of emergencies at the initial stages of the heavy coolant application were analyzed, and the set of methods and means for prevention of similar accidents was substantiated.

The heavy coolant technology preventing reoccurrence of the above-mentioned events has been developed in Russia by the end of the XX century. The technology provides for implementation of the measures aimed to reduce or completely eliminate the possibility for accumulation of impurities and slag. Special-purpose filters are developed for continuous purification of the coolant from any metal impurities. In case of any emergency hydrogen regeneration is carried out upon appearance of the first signs of lead oxide-based slag formation in order to purify the coolant and to clean the circulation circuit surfaces. Protection of steels against corrosion is performed in accordance with the proven oxygen technology. In this case, the optimal oxygen concentration for formation of the anti-corrosion oxide film on the steel surface is continuously maintained in the coolant.

It should be noted that due to application of the advanced coolant technology methods and means no negative situations as happened in 1968 have been registered on any nuclear-powered submarines with the lead and bismuth coolant [33].

Heavy coolant technology. In general, the integral problem related to maintenance of the required coolant quality, the primary circuit purity and corrosion protection of steels has been solved at present. The set of measures associated with this problem was called "the heavy coolant technology."

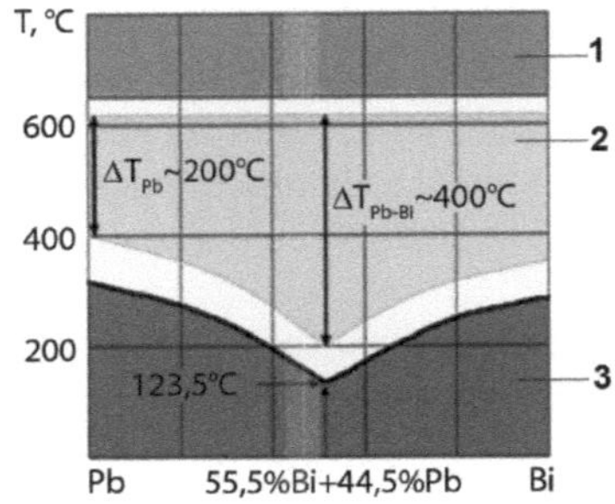

FIGURE 6.1 Equilibrium diagram and operating temperatures of Pb and Pb-Bi smelts as coolants: 1 – unassimilated zone; 2 – operating zone of the coolants; - transient zone; 3 – solid state zone.

Two main objectives of the heavy liquid-metal coolant technology for long-term operation (several dozens of years with the reactor plant operation with the power of up to 100%) may be specified:

1) Purity maintenance for the coolant and the circulation circuit surfaces in order to ensure the design thermal and hydraulic characteristics [40];
2) Prevention of corrosion and erosion of the structural materials.

The equilibrium diagram of the lead and lead-bismuth coolant is shown in Fig. 6.1. Coolants have the common upper operating temperature limit of ~ 650°C. Corrosion resistance of the steels used in the reactor core is justified at this level. In this case the introduced restrictions are related not to the average mixed temperature of the coolant at the reactor core outlet but to the temperature of individual "hot spots" on the fuel rod surfaces.

The lower operating temperature limit of the coolants has significant differences due to the different melting temperatures: ~123.5°C for lead and bismuth, ~327°C for lead. Assuming that the operating temperature of the coolants must exceed the melting point by 70–100°C, the minimum permissible coolant temperature is determined: ~ 400°C for lead, and ~ 200°C for lead and bismuth. For the same reason the operating temperature range of the lead and bismuth coolant is ~200°C (650–200°C = 450°C) wider than the operating temperature range of the lead coolant (650–400°C = 250°C).

It should be noted that the main processes of the lead-bismuth and lead coolant technology have common physical and chemical nature, as lead demonstrates the predominant chemical activity in Pb-Bi. Taking into account the fact that the structural materials for both coolants are identical the coolant technology methods and means are also similar in their action principle.

This brings up the question of whether transition to the lead coolant with the melting temperature which is 200°C higher is necessary if the lead and bismuth coolant is applied, and the acquired set of measures for the lead and bismuth coolant technology is also applicable to the lead coolant.

Two problems of the lead and bismuth coolant are usually specified as the reasons for selection of lead.

Firstly, it is the problem of potentially hazardous radioactive polonium (^{210}Po) generated mainly by bismuth irradiated with neutrons and the problem of its safe confinement. Almost all ^{210}Po is contained in the coolant in the form of plumbum polonide PbPo. Less than 1% of polonium is converted into the gaseous phase (aerosols), so it can be removed from the circuit through the use of the relevant filters. This problem is not so urgent under normal operation conditions and with the leak-tight circuit. The situation changes in case of any intercircuit leaks, loss of the primary circuit integrity in the course of planned maintenance, and refueling or emergency spillages of the radioactive coolant in the attended rooms.

The second problem is related to the high cost of bismuth and its limited resources that becomes an economic constraining factor in case of widespread implementation of nuclear power installations with the lead and bismuth coolant. This particular problem is determinative for selection of lead as coolant instead of Pb-Bi for development of large-scale nuclear power with power units of hundreds and thousands MW.

Comparison of coolants. Let us compare different coolants (water, sodium and heavy liquid metal or alloy) from the viewpoint of the potential energy amount related to the coolant which can be converted into the kinetic energy of structural material missiles and radioactive substances under certain conditions (Table 6.1). In the final count, the main potential danger of nuclear power installations is associated with the amount of radioactivity engaged in radiation damage of the humans and biosphere in case of any emergency release of all potential energy related to the coolant in any way. According to the definition, the thermodynamic risk of potential energy release is directly proportional to the energy itself.

It can be seen from the table that the relative thermodynamic risk for the water coolant is 20 times higher, and for the sodium coolant – 9 times higher than the risk in case of the heavy liquid-metal coolant use. Of course, the thermodynamic risk is rather a conditional value, and the risk determined in the PSA as the product of the potential damage caused by any accident and the probability of its occurrence should be used for adequate comparative assessment. But in the first approximation the thermodynamic risk provides rational and clear presentation of the relative danger of different coolants.

So, selection of lead as the coolant for nuclear power installations is due to:

- Availability of the substantiated lead coolant technology, that is, the set of measures and means for maintenance of the required coolant quality and purity of the primary circuit in the course of operation;
- Low potential energy related to possible chemical reactions with involvement of lead;

TABLE 6.1 Potential energy related to the coolant use.

Parameter	Water P = 16 MPa T = 300°C	Sodium P = 0.7 MPa T = 500°C	Lead and bismuth, lead P = 0.1 MPa T = 500°C
Thermal energy of the coolant, GJ/m^3	0.9	0.6	1.1
Potential energy of interaction, GJ/m^3			
Between the coolant and the fuel rod cladding with hydrogen release	11.4*	None	None
Between the coolant and water with hydrogen release	–	5.1	–
Between hydrogen generated in the course of reactions with the coolant and air	9.6	4.2	–
Stored potential			
Energy related to the coolant, GJ/m^3	21.9	9.9	1.1
Relative thermodynamic risk of the coolant use	20	9	1

**Steam-zirconium reaction.*

- Negligibly low moderating capacity of the lead nuclei solving the problem of the positive void reactivity effect of the coolant, on the one side, and enabling to use a wide grid in the core, on the other side, thus providing for the efficient natural circulation of the coolant in the core with simultaneous reduction of the coolant velocity (almost twice in comparison with the sodium coolants in the BN type reactors) and flow resistance of the circulation circuit, and consequently reduction of power consumption for the coolant pumping;
- Market availability of lead in comparison with bismuth, availability of the domestic production facilities and significantly lower cost (~4 USD/kg for lead, ~20 USD/kg for bismuth);
- Low activation of lead nuclei in the neutron flux (particularly with minimum generation of ^{210}Po) enabling to replace the vessel design with the pool one characterized by higher thermal capacity and to install the equipment in the concrete cavity lined with steel or iron compatible with lead, thus reducing the cost of the reactor plant and enhancing safety of the reactor in any transient and emergency processes due to heat retention of the circuit;

TABLE 6.2 Lead and sodium properties.

Property	Unit of measurement	Lead	Sodium
Melting temperature	°C	327.4	98.0
Boiling temperature	°C	1745	883.0
Fusion heat	kJ/kg	24.7	114.8
Heat of evaporation	kJ/kg	856.8	3871
Density (450°C)	kg/m^3	10520	845
Specific heat capacity (450°C)	kJ/(kg·K)	0.147	1.269
Volumetric heat capacity (450°C)	$kJ/(m^3 \cdot K)$	1546	1072
Thermal conductivity (450°C)	W/(m·K)	17.1	68.8
Kinematic viscosity (450°C)	m^2/s	$1.9–10^{-7}$	$1.4–10^{-7}$
Prandtl number Pr (450°C)	–	0.0174	0.0048
Volumetric change in the course of melting	%	+3.6	+2.65
Thermal neutron capture cross-section	barn	0.17	0.49

– High boiling temperature (~ 1745°C) of the lead coolant preventing any accidents related to departure from nucleate boiling (the lead boiling temperature in the reactor can reach 2300°C due to higher pressure in the core).

Some thermal, physical, and neutron-physical properties of the main liquid-metal coolants are presented in Table 6.2.

6.6 Primary circuit air heat exchanger for residual heat removal

The emergency heat removal system (EHRS) shall perform the double function. Firstly, it shall maintain the core integrity and operability in case of any initiating events resulting in loss of heat removal, and secondly, it shall perform the accident management function in case of any severe accidents (including the ones initiated by catastrophic external impacts) in order to reduce the core damage degree and to guarantee compliance with the nonresettlement criterion.

At present the problem still remains urgent as confirmed by the accident at Fukushima Daiichi NPP; in this case its relevance and complexity increase with the reactor capacity growth, and the possible engineering solutions usually represent the attempts to find a compromise between reliability and technical efficiency of the corresponding system and the expenditures for its implementation. The term "technical efficiency" mainly means the system capability to remove residual heat with the smallest system size possible.

Main components of reliability:

- Activation and operation are based on the passive principles;
- Independence;
- Structural diversity principle aimed to prevent the possibility for common cause failures;
- Self-protection in case of any accidents (maintenance of operability in case of any additional failures);
- Substantiation of the design solutions including justification of the parameters by calculations through the use of the advanced computational means, experimental substantiation through the use of special-purpose benches and consideration of the international experience.

On the other side, modern trends in nuclear power are aimed at reasonable reduction of expenditures, particularly for safety, but without any impairment thereof.

A wide range of possible engineering solutions for this task is known in the global and national practice. The problem of the search for the optimal emergency heat removal system design with regard to FNRs cooled with the liquid-metal coolant has a long history both in Russia and abroad. Residual heat removal from the reactor may be arranged in different ways [112], particularly:

a) Via the main heat removal system (the system is not independent):
   - With natural circulation in the primary circuit,
   - Through the use of auxiliary pumps (usually electromagnetic ones);

b) Via the separate independent heat exchange loop:
   - Through the use of emergency heat exchangers submerged into the upper reactor chamber (DRACS – Direct Reactor Auxiliary Cooling System),
   - Through the use of emergency heat exchangers installed within the intermediate heat exchanger volume (PRACS – Primary Reactor Auxiliary Cooling System),
   - Through the use of the loop "suspended" to the secondary circuit (IRACS – Intermediate Reactor Auxiliary Cooling System);

c) By arrangement of the reactor vessel air cooling (RVACS – Reactor Vessel Auxiliary Cooling System), blow-off of the primary and secondary circuit pipelines and the steam generator (this EHRS type can be applied for the reactors of limited power, or as an additional one).

While the IRACS scheme was most frequently used in the early designs of fast reactors, the scheme with submersible emergency heat exchangers in accordance with the DRACS option is preferred for the majority of the modern foreign reactor designs, as it is completely independent and based on the passive principles. Additional expenditures for development, substantiation and operation should be classified as drawbacks. The system of DRACS type is preferred in some foreign projects – the European Fast Reactor (EFR), the Chinese Experimental Fast Reactor (CEFR), the Indian demonstration reactor (PFBR), the Korean demonstration reactor (KALIMER) and the Japanese demonstration reactor (DFBR).

The Korean demonstration reactor KALIMER-600 [57] is a notable example of a foreign reactor where the structural diversity principle is implemented in the design for enhancement of its reliability and safety. The possibility for parallel residual heat removal via two independent systems operating in accordance with different principles is provided in this reactor. Two DRACS loops are arranged in accordance with the passive principle (emergency heat exchangers are submerged into the upper chamber, and heat removal in all circuits is performed due to natural circulation), and two IRACS loops are "suspended" to the secondary circuit (in parallel with the SG) and have the active component (a compressor) for arrangement of circulation via the air duct.

The third generation domestic FNRs BN-600 and BN-800 are equipped with the emergency heat removal systems of IRACS type. In the prototype reactor BN-350 residual heat removal was arranged via the main heat removal system but through the use of auxiliary electromagnetic pumps.

In the innovative Russian fast reactor designs BN-1200 and BREST-OD-300 heat removal from the core is also arranged directly via the primary circuit as in the majority of the advanced foreign designs, but in both cases there are differences from the "classic" variant of the DRACS type system.

A three-circuit four-loop "sodium – sodium – air" EHRS with straight-tube emergency heat exchangers located in the upper reactor chamber like in the DRACS system is provided for residual heat removal in BN-1200 reactor (Fig. 6.2). The difference resides in the fact that the emergency heat exchanger outlet is not open to the upper chamber, but connected to the pressure chamber via the pipeline with a hydraulic suspended check valve. In case of the flow rate decrease in the primary circuit up to a certain setpoint, the valve is opened in the passive way, and natural circulation via the line "emergency heat exchanger – pressure chamber – the core" is established.

Selection of this emergency cooldown system design option was the result of comparative analysis of various options from the viewpoint of the entire set of requirements imposed on the EHRS [142]. Particularly, in accordance with the comparative calculations [142] it is superior in efficiency as compared with the DRACS option. This design solution is based on the computational and experimental substantiation [189].

High efficiency of the EHRS is confirmed by the fact that the maximum cladding temperature T_{Max} is < 530°C with all four emergency heat exchangers in operation, while the operational limit is 650°C. The fact that cooldown is provided in the course of the EHRS operation with reduced strength gives

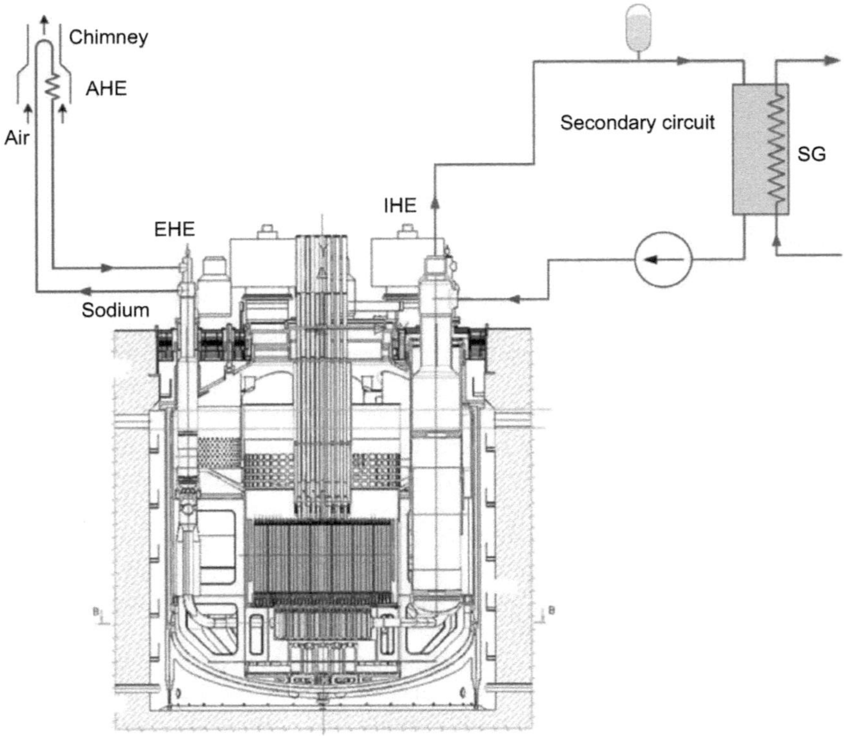

FIGURE 6.2 Heat removal system of BN-1200 reactor: AHE, air heat exchanger; EHE, emergency heat exchanger; IHE, intermediate heat exchanger.

evidence of high self-protection reserves against additional failures. T_{Max} is <555°C even with one emergency heat exchanger (out of four) in operation.

The fact that heat is removed with sufficient efficiency, and the fuel rod temperature remains within the permissible limits even in case of failure of all four check valves to activate stands out. The point is that there are two more heat transfer mechanisms in the emergency heat exchanger apart from direct sodium circulation via the emergency heat exchanger line, namely: thermal conductivity through the emergency heat exchanger walls and due to natural circulation in the area of inlet apertures. As it follows from the stated results, this mechanism can be efficient enough.

Emergency cooling systems of fast reactors. It is demonstrated above that the problem of residual heat removal for high-capacity sodium-cooled fast reactors is being solved successfully. Nevertheless, let us consider the potential ways to improve the emergency cooldown schemes for sodium-cooled reactors.

Refusal of a check valve and actual return to the "classic" scheme with emergency heat exchangers submerged into the upper chamber provides such a possibility for a high-capacity sodium-cooled reactor. The preliminary calculation analysis showed [189] that the "classic" scheme slightly inferior in heat removal efficiency also enables to ensure heat removal from the core, though with a bit higher level of temperature.

The emergency heat removal system of BREST-OD-300 reactor is a double-circuit one ("lead – air") and includes four independent loops. The primary circuit of the emergency core cooling system (ECCS) is a lead loop arranged inside the reactor and connected in parallel with the SG. The secondary circuit is an open air loop with gate valves for the air flow rate control; in case of any temperature increase they are opened in the passive way. Heat is transferred from lead to air via heat exchangers of the Field tube type. Movement of coolants in both cases is arranged due to natural circulation.

In the cooldown mode, the hot coolant from the core is supplied to the cooldown channels via the outlet nozzles, the steam generator and the primary circuit circulation pump. Here the coolant flows past the outer surfaces of the Field channels in the course of its downward movement, gives up the heat and then is supplied to the core inlet via special-purpose channels.

The complex branched coolant flow path inside the reactor and availability of large void spaces where stratification is possible in the cooldown mode should be mentioned as peculiarities of the selected cooldown scheme. Difficulty of replication for high-capacity reactors due to significant increase of the volume occupied by the cooldown channels inside the reactor because of the refrigeration capacity growth should be noted as the drawback of this design.

In general, the ECCS of BREST-OD-300 reactor is innovative in its design and has no national or foreign counterparts that impose special requirements for its substantiation.

The substantiation calculations were performed through the use of 1D engineering thermohydraulic codes RELAP-5 and DINAR; it was demonstrated that the temperatures in the reactor remained within the permissible limits upon activation of the cooldown mode. Within the framework of the self-protection reserve analysis it was demonstrated that the temperatures in the reactor also remained within the permissible limits even upon a failure of two ECCS loops out of four. The calculation analysis of transition into the cooldown mode performed through the use of SVIR 3D thermohydraulic code confirmed that even with due regard for stratification phenomena in the large primary circuit spaces the temperature of fuel rod claddings did not exceed the safe operation limits within the entire transient process duration.

The ECCS experimental substantiation program is focused on operability justification for the most important ECCS loop components. Experimental research of heat transmission in the "lead – air" heat exchanger of Field type confirmed the characteristics assumed for the design.

In general, the set of the performed calculations and experimental works confirms sufficient efficiency of the ECCS of BREST-OD-300 and its capability to perform its main function, particularly to remove residual heat from the core in case of any accidents without unacceptable overheating of the fuel rods even upon failure of two ECCS loops out of four.

Thus all emergency cooldown schemes proposed for solution of the residual heat removal problem in the advanced Russian fast reactors with liquid metal cooling have a certain degree of innovation in their design (the scheme with a check valve for BN-1200, the scheme for a high-capacity lead-cooled reactor, the double-circuit scheme with Field tubes for BREST-OD-300 medium-capacity reactor). Innovation imposes strict requirements for substantiation of the adopted solutions.

Calculations confirming the capability of the selected emergency cooldown system design options to provide residual heat removal without any exceedance of the permissible temperature level in the reactor were performed for both projects under consideration. Experimental programs aimed to confirm operability and efficiency of the individual cooldown system components as well as the entire system for BN-1200 have been developed and are being implemented for BN-1200 and BREST-OD-300 projects.

6.7 Reactor designs preventing heat removal accidents

All severe heat removal accidents occurred because of the primary circuit integrity loss outside the reactor vessel. In 1960s, N.A. Dollezhal proposed to use a single-block reactor design for nuclear-powered submarines particularly to prevent such accidents. In this design all primary circuit components (the core, steam generators, coolant pumps, pressurizers, etc.) are installed in the single casing of the steam generating unit, and no coolant loss occurs in case of any damage of the primary circuit wall. The integral reactor plant designs as they are called at present implemented with water coolant have been further embodied in the designs of plants with liquid-metal coolants (BREST-OD-300 and BN-1200) where the advantages of this engineering solution are enhanced with the above-mentioned advantages of the coolants.

Chapter 7

Codes for development and safety analysis of reactor plants

7.1 Design codes

Brief characteristics of the codes initially used for the design safety analysis of BREST-OD-300 and BN-1200 reactor plants are given in Tables 7.1 and 7.2. They were called design codes as they were developed prior to implementation of the Federal Special Program "New Technology Platform for Nuclear power" (2010) and included into this program due to the necessity of verification and validation.

At present all listed codes are verified, and the majority of them are validated by Rostechnadzor.

7.2 New generation codes

The Russian Federation as the world leader in the area of nuclear power technologies with fast reactor plants shall have its own software for safety analysis at all stages of the NPP lifecycle, including RW and SNF handling, with due regard for the achievements in the area of theoretical knowledge, experimental measurements, and the capacities of computational tools. Besides, the innovative plants developed within the framework of Proryv Project have no counterparts in the global practice and require improvement of the federal regulations and rules for these facilities. New generation codes shall be developed in order to cope with the intrinsic difficulties of scientific and technical nature in compilation of the regulatory documents.

The term "new generation code (software tool)" means commercialized alienable software product with the basic characteristics specified in Tables 7.1 and 7.2.

Main characteristics of the new generation codes:

- Usage of physical models providing for higher accuracy of the calculation results and covering greater number of phenomena as compared with the previous design codes;

Closed Nuclear Fuel Cycle with Fast Reactors: Handbook of Russian Nuclear Power.
DOI: https://doi.org/10.1016/B978-0-323-99308-1.00006-4

TABLE 7.1 Safety analysis design codes for BREST-OD-300.

Code name	Code purpose
FACT-BR	Calculation of the neutron and physical characteristics of BREST-OD-300 reactor and fuel burnup computational modeling
MCU-BR	Performance of precise calculations based on the Monte-Carlo method for the NPC of fast spectrum nuclear reactors with due regard for changes in the nuclide composition of fuel, materials of the control rods, the reflector, etc.
CASCADE/KATRIN	Calculation of the radiation parameters in 3D/2D geometry (the neutron flux density, the activation reaction rates, DDIR, neutron and photon dose rates, neutron and photon energy emission) for the reactor internals, the reactor vessel and biological shielding of BREST with the predetermined distribution of neutron and photon sources. Both specific and integral (by the volume of geometrical zones) values of the above-mentioned radiation parameters are calculated. CASCADE SW enables to evaluate 3D/2D effects of neutron and photon transfer that cannot be calculated through the use of 1D programs.
TARUSA	Calculation of accumulation and transfer of radioactive and stable actinide fission products in the sequential chain of zones under normal and emergency operation of the nuclear reactor with random independent time variation of the reactor power, the neutron flux density and all mass transport parameters that can be set individually for different chemical elements or groups of elements.
PUCHOK-ZhMT	Calculation of the local thermal and hydraulic parameters of the LMC in case of steady-state flow in the rod-type FNR fuel assemblies and determination of the maximum cladding and fuel rod temperature values, the azimuth irregularity of the cladding temperature and also the flow resistance of the fuel rod section on the basis thereof.
MASKA-LM	Modeling of 3D processes of heat and mass transport, generation and interaction of impurities in the lead coolant circuits: fluid dynamics of the lead coolant; heat transfer and heat exchange in the reactor; surface and volumetric sources of impurities; interaction between the impurity components; diffusion and convective transport of impurities; deposition of impurities; reduction of oxides and coolant purification.

(*continued on next page*)

TABLE 7.1 Safety analysis design codes for BREST-OD-300—cont'd

Code name	Code purpose
Module for modeling of the processes in case of any SG tube rupture in the lead coolant	Models enabling to predict the real accident development pattern, including investigation of the possibility for steam ingress into the reactor core and the possibility for breakage of the adjacent tubes because of the hydrodynamic impact
PLATFORM-BREST	Integrated software packages performing both neutron and physical and thermohydraulic calculations with different pre- and postprocessors as well as visualization systems.
FACT-BREST Integral Dynamic Code	3D neutron and physical calculation of steady states, fast and slow transient processes in the reactor core with due regard for functioning of the neutron power control systems and the fuel and coolant temperature feedbacks.

Note. DDIR, damaging dose intake rate; FNR, fast neutron reactor; LMC, liquid-metal coolant; NPC, neutron and physical characteristics.

- Application of the advanced numerical calculation methods and programming technologies;
- Use of the material property databases based on the state-of-the-art experimental and/or calculation data;
- The possibility to perform efficient calculations through the use of high-performance computational systems;
- Availability of the user-friendly interface;
- Automated link to the design model of the item (for the 3D new generation codes).

Intensive development of the new generation codes for reactor plant also takes place abroad as evidenced by significant increase in the number of studies dedicated to development and verification of the integral codes for safety analysis and substantiation of fast reactor with liquid-metal coolant [48,115,120,141,144,150,165,219,221]. Modeling is carried out in 1D, 2D and 3D approximation depending on the code development objective and stage. Gradual complication of models and interactions between the modeled processes takes place in the course of code development and verification.

Requirements for the new generation computer codes within Proryv Project Framework. It is necessary to develop the system of codes for analysis of the processes taking place at a NPP, including the reactor plant, the NPP equipment and rooms, the fuel fabrication/refabrication module (fuel production and refabrication), the site territory and the adjacent area, in consideration of the effect of any emergency and/or regular releases and discharges on the personnel, the public and the environment.

TABLE 7.2 Safety analysis design codes for BN-1200.

Code name	Code purpose
CONSYST/BNAB-RF	Constant support system with BNAB-RF library
TRIGEX	Package of programs for calculation of the main NPC of fast reactors and their models with the zero-power critical assemblies under the predetermined values of dimensions, compositions and temperature of the structural components for these plants.
JARFR	Package of programs for calculation of the main NPC of fast reactors and their models with the zero-power critical assemblies under the predetermined values of dimensions, compositions and temperature of the structural components for these plants.
MMK	Multi-group code of the Monte Carlo method for calculation of the following functions: the effective multiplication factor, the reaction rates and balance tables, the power density field, DDIR, burnup calculations, lifetimes, the effective delayed neutron fraction, reactivity coefficients, sensitivity coefficients, calculation errors.
ALPHA-M	Modeling of the processes for release of gaseous, volatile and solid fission products from fuel, their migrations in the primary circuit coolant, release of corrosion products from steel and formation of deposits on the equipment surfaces
VYBROS-BN	Calculations for assessment of the radiological consequences of persistent and nonrecurrent releases from NFs both under normal operation conditions and in case of any nuclear and radiation accidents.
DINRUN	Modeling of the transient and accident processes in FNRs related to emergency heat removal, substantiation calculations and optimization of the EHRS characteristics and their control algorithms
ANPEX	Calculation of reactivity-related accidents in FNRs caused by the uncontrolled reactor runaway with prompt neutrons
INTERACT	Calculation of the thermal interaction processes between melted fuel and sodium, pressure changes and sodium movement in the reactor due to this interaction.
BRUT	Calculation of severe accidents with the core melting in sodium-cooled FNRs and determination of the following parameters: depth of penetration for the structures; sodium temperature; temperature of the heat-generating layer.
DINROS	Software package for calculation of the transient and accident processes in a multi-loop three-circuit nuclear power installation based on a FNR with single-phase sodium coolant
BOS-TWC	Calculation of accident processes in the reactor accompanied with sodium boiling in the core

(*continued on next page*)

TABLE 7.2 Safety analysis design codes for BN-1200—cont'd

Code name	Code purpose
TWOCOM	Calculation of operational occurrences related to loss of integrity of fuel rod claddings and release of GFP into sodium and also calculation of sodium boiling processes in the core
GRIF	Single-phase integrated thermohydraulic code for calculation of the dynamics of thermal and hydraulic parameters in a nuclear reactor with liquid-metal coolant both in steady-state and transition modes
COREMELT	Software package for combined multi-dimensional calculation of neutron, physical and thermohydraulic processes in the reactor based on the multi-component, multi-rate, thermally nonequilibrium model
TR-BN	Calculation of the static parameters (temperature and flow rates) of the reactor plant with a three-circuit heat transfer scheme from the reactor to the turbine generator under forced circulation of the media
RATO	Calculation of a 2D shell-and-tube heat exchanger
UMB BK UPAKS	Calculation of damages accumulated in the structural materials of the primary and secondary circuit equipment and pipelines of BN-1200 reactor plant in the course of operation
DRAGON	Numerical modeling of temperatures and SSC in the rod-type cylindrical FNR fuel rods with dense U-N and mixed U-P-N fuel and LMC operated in quasi-steady modes

Note. GFP, gaseous fission products; NF, nuclear facilities; SSC, stress-strain condition; U-N, Uranium Nitride; U-P-N, Uranium Plutonium Nitride.

The term "system" means the set of codes containing the required physical and mathematical models for computational support of design, engineering, operation and safety analysis of reactor plants and their fuel cycles complying with the uniform requirements for the calculation arrangement procedure, the format of the source data entry and output of the calculation results, using the common database on the properties of materials and coolants.

A new generation code system shall model the processes taking part at a NPP both under normal operation conditions and in case of any operational occurrences including accidents (design basis and beyond design basis ones), particularly severe accidents with low probability: positive reactivity insertion (UTOP), loss of power supply (with subsequent loss of forced coolant circulation) with failure of the reactivity control devices (ULOF), clogging of the fuel assembly flow area (TIB). In this case the codes shall provide for modeling of the determinative physical processes thus enabling to use them not only for the particular plants but also for different plants of the same type. Besides, the possibility for automated mass calculations with uncertainty and sensitivity analyses shall be implemented.

One of the main requirements for the advanced computer codes prescribes their integrity in order to provide the opportunity for interrelated end-to-end calculations of different NPP operation modes, from the behavior of a fuel pellet to the effect on the environment and the public. In order to solve this problem according to the project "New Generation Codes" individual modules intended for modeling of certain groups of physical processes are developed at first, and then they are united into integral software packages. In this case it is important to take into account the requirement for performance of calculations within the time period acceptable for engineers, designers and operating organizations, including multivariant calculations for uncertainty and sensitivity analyses.

The system of computer codes shall be aimed to the possibility for its application independent from the developers, particularly by any supervisory authorities for safety analysis and assessment of compliance with the regulatory requirements for the NPP safety.

In accordance with the peculiarities of NPPs with sodium- or lead-cooled reactor plants special attention in development of computer codes shall be paid to modeling of the following processes:

- Behavior of mixed uranium-plutonium nitride fuel with due regard for time-dependent evolution of the nuclide composition, release of gaseous fission products, thermochemical transformations, and changes of the fuel structure;
- Oxidation and reducing reactions and diffusion processes on the fuel rod claddings, the pipeline walls and other surfaces of the primary circuit of the lead-cooled reactor plant, including formation of the oxide layer, dissolution of plumbum oxide particles, changes in oxygen activity, transport and deposition of polydisperse corrosion product particles in the coolant flow, interaction between fission products and impurities in the coolant, formation of deposits;
- Heat and mass transport in liquid-metal coolant including formation of dead zones;
- Heat and mass transport in the gas cavities filled with inert gas (argon) with due regard for dissolution and transportation of gas with liquid-metal coolant;
- Flow of the coolant with gas phase inclusions, for example, in case of any intercircuit leaks of the steam generator and severe accidents with breakdown of fuel rods;
- Dynamics of the heat emission field across the core (with due regard for uncertainty of the fuel composition supplied from the closed NFC to the reactor);
- Behavior of the neutron field in case of any cavity formation in the core or on its boundaries;
- Residual heat emission in the fuel-containing spaces;
- Attenuation of ionizing radiation determining the mass and dimensional parameters of protection and radiological characteristics of the structural materials;

- Thermomechanical parameters of fuel rods and fuel assemblies with due regard for peculiarities of new structural materials;
- Sodium fires;
- Behavior of sodium aerosols and fission products.

The critical operations of MNIT (or MOX) fuel fabrication modules, SNF processing with the combined technology and RW handling shall be selected for mathematical description of the processes at the marginal stages of the fuel cycle. In modeling of these processes it is necessary to take into account the impact of various factors - from the operating mode parameters (temperature, pressure, the working medium flow rate, etc.) and the material flow characteristics (composition, density, phase state, etc.) to the structural parameters of the equipment where these processes are implemented.

Composition and development status of the new generation code system within the framework of Proryv Project. The developed codes for modeling of physical processes and phenomena are presented in Table 7.3. The most advanced foreign counterparts enabling to align with the global trends in the area of computer modeling are defined for each of them.

The list of new generation codes provided in Table 7.3 enables to model the reactor plant behavior, the heat and mass transport and fission product dissemination processes in the NPP rooms and the environment, as well as the closed NFC processes to the full extent, to solve the task of its optimization, to perform the probabilistic safety analysis at all lifecycle stages of nuclear installations and other nuclear facilities.

The advanced software product CRISS 5.3 for modeling and analysis of safety systems and nuclear installations in general in the course of probabilistic safety analysis at all lifecycle stages of nuclear installations and other nuclear facilities was developed in the project "New Generation Codes" within the framework of Proryv Project. The code does not impose any restrictions on the size of logic and probabilistic modes, enables to analyze complex models of large size with high speed, eliminates any errors in consolidation of individual models of systems and emergency sequences into the integrated model of the nuclear installation and enables to use universal models of common cause failure consideration for highly redundant systems.

The codes for description of the nuclear fuel cycle closing processes are united into a separate group. In particular, VISART software package (Virtual Radiochemical Technology Plant) performs balance calculations in order to determine flow rates and characteristics of all process streams, accumulation of fissionable and nuclear materials in the production units and intermediate storage tanks as well as to plot the equipment operation cyclograms. In accordance with the system modeling principles equipment or equipment groups are considered as parts of a complex system, and the connections between them - as substance streams [35]. A model (balance module) is assigned to each equipment unit included into the system and enables to determine the flow parameters at the

TABLE 7.3 Modeling codes for individual groups of physical processes and phenomena.

Code and developing organization	Brief characteristics	Development status	Most advanced foreign counterparts
MCU-FR (Kurchatov Institute)	Neutron and physical code based on the Monte Carlo method for calculation of the reactor NPC, protection and the CNFC facilities	Development completed. Verification is in progress. Submitted for validation	MCNP-5, 6, KENO (USA)
ODETTA (the Nuclear Safety Institute of the Russian Academy of Sciences, Keldysh Institute of Applied Mathematics)	Neutron and physical code based on kinetic approximation on the basis of the finite element methods and discrete ordinates for calculation of the core, protection and the CNFC facilities	Verification is in progress. Submitted for validation for the protection purposes	ATTILA (USA)
NDP-ACE (Kurchatov Institute)	Nuclear data processing program	Under development	NJOY (USA)
HYDRA-IBRAE/LM (the RAS Nuclear Safety Institute of the Russian Academy of Sciences, Afrikantov OKBM JSC, NIKIET JSC, National Nuclear Research University MEPhI)	Systemic thermohydraulic code enabling to model all coolant circulation circuits of NPPs with fast neutron reactor plants, including the porous body models and CDF-modeling of individual RP components	Validation of the first version (without multi-dimensional models)	RELAP5-3D (USA)

(continued on next page)

LOGOS (FSUE RFNC VNIIEF)	CFD-code with RANS-models of turbulence for detailed investigation of the local heat and mass transport processes in the RP equipment components and rooms of NPPs with fast neutron RPs	Submitted for validation	STAR CCM+ (Germany), ANSYS FLUENT, ANSYS CFX (USA), OpenFOAM (France)
CONV-3D (Nuclear Safety Institute of the Russian Academy of Sciences, Keldysh Institute of Applied Mathematics)	CFD-code based on DNS approximation for detailed investigation of the local heat and mass transport processes in the RP equipment components and NPP rooms	Verification, submitted for validation	Nek5000 (USA), Trio (France)
BERKUT (Nuclear Safety Institute of the Russian Academy of Sciences, VNIINM JSC, IPPE JSC)	Code for substantiation of fuel rod behavior under normal and emergency operating conditions of fast reactors	The first version (engineering models of fuel behavior) is submitted for validation.	CEPTAR (Japan)
–	Thermomechanics: Behavior of the core components and the support grid, the pressure vessel, pipelines, valves in the temperature and radiation field with due regard for changes in the properties of structural materials over time, strength calculations	Under development	MSC.MARC (USA)

(*continued on next page*)

TABLE 7.3 Modeling codes for individual groups of physical processes and phenomena—cont'd

Code and developing organization	Brief characteristics	Development status	Most advanced foreign counterparts
KUPOL-BR (IPPE JSC)	Modeling of the fission product dissemination processes as well as heat and mass exchange in the NPP rooms	Verification, the first version is submitted for validation	COCOSYS (Germany) CONTAIN/LMR (USA)
ROM (Nuclear Safety Institute of the Russian Academy of Sciences)	Assessment of radiation situation in the course of atmospheric transport	Submitted for validation	MINERVE-SPRAY (France), ADPIC (USA)
ROUZ (Nuclear Safety Institute of the Russian Academy of Sciences)	Three-dimensional calculation of radiation situation at the site	Submitted for validation	FEM3MP, MISCAM (USA)
Sybilla (Nuclear Safety Institute of the Russian Academy of Sciences)	Code for calculation of exposure by water	Validation is near completion	PC-CREAM 08 (England)
GeRa (Nuclear Safety Institute of the Russian Academy of Sciences, the RAS Institute of Computer Modeling)	Safety analysis for disposal of all types of conditioned radioactive waste from the CNFC process stages	The first version is submitted for validation	Processing Modflow, TOUGH2 (USA), FEFLOW (Germany)
CRISS 5.3 (Afrikantov OKBM JSC)	Probabilistic safety analysis and risk assessment for power units	Validation is near completion	Risk and Reliability (R&R) Workstation, Systems Analysis Programs for Hands-on Integrated Reliability Evaluations (SAPHIRE), WinNUPRA (USA), RiskSpectrum (Sweden), FinPSA (Finland), RiskA (China)

(*continued on next page*)

VISART (FSUE Zababakhin RFNC VNIIEF)	Calculation of material balance and nuclide flows for the process diagrams of the marginal NFC stages and individual production sites, particularly dynamic calculations with due regard for capacity of the equipment, transportation lines and volumes of storage tanks	Under development	SuperPro Designer, ShedulePro, EnviroPro Designer, Aspen Plus, Aspen Plus Dynamics, Aspen Custom Modeler, CHEMCAD Steady State, CC-Batch, CC-Dynamics, CC-Therm, CC-Safety Net, CC-Flash (USA)
TP CODE (Tomsk Polytechnic University, VNIINM JSC)	Code for simulation of the process diagram operation for a plant or individual production sites on the real-time basis in order to investigate operability, controllability and optimization of the processes, units and facilities as well as control and instrumentation systems for both individual components of the diagram and production lines	Under development	PAREX (France)

equipment outlet depending on the flow parameters at the equipment inlet. Balance modules are interconnected in accordance with the scheme set by the user. For dynamic balance calculations VISART code is supplemented with the functions for consideration of the equipment capacity, transportation lines, volumes of storage tanks, and duration of individual operations. Calculations with the use of this code enable prompt detection of any mismatches in the equipment operation (especially for the equipment operated in the periodic mode) and optimization of the production process, as well as assessment of the maximum simultaneous storage of nuclear materials in tanks and equipment units.

GeRa calculation code is developed to solve the set of tasks related to substantiation of radiation equivalence, selection of the places for subsurface and underground isolation of radioactive waste and disposal safety analysis [35,44]. The concept of its development provides for consideration of the maximum possible number of physical and chemical processes significant from the viewpoint of contamination spreading in the underground hydrosphere: saturated filtration (in the subartesian and phreatimetric conditions), saturated and unsaturated filtration, advective, diffusive, and dispersive transport with due regard for radioactive decay, including calculation of the radioactive transformation chains and sorption equilibrium, particularly with the variable distribution factor, transport with detailed calculations of chemical reactions in the water-ground system and density convection.

The new generation codes are already used to solve the applied problems within the framework of Proryv Project. Material flow balance calculations and equipment operation cyclograms developed through the use of VISART code are used to prepare the source data for design of the process lines and equipment of SNF processing and fuel fabrication modules. Dynamic calculations enable to optimize the number of equipment units for each operation in order to ensure smooth functioning of the entire process line, the volume of intermediate storage tanks for accumulation of products in case of different capacity of the adjacent equipment units. Dimensions of the equipment units or structural components and their service life characteristics (for example, frequency of filter replacement in the gas treatment system) are determined through the use of physical and chemical models of individual processes.

Assessment of the exposure doses for the public in case of certain accident scenarios at the designed reactor plant BREST-OD-300 was performed with the use of ROM code. It was demonstrated that there was no need for evacuation of the population in the scenarios under consideration. ROM code with the Lagrange trajectory model enabled significant reduction of the calculation conservatism in comparison with the applied design codes.

The integral codes for safety analysis of NPPs with fast neutron reactor plants currently under development are presented in Table 7.4.

TABLE 7.4 Composition and development status of new generation integral codes.

Code (developing organization)	Brief characteristics	Development status	Most advanced foreign counterparts
SOCRAT-BN (the RAS Nuclear Safety Institute of the Russian Academy of Sciences, Africantov OKBM JSC, TRINITI JSC)	Integral computer code for safety analysis of NPPs with sodium-cooled reactor plants	Validation of the first code version is near completion. The second version is submitted for validation	SIMMER-III (Japan), SAS4A (USA)
EUCLID (Nuclear Safety Institute of the Russian Academy of Sciences, Afrikantov OKBM JSC, NIKIET JSC, National Nuclear Research University MEPhI)	Universal integral computer code for safety analysis of NPPs with fast neutron reactor plants	The first version is submitted for validation	SAS4A/SASSYS-1 (USA), SIMMER IV (Japan)
–	Precision thermohydraulic, thermomechanical, neutron and physical calculations of the fast reactor core characteristics.	Commencement of development in 2019	ASTERIA-FBR (Japan)

The new generation integral codes presented in Table 7.4 enable to model behavior of fast neutron reactors plants developed within the framework of Proryv Project to the full extent.

As technology unification and compatibility of the code system components along the entire chain from modeling of the processes in the core to the impact on humans and the environment on the basis of advanced computational technologies are taken as the principle for the approaches adopted for development of the new generation codes development of the integral codes is considerably simplified. In particular, EUCLID integral computer code was developed by consolidation of individual new generation codes within a short period of time. The code uses advanced physical and mathematical models, for example, neutron and physical processes are described not only in the diffusive, but also in the kinetic approximation. The code includes the models enabling to describe evolution of the mixed uranium-plutonium nitride fuel micro-structure, intra- and intergranular transport of fission products and release into open porosity, thermal and chemical mutual transformation of fission products and their distribution by phase states. The Lagrange trajectory transport model is used to assess the radiation situation around the NPP in case of any emergency or regular releases of radioactive substances into the atmosphere (in contrast with Gaussian models widely used in the previous design codes) taking into account real three-dimensional distribution of the meteorological parameters within the boundary atmospheric layer.

Precision software tools MCU-FR and CONV-3D are not included into EUCLID integral code and are used independently for verification and adjustment of the models based on simplified methods. In particular, CONV-3D new generation scalable CFD-code on the basis of DNS approximation is used to calibrate RANS-models of turbulence and to obtain the closing relations for systemic thermohydraulic codes as well as for cross-verification and testing of the above-mentioned codes [82,151]. Automated grid generation from solid-body CAD formats is provided in CONV-3D code.

New generation integral codes as well as the codes for modeling of individual groups of physical processes and phenomena are already used to solve the applied problems within the framework of Proryv Project. In particular, integrated residual risk accident calculations for a NPP with BN-1200 reactor plant (insertion of positive reactivity with failure of all active and passive reactor protection control systems) were performed with the use of SOCRAT-BN/B2 code. Thermal and hydraulic processes in the circuits including sodium boiling, neutron and physical processes in the core, accumulation and release of fission products from fuel into the gas gap, thermal and mechanical aspects of fuel and fuel rod cladding, physical and chemical interaction between fuel and coolant, release and dissemination of fission products in the primary circuit, release of fission products into the NPP rooms and the environment, transport in the environment, and exposure doses for the public are modeled. It is demonstrated that any release of fission products into the environment does not result in the public exposure doses requiring evacuation or resettlement.

Modeling of the operation mode caused by the initiating event with rupture of steam generator tubes was performed for the designed reactor plant with lead coolant and mixed uranium-plutonium nitride fuel through the use of EUCLID/VI code. It was demonstrated that steam and water mixture did not enter the core. Absence of the required models in domestic and foreign codes prevented from performance of such calculations in the past.

Approaches to verification and validation of new generation codes. The capability of a software tool to model the processes and phenomena taking place in certain facility operation modes with the definite error evaluated on the basis of validation and verification calculations shall be confirmed in order to apply computer codes for safety analysis of any nuclear facility.

For this purpose, the verification matrix containing the list of the nuclear facility operation modes modeled by the computer code, the list of processes and phenomena to be modeled with their ranking in accordance with the degree of significance and the experimental data sufficient for the computer code validation and compilation of the verification report in compliance with the requirements of Rostechnadzor (RD 03-34-2000) shall be developed at the early stages.

Verification is carried out by stages on the basis of analytical tasks and computational solutions, experiments on individual phenomena and integrated experiments. It should be noted that due to innovation and absence of any counterparts for the plants developed within the framework of Proryv Project verification of codes at nuclear facilities can be performed only subsequent to completion of their construction and commissioning.

Users independent from the developers shall be inter alia engaged in verification works. It will enable to control alienability of the computer codes under development at the early stages.

In case the experimental data are not sufficient for the code validation and substantiation of calculation errors for individual parameters the requirements for additional experimental research shall be compiled, and the objectives for performance of calculations with the use of reference (precision) or validated codes for these parameters shall be stated. It should be noted that development of precision computer codes such as the mechanistic fuel module of BERKUT fuel rod code and CONV-3D CFD-code of DNS class requires special-purpose experimental research to obtain individual microscopic parameters or local flow characteristics respectively.

As the requirement for compliance with the international level is imposed on the new generation codes participation in international projects and comparison of the calculation results with foreign counterparts on the basis of benchmark tasks or experimental data is an important stage.

Approaches to team development of the new generation code system. Development of computer codes complying with the requirements of the modern software development methodology defines the necessity to use the lifetime management means providing the unified repository for the data on the computer code, systems for management of the requirements, version and testing control, arrangement of coordination between the developers and users. For this purpose, the computer code development management system is developed and implemented in the project "New Generation Codes" within the framework of Proryv Project, and the set of procedures followed at all stages of the computer code development from compilation of the requirements to its implementation for commercial use is established [44].

Iterative development takes place in implementation of the computer code development technology when the development phases are repeated at each iteration, the development priorities are re-evaluated upon its completion, and then the next iteration begins. Software is tested and partially verified at each iteration. When this approach is applied a version of the ready-to-use product is obtained upon completion of each iteration. Further expansion of the functional capabilities takes place at the next iteration.

Thus, arrangement of nuclear power on the basis of the new technology platform based on fast reactors with the closed NFC necessitated development of the new generation code system taking into account the peculiarities of the designed innovative facilities. Development of the fundamentally new codes, for example, for description of the NFC closing processes, was also necessary together with development of software for traditional groups of physical processes (particularly for thermal reactors) taking into account the peculiarities of NPPs with fast neutron reactor plants (new coolants, structural materials, fuel, SNF handling approaches, etc.).

The system of computer codes with different degree of complexity enabling to solve the main safety analysis problems for NPPs with fast reactors and the closed NFC was successfully developed in the project "New Generation Codes" within the framework of Proryv Project for the period of six years. Both integral codes compiled due to connection of the individual new generation code system components and independent precision computer codes intended inter alia for partial replacement of experimental research were developed.

Development of the new generation codes became possible due to consolidated efforts of Rosatom organizations, the Federal Agency for Scientific Organizations, national centers and higher education institutions; their efficient joint works were performed with the use of the team computer code development management system.

Calculations performed by now with the use of the new generation codes enable to make the conclusion that substantiation calculations are carried out at a totally new level both due to elimination of excessive conservatism and consideration of the mutual impact of processes, and performance of more calculations, particularly with uncertainty and sensitivity analysis.

Chapter 8

SNF and RW handling as a risk factor for the public

Unique high-specific energy release in nuclear fuel results in relatively low amounts of generated waste with regard to mass characteristics. It is enormously lower as compared with traditional hydrocarbon, let alone coal-fired power generation. CO_2 emissions alone amount to millions of tons, ash waste of coal-fired TPPs are massive, while approximately 370 thous. tons of irradiated nuclear fuel (SNF) are accumulated worldwide for the entire period of nuclear power existence. In this case the total volume of SNF is much less than for ash waste of the same weight due to high specific density. The amount of SNF in Russia is near 25 thous. tons [85].

Maybe, that is why it took some time for the scientific community to recognize the SNF problem as one of the most urgent for nuclear power technologies. It was expected that sooner or later it would be possible to bury SNF in reliable underground repositories. Many nuclear countries supposed to follow this path. One has only to recall the SNF disposal projects in Gorleben (Germany), Yucca Mountain (USA), Östhammar (Sweden), or Onkalo (Finland). But solution of this problem is being constantly postponed in all countries without any exceptions, increasingly passing on to the next generations. And the requirements for the disposal conditions are getting more and more stringent. While safety substantiation for 10 thous. years was required in the USA in the past, nowadays safety should be substantiated for 1 mln. years. It has already become one of the most significant factors of public distrust of nuclear technologies.

It is obvious that the SNF and RW problem is first of all related to the nuclear, radiological, toxicological and environmental aspects of their disposal safety assurance within a long period of time (hundreds of thousands years) rather than to their amount. It is almost impossible to substantiate safety of any technical facilities or techniques for such a long perspective beyond reasonable doubt. In this case it should be particularly noted that the proposed proofs shall be addressed not only and not so much to the scientific and technical community as to the general public because their consent is necessary to make any decision in the area of environmental protection.

Closed Nuclear Fuel Cycle with Fast Reactors: Handbook of Russian Nuclear Power.
DOI: https://doi.org/10.1016/B978-0-323-99308-1.00038-6

Taking this into account the way to solve the problem not by arrangement of multiple reliable safety barriers in the radioactivity spreading paths but by its hazard reduction down to the permissible level was described in the studies [31,39,126,193]. These ideas are consistently developed and particularized in these studies facilitating their technical implementation in the nearest future.

8.1 Radiation-equivalent RW management principle

The fundamental provisions of the proposed RW handling principle are the following:

- Recognition of the difficulties in arrangement of long-term repositories with nuclear- and radiation-hazardous substances maintaining the safety barriers and preventing any unauthorized entry of nuclear- and radiation-hazardous substances into the human habitat (except for professional activities of nuclear specialists having voluntarily selected their line of work) and also eliminating illegal proliferation of nuclear materials;
- In view of this – elimination of the possibility for disposal of the main nuclear fuel components (enriched uranium and plutonium), and consequently
- The need for SNF processing and return of fissionable materials to the fuel cycle;
- Reduction of hazard for the generated radioactive waste to the acceptable level within the time period permitting their controlled storage in the area of professional nuclear human activities with the possibility to substantiate sufficient reliability of the safety barriers;
- Selection of the biological hazard level of extracted fuel raw materials (uranium and thorium together with their related decay products) as the hazard acceptability criterion for the radioactive waste returned into the environment (outside the professional activities of nuclear specialists);
- In assessment of the relative hazard of waste and raw materials it is necessary to take into account the difference not only in their composition but also in the migration capacity in the disposal location environment.

Thus the natural radiation balance is maintained within the framework of the radiation and migration equivalence concept for fuel raw materials and RW based on the acceptability of RW disposal only subsequent to the balance between (potential) biological hazard of buried radioactive waste ($G_w = G_w{}^{(sp)} \cdot m_w$) and potential biological hazard of the relevant amount of extracted fuel raw materials ($G_{rm} = C_{rm}{}^{(sp)} \cdot m_{rm}$). It is important to note that this balance shall be achieved within the acceptable controlled storage time limits ($t_{cst} < 400$ years).

The basic radiation and migration equivalence condition for RW disposal apparently has the following form:

$$G_w{}^{(sp)} \cdot m_w / {}_{rm}{}^{(sp)} \cdot m_{rm} \leq 1.$$

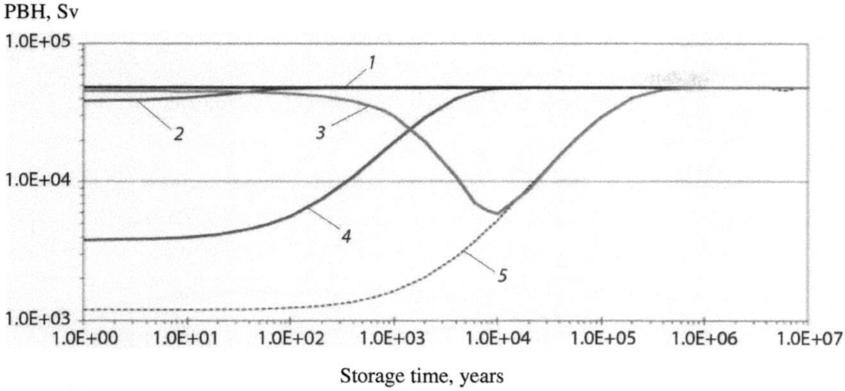

FIGURE 8.1 Potential biological hazard of 1 ton of natural uranium with and without decay products: 1 – uranium + thorium + radium + lead; 2 – uranium + thorium + radium; 3 – uranium + radium; 4 – uranium + thorium; 5 – uranium.

Biological hazard. It may be assumed in the first approximation that biological hazard of the i-th radioactive nuclide G_i for humans is directly proportional to:

- Its radiation activity (A_i);
- Its migration activity (B_i), that is, the probability of migration from the geosphere to the biosphere;
- Intensity of its contact with the human biological tissues (C_i);
- Its capability to accumulate in them (D_i);
- Energy released in the course of the radioactive nuclide decay (H_i);
- And its destructive biological effect (F_i).

So:

$$G_{ii}\tilde{} \cdot {}_i \cdot C_i \cdot D_i \cdot {}_i \cdot F_i = {}_i \cdot {}_{i(rad)} \cdot {}_{i(mig)},$$

where $K_{i(rad)} = C_i \cdot D_i \cdot H_i \cdot F_i$ is to be called the radiation toxicity coefficient of the i-th nuclide, and $K_{i(mig)}$ – its migration coefficient.

The annual intake limits or dose factors may be used as the basis for numerical evaluation of the radionuclide toxicity for the public having no professional contact with radiation sources.

Mining of uranium involved into the fuel cycle is usually accompanied with extraction of ^{230}Th and ^{226}Ra – long-lived products of ^{238}U decay as well as ^{231}Pa – a long-lived product of ^{235}U decay. Potential biological hazard of 1 ton of natural uranium with due regard for the daughter products is significantly higher than for uranium itself (Fig. 8.1) and amounts to 48,230 Sv (1 ton of natural uranium contains 0.36 g of ^{226}Ra (T = 1600 years) and 16.9 g of ^{230}Th ($T_{1/2} = 7.51 \cdot 10^4$ years). This value is assumed as the standard for radiation equivalence assessment.

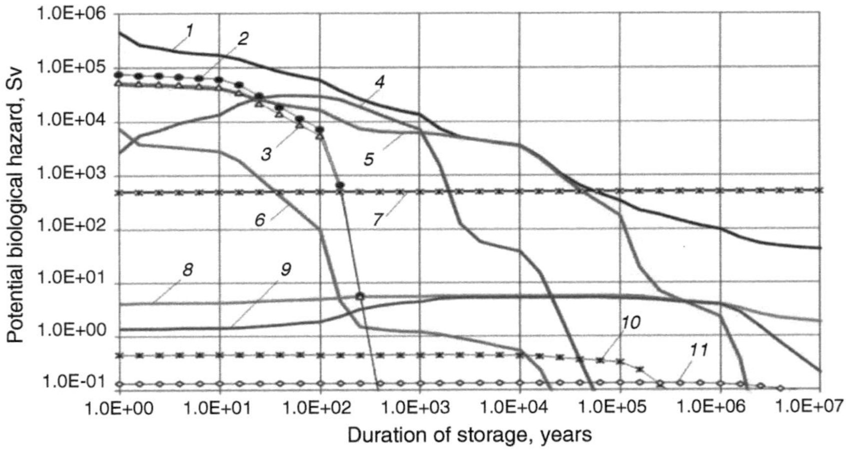

FIGURE 8.2 Contribution of individual elements and nuclides into potential biological hazard of 1 kg of VVER-1000 SNF (initial enrichment by ^{235}U – 4.4%, fuel burnup depth – 40 MW·day/kg of uranium): 1 – irradiated fuel; 2 – $^{90}Sr + ^{90}Y$; 3 – $^{137}Cs + ^{137m}Ba$; 4 – all americium; 5 – all plutonium; 6 – all curium; 7 – natural uranium, 8 – all uranium, 9-^{237}Np;10-^{99}Tc; 11-^{129}I.

At present long-lived decay products are not involved into the fuel cycle and shall be considered as RW subject to disposal right after extraction. Within the framework of the concept under consideration these products of natural uranium decay shall be involved into the fuel cycle and burnt (transmuted).

The results of the radiation component assessment for biological hazard of various radionuclides contained in spent nuclear fuel of a thermal (Fig. 8.2) and fast (Fig. 8.3) reactor are given below. It can be seen from these results that in case of plutonium disposal in SNF radiotoxicity of the waste is considerably higher than the corresponding value for natural uranium for dozens of thousands years. It is obvious that plutonium disposal is unacceptable for maintenance of the natural radiation balance. This problem is comfortably solved within the framework of the closed nuclear fuel cycle (CNFC) by complete plutonium combustion in the course of repeated recycling.

Nevertheless, the radiation-equivalence principle in handling of fissionable materials imposes additional restrictions for practical implementation of the CNFC. The problem is that complete prevention of any fissionable fuel material ingress into RW in the course of radiochemical SNF processing and subsequent fuel re-fabrication is impossible. Some part of any nuclides will somehow enter RW in the form of losses. Then the equation for RW hazard calculation may be presented as the component-wise sum:

$$G_w \cdot m_w = \sum_i \cdot K_{i(rad)} \cdot K_{i(mig)} \cdot \xi_i \cdot m_i$$

and the condition for radiation and migration balance of the disposed RW (index i) and extracted uranium with its daughter decay products (index j(U))

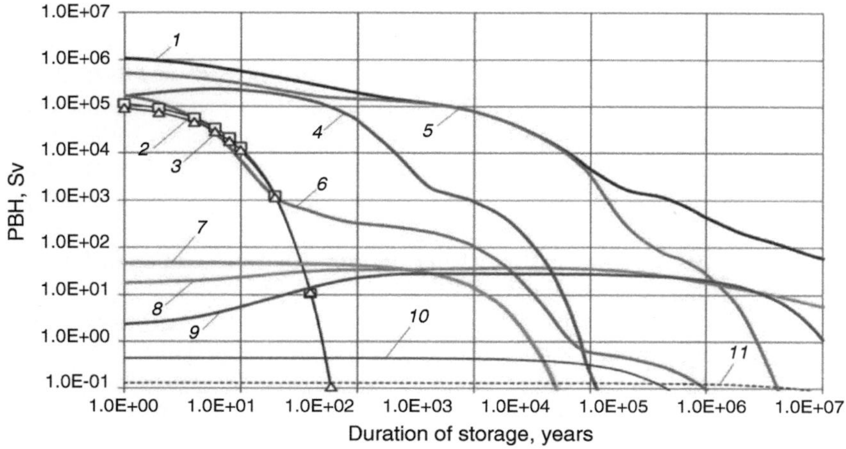

FIGURE 8.3 Contribution of individual elements and nuclides into potential biological hazard of FNR SNF (rated for 1 kg of irradiated actinides (1.06 kg of nitride fuel): 1 – irradiated fuel; 2 – $^{90}Sr + ^{90}Y$; 3 – $^{137}Cs + ^{137m}Ba$; 4 – all americium; 5 – all plutonium; 6 -all curium; 7 – ^{14}C; 8 – all uranium, 9 – ^{237}Np; 10 – ^{99}Tc; 11 – ^{129}I.

may be presented as follows:

$$\frac{\sum i \cdot Ki(rad) \cdot Ki(mig) \cdot \xi i \cdot mi}{\sum j(U) \cdot Kj(U)(rad); \cdot Kj(U)(mig); \cdot mj(U)} \leq 1$$

Calculations show that the radiation and migration balance condition requires limitations with regard to the relative proportion of actinide losses, primarily plutonium isotopes, not exceeding 0.001 (or 0.1%).

As it follows from the calculation data, americium, heat-emitting fission isotopes (^{90}Sr-^{90}Y, ^{137}Cs-^{137m}Ba) are the next significant nuclides determining long-term activity, then – curium, uranium, neptunium, ^{14}C (for fast reactor with nitride fuel), and then ^{99}Tc and ^{129}I. The most significant fission fragments are not critical for radiation equivalence achievement due to their relatively short half-life.

But destruction of so-called minor actinides (MAs) through their transmutation is necessary. The MA transmutation problem is discussed by the scientific community in sufficient detail. But the approach under discussion provides a new methodology for development of particular practical recommendations and creates the required scientific and technical basis for development of well-reasoned numerical criteria. On the basis of the obtained data the following recommendations on the need for transmutation of SNF processing products may be compiled:

- Disposal of plutonium is prohibited, and it should be returned into the cycle (as provided in the CNFC concept) with the minimal losses not exceeding 0.1%;

- Americium is certainly subject to transmutation with the losses in processing not exceeding 1% in the first approximation;
- Curium shall be eliminated from disposed radioactive waste, but its transmutation is not mandatory – storage for about 70–100 years with subsequent return of its decay products (plutonium isotopes) to the cycle is sufficient;
- Neptunium and uranium are not subject to complete mandatory transmutation, although it is desirable in cases when it does not cause any severe technology problems;
- It is feasible to extract fission fragments from RW and send them for controlled storage with subsequent disposal;
- Transmutation of long-lived ^{99}Tc and ^{129}I is not mandatory from the viewpoint of radiation equivalence assurance, although it remains possible.

So, the radiation equivalence ideology for handling of fissionable materials indicates the ways to solve the postponed problem related to SNF accumulation and practical implementation within the CNFC framework clearly enough and sets the following technical and technology problems:

- The technology for coextraction of thorium and radium together with uranium;
- Fine purification of radioactive waste from any actinides and long-lived fission products;
- Transmutation of americium with separation of curium as well as neptunium and uranium where necessary;
- Long-term controlled storage of waste prior to their final disposal;
- RW disposal in the uranium mining areas in the course of their rehabilitation (according to the local balance requirement).

8.2 Transmutation of minor actinides

Prior to discussion of transmutation as one of the essential components of the concept the authors consider it useful to remind that this problem ranks the second in significance that is frequently forgotten. Elimination of the dominant element (plutonium) from radioactive waste by its repeated recycling in the CNFC is the first and the most important stage.

The initial physical fact distinguishing the approach to recycling of the basic actinides (uranium and plutonium) from the approach to recycling of minor actinides is related to considerably lower nuclear properties of the latter. Typical values of the average neutron cross-sections in the fast reactor spectrum for uranium, plutonium and MA isotopes are presented in Table 8.1. Attention should be paid to dominance of radiative neutron capture over fission (α value) for the main MAs under discussion: ^{237}Np, ^{241}Am, and also ^{243}Am. While for the main fuel isotopes this value is less than 0.3, for MAs it is higher by an order (more than 3).

TABLE 8.1 Typical average cross-section values of neutron reactions in the fast neutron spectrum.

Basic nuclides	σ_c	σ_f	α	Minor nuclides	σ_c	σ_f	α
^{235}U	0.47	1.79	0.27	^{237}Np	1.35	0.38	3.55
^{236}U	0.40	0.11	3.65	^{241}Am	1.58	0.33	4.82
^{238}U	0.25	0.04	5.53	^{242m}Am	0.34	3.74	0.09
^{238}Pu	0.50	1.16	0.43	^{243}Am	0.93	0.26	3.61
^{239}Pu	0.44	1.76	0.25	^{242}Cm	0.46	0.80	0.58
^{240}Pu	0.48	0.40	1.21	^{243}Cm	0.21	2.50	0.09
^{241}Pu	0.37	2.37	0.16	^{244}Cm	0.48	0.46	1.03
^{242}Pu	0.40	0.28	1.41	^{245}Cm	0.29	2.54	0.11

TABLE 8.2 Balance of neutron absorption and production processes.

Reactor	Neutron absorption	Production of fission neutrons	Balance
		^{237}Np nuclide	
Fast	2.1	2.9	0.8
Thermal	3.76	2.93	–0.8
		^{241}Am nuclide	
Fast	2.16	3.2	1.0
Thermal	3.5	2.95	–0.7

That is why MA transmutation is accompanied with much more complex nuclide transition patterns and longer chains of successive neutron capture (see the diagrams in Figs. 8.4 and 8.5, Table 8.1). And insertion of minor actinides into the core negatively affects the neutron balance. But it was successfully demonstrated in the study [27] that the total neutron balance in a fast reactor was evolving to the benefit of fission neutrons taking into account multiple inter-nuclide transitions (according to the presented patterns), and the total amount of these neutrons exceeded the amount of absorbed ones (Table 8.2). A similar calculation with regard to thermal reactors gave a negative result thus supporting selection of a fast reactor for MA transmutation.

But arrangement of efficient transmutation in the fast neutron spectrum also necessitates search for the optimal solutions. It is shown in the study [54] that a single MA recycle brings only harm. And this presents very serious problems in case of any attempts for MA burn-out in many alternative systems aimed at instant transformation of MAs into the form suitable for direct disposal (for example, so-called ROX matrices).

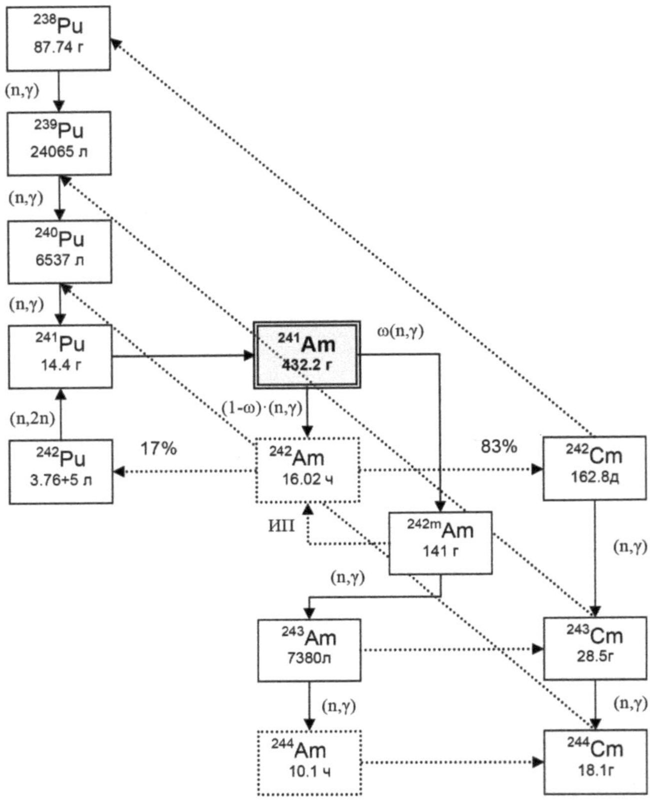

FIGURE 8.4 Pattern of the basic nuclide transitions during irradiation of ^{241}Am.

In case of repeated americium recycling (not synchronized with plutonium recycling) great burnup depth is necessary (more than 90%) so that the composition of the resulting nuclide mixtures with due regard for daughter decay products should be much less active. Otherwise americium irradiation brings only harm. And finally, it is shown in the same study that common recycling of MAs with the basic fuel increases efficiency of transmutation significantly. From the physical viewpoint it is easily understood from the diagrams of inter-nuclide transitions during americium and neptunium irradiation (see Figs. 8.4 and 8.5). In both cases irradiation results in accumulation of plutonium isotopes, first of all – ^{238}Pu, with subsequent generation of ^{239}Pu. In case of curium production one may remember that the basic curium isotopes with relatively short half-lives (28.5 years for ^{243}Cm and 18.1 years for ^{244}Cm) decay into $^{238\text{-}240}$Pu isotopes.

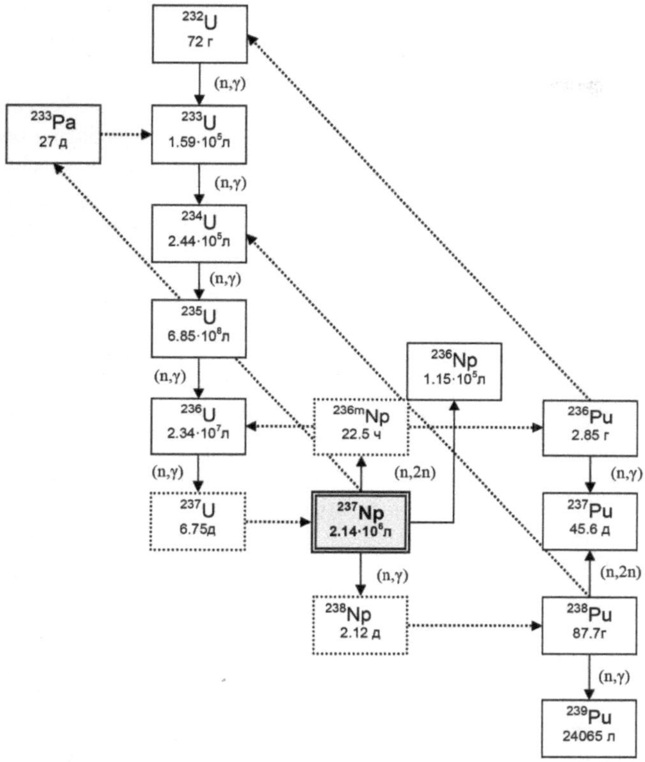

FIGURE 8.5 Pattern of the basic nuclide transitions during irradiation of ^{237}Np.

Taking into account analysis of the obtained results the proposed concept for radiation-equivalent handling of fissionable materials provides for clear and comprehensive logic with regard to MA transmutation approaches.

First, there is no sense in MA burn-out forcing with the rate considerably higher than for the basic plutonium (or rather mixed) fuel.

Second, in the course of common MA recycling with the basic fuel in the CNFC americium and neptunium nuclides transform into fuel components quickly enough and thus cease to be minor actinides in the narrow sense. In some way they perform approximately the same function as ^{238}U – additional breeding of plutonium. In case of MA burn-out separately from plutonium we have to breed plutonium from ^{238}U in addition to MA burn-out and production of plutonium within MAs thus deteriorating the activity balance in the system. And (separate) MA burn-out to fission fragments with significant burnup is getting a much more difficult task in itself.

Thus, the so-called homogeneous approach to MA transmutation in a fast reactor seems the most logical. In this reactor minor actinides are automatically mixed with the basic fuel in the course of transmutation and transformation into plutonium, and their quantity stabilizes at the low level of approximately ~0.4–0.5% (in total) quite rapidly during recycling. It can be seen from Table 8.1 that the amount of minor actinides in case of repeated recycling in the fast reactor fuel is significantly lower than their accumulation during one irradiation cycle in a VVER reactor. The difference between FNR types (with lead and sodium coolants) is not essential from the viewpoint of MA transmutation with comparable heat density in the core (accumulation in BREST-OD-300 is slightly lower due to a harder neutron spectrum). Such amounts of minor actinides in the fuel do not pose any problem for the physical characteristics of the fast reactor core.

However, two problems are possible from the technology point of view. The first one is related to potential volatility of americium in the course of high-temperature sintering of fuel pellets. The second one is related to considerably poorer radiation properties of refabricated fuel.

Calculations show that involvement of ^{237}Np into the fuel cycle increases the content of ^{238}Pu almost tenfold in comparison with equilibrium one and causes growth of the dose rate from exposure to neutron radiation. It also results in increase of ^{232}U concentration in the fuel up to ~4–10^6%. In this case the dose rate from γ-radiation of this isotope and its decay products becomes comparable with the aggregate dose rate from all regenerated mixed uranium-plutonium fuel with plutonium of energy-grade isotope composition.

Involvement of americium results mainly in significant increase of the γ-radiation dose rate from regenerated fuel (10–20 times). In this case the main hazard is not due to the basic isotope ^{241}Am but to ^{238}Np – the decay product of ^{242m}Am isotope containing the lines with γ-quantum energy of ~1030 keV.

And finally, presence of $^{242\text{-}244}Cm$ isotopes in the fuel causes increase in heat emission of regenerated fuel (~7 times) and neutron radiation power – by two orders (up to 500 times).

Based on the above-mentioned evaluations involvement of curium into the fuel cycle seems highly undesirable, particularly taking into account relatively short $T_{1/2}$. Involvement of neptunium does not cause any significant problems from the viewpoint of radiation situation, and thus it is reasonable.

In general, involvement of americium does not seem too problematic either from the viewpoint of radiation situation although it is more significant in comparison with neptunium. But taking into account its potential (although not sufficiently proven yet, as single experiments performed do not confirm increased volatility of americium) volatility it is reasonable to consider an alternative (so-called heterogeneous) option of its transmutation.

Taking into consideration the selected strategy of americium transmutation harmonization and fuel recycling within the CNFC in the same reactor it is reasonable to choose uranium as the fuel matrix thus enabling to maintain the BRC in case of any nonstandard fuel location in the core. Development of the options for heterogeneous positioning of americium burnup elements: inside fuel

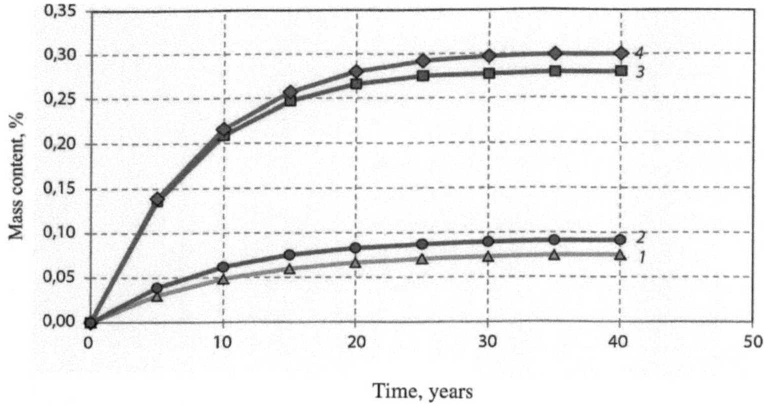

FIGURE 8.6 Accumulation of americium and neptunium in the FNR fuel in case of homogeneous transmutation: 1 – neptunium (BREST); 2 – neptunium (BN); 3 – americium (BREST); 4 – americium (BN).

rods, inside the cassettes and/ or cassette by cassette – requires analysis and optimization. However, calculations show that it is possible to reach the same transmutation degree as in case of the homogeneous approach with the number of americium burnup elements of less than 10% of the total number of fuel rods with the initial americium concentration of about 5–10%.

Composition of the fuel with americium has peculiar nature (see Table 8.1) with a large fraction of ^{238}Pu isotopes; it is the consequence of americium transmutation on the one side, and provides the composition of plutonium mixture obviously unsuitable for military application on the other side. In this case no doubts from the viewpoint of nonproliferation arise for application of the uranium matrix.

It should be noted that the heterogeneous americium transmutation pattern shall be synchronized with the basic fuel recycling pattern in order to enhance its efficiency and also to eliminate excessive accumulation of actinides as described above. In other words, americium is extracted each time in the course of the basic fuel mass recycling and concentrated in the americium burnup elements. In this case plutonium generated in the americium burnup elements shall be extracted from them and added to the basic fuel.

8.3 Transmutation nuclear fuel cycle

On the basis of the proposed concept and the results of analysis described above, the transmutation fuel cycle scheme presented in Fig. 8.6 and Table 8.3 may be offered for implementation. The transmutation closed nuclear fuel cycle based on the principle of radiation-equivalent RW handling is presented in Fig. 8.7.

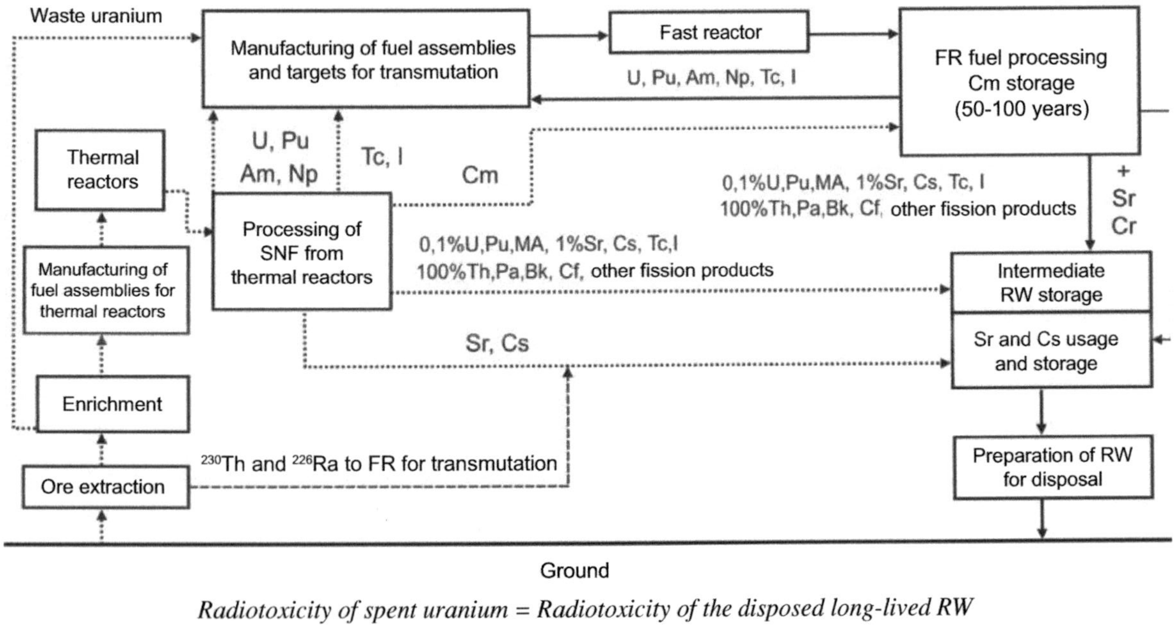

FIGURE 8.7 Transmutation CNFC implementing the principle of radiation-equivalent handling of fissionable materials.

TABLE 8.3 Composition of the trans-uranium component in VVER fuel (after storage for 10 years) and fast reactor fuel in case of homogeneous and heterogeneous transmutation schemes, %.

Nuclide	VVER	FNR	
		Homogeneous	Heterogeneous
^{238}Pu	2.07	1.03	11.45
^{239}Pu	52.9	62.53	40.12
^{240}Pu	18.82	27.75	4.80
^{241}Pu	8.03	3.25	0.15
^{242}Pu	4.49	2.15	2.94
^{237}Np	6.16	0.60	–
^{241}Am	6.29	1.79	26.89
^{242m}Am	0.01	0.07	1.00
^{243}Am	1.01	0.82	12.65
^{244}Cm	0.2	–	–
^{245}Cm	0.02	–	–
Fraction in the fuel, %	1.2	13.5	4.8

So, the transmutation nuclear fuel cycle (see Fig. 8.7) based on the principle of radiation-equivalent RW handling enables fundamental solution of the two interrelated and constantly postponed problems of nuclear power – SNF and RW problems. In case of its technical implementation it opens the door for essentially zero-waste power engineering without any changes of the environment.

The described principle provides a working quantitative criterion for scientifically grounded selection of the methods and engineering solutions for the CNFC implementation.

The main features of the conceptual approach to practical implementation and demonstration of the technology CNFC approaches proposed within the framework of Proryv Project include the following provisions:

- Repeated recycling of plutonium in order to eliminate its disposal within RW with the losses exceeding 0.1%;
- Homogeneous transmutation of neptunium together with the basic mixed uranium-plutonium fuel;
- Homogeneous or heterogeneous americium transmutation in the fast reactor core harmonized with the basic fuel recycling;
- Extraction and controlled storage of curium till its decay into plutonium with subsequent return of the latter to the CNFC.

Chapter 9

Radiation and radiological equivalence of radioactive waste in two-component nuclear power engineering

The share of electricity produced by nuclear power plants (NPP) in the world total electricity production is very small. Though the nuclear fuel is characterized as the fuel with the higher specific energy per a unit mass, the problems related to the management of spent nuclear fuel (SNF) in operating thermal neutron reactors (thermal reactors) delay the development of nuclear energy. Inevitable long-term controlled storage of the SNF from thermal reactors and disposal of long-lived high-level radioactive waste (LHLW) lead to accumulation of significant amounts of SNF and LHLW that in turn increases potential threat of nuclear energy to humans and the environment.

Development of two-component nuclear power system based on closed nuclear fuel cycle with thermal and fast neutron reactors (fast reactors) operating in the closed nuclear fuel cycle (CNFC) makes possible significant reduction of the nuclear energy danger.

To secure an acceptable level of protection of human health and the environment, International Atomic Energy Agency (IAEA) pays great attention to the research on management of radioactive waste (RW), formed during the CNFC, as part of the International Project on Innovative Nuclear Reactors and Fuel Cycles (INPRO). The Generation IV International Forum indicated the importance of the fast reactors development (particularly, fast reactors with lead coolant) and the relevant CNFC technologies for the nearest 15–25 years [24].

The amounts of SNF and LHLW will be significantly reduced due to the implementation of the CNFC, but it will fail to arrange zero-waste process, therefore geological disposal of RW will remain necessary.

Current large-scale nuclear power systems based on CNFC are being developed. To solve the problem of RW management the new technology has been developed, it is based on the principle of equivalence. The time of the equivalence achievement depends on composition and the amount of radioactive waste (and

Closed Nuclear Fuel Cycle with Fast Reactors: White Book of Russian Nuclear Power.
DOI: https://doi.org/10.1016/B978-0-323-99308-1.00007-6

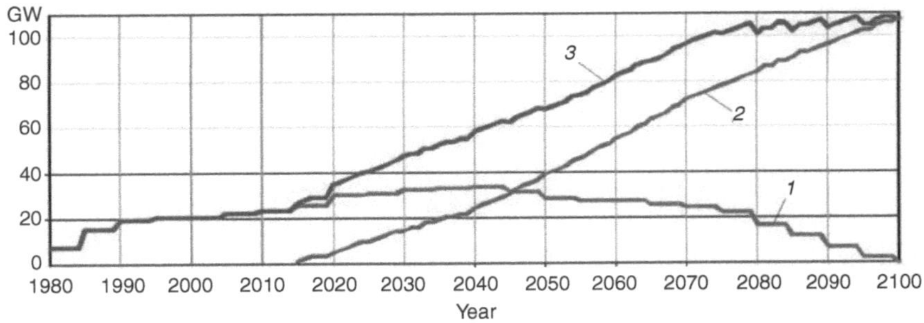

FIGURE 9.1 Capacities of thermal and fast reactors in the nuclear power system: 1 – thermal reactors; 2 – fast reactors; 3 – all nuclear power engineering.

its disposal). The equivalence is achieved if hazardous effect of radioactive waste is equivalent to or less than effect of natural raw uranium [46,106,110].

9.1 Equating lifetime radiation risks of possible cancer from RW and natural raw materials

Committed effective doses [46] can be used as a basis for the assessment of potential hazard to human health (in case of the radiation equivalence principle implementation) or, more correctly, by estimating lifetime attributable risk (LAR) of cancer from radionuclides entering the human body. For estimating LAR radiation risk models recommended by the International Commission on Radiological Protection (ICRP) [78] (in case of the radiological equivalence principle implementation) are used.

According to computationally modelled scenario development of the nuclear energy in Russia in the XXI century the thermal reactors operating on enriched uranium fuel will operate until the adopted in the model natural uranium resource of 540 thousand is exhausted. All irradiated fuel from thermal reactors will be processed. Plutonium, minor actinides, and some long-lived fission products will be extracted from it. The line of fast reactors will be developed on the basis of extracted plutonium. The production output is assumed to be annually, if plutonium amount to be sufficient. Thermal VVER and RBMK and fast reactors, such as BREST-1200, were included in the model. The nuclear power capacity will reach 47.3 GW by 2030 and 107 GW by 2100. To that time fast reactors only will operate (Fig. 9.1) in the closed fuel cycle. In processing of irradiated fuel from thermal and fast reactors the content of uranium, plutonium, minor actinides, strontium, cesium, technetium and iodine in long-lived radioactive waste should not exceed of 0.1% of their amount in the processed fuel. It is assumed that RW batches removed from reactors will be stored during some time; their nuclide composition has changed due to radioactive decay. About 7523 tons of long-lived radionuclides are assumed to be accumulated in the nuclear power system by 2100.

Potential biological hazards (PBH) of radioactive waste (RW) and natural raw material (uranium) consumed by nuclear power are different. The objective of the study was to identify the time when equivalence between hazardous biological effects of RW and the natural raw material has been achieved. The problem solution is to identify the time for achievement of the equality of committed effective doses (radiation equivalence) and the time for achievement of lifetime radiation risks equality (radiological equivalence).

The effective dose concept was developed by the ICRP for the purposes of radiation protection management. The limits, restrictions and reference levels within the system of radiation protection rules and regulations are expressed in the effective dose terms.

The effective dose *E* is determined as the weighted average value of equivalent doses in organs and tissues:

$$E = \sum \omega'_T \cdot \cdot H_T = \sum \omega'_T \cdot \cdot \sum \omega'_R \cdot \cdot D_{TR}$$

where H_T is the equivalent dose to any organ or tissue T; $\acute{\omega}_T$ is the relative (nondimensional) weight of radiation damage to organ or tissue *T* (the weighting tissue coefficient), in this case $\Sigma\acute{\omega}_T = 1$, and the sums are taken for all organs and tissues of the body with determined radiation damage values; $D_{T,R}$ is the average absorbed radiation dose *R* to the organ or tissue *T*; $\acute{\omega}_R$ is the weighting coefficient for radiation *R*; the effective dose unit (J/kg) has the special name "Sievert" (Sv).

The effective dose *E* shall be calculated with the use of some assumptions:

1. Tissue coefficients $\acute{\omega}_T$ are calculated for nonrecurring uniform exposure of a human body with averaging in accordance to sex and age in the model of European-American-Asian population.
2. To calculate tissue coefficients $\acute{\omega}_T$ the ICRP considers radiation related harm to the health as the weighted average of radiation-related cancer incidence, mortality rate, reduction of life expectancy, and hereditary effects of radiation estimated through expertise.
3. Though ICRP estimated radiation-related harm to the health for 14 various organs and tissues, the equation for *E* includes four different rounded $\acute{\omega}_T$ quantities only.
4. In calculating the committed effective dose (CED) the aggregate equivalent dose H_T from radionuclide transfer to the human body expected in the course of lifetime is assigned to a year of this radionuclide intake.

That is the ICRP points out [78] that organ/tissue doses rather than effective doses are required to assess the probability of cancer induction in the irradiated persons.

The LAR [78] of cancer incidence representing the expected lifetime number of radiation-induced cancer cases in the irradiated group of people scaled to its size is adopted for measurement of potential biological hazard in assessment of radiological equivalence.

TABLE 9.1 Ratio of lifetime radiation risks per 1Sv of CED of cancer incidence from internal exposure of the USA population to 238U and 241Am.

Radionuclide	Dose coefficient [175], Sv/Bq	Disease incidence risk [136], 1/Bq	Disease incidence risk, 1/Sv
^{238}U	$4.50 \cdot 10^{-8}$	$2.34 \cdot 10^{-9}$	0.0520
^{241}Am	$2.00 \cdot 10^{-7}$	$3.63 \cdot 10^{-9}$	0.0182
Ratio of risks at 1 Sv from ^{238}U and ^{241}Am			2.87

The ratio of LARs for two different radionuclides calculated with the use of the initial annual risk models for the particular population differs from the relevant CEDs ratio.

Evaluation of lifetime radiation risks of cancer incidence following internal exposure of the USA adult population to ^{238}U and ^{241}Am may serve as an example [136,175]. The data presented in Table 9.1 show that the radiation-related cancer risk from ^{238}U is 2.87 times higher than from ^{241}Am with the same effective dose of 1 Sv.

In this case the CED of 2.87 mSv from ^{241}Am corresponds to the same lifetime radiation risk of cancer from 1 mSv of ^{238}U.

As far as the problem of radiation and radiological equivalence is concerned this means that the estimated RW storage period prior to their disposal calculated on the basis of lifetime attributable risks ratio may differ from the ratio calculated on the basis of CED values.

Potential biological hazards in terms of the CED from the significant RW radionuclides for their different storage periods and the CED from natural uranium isotopes are used as the source data for evaluation of radiation and radiological equivalence. These data were obtained in the modeled scenario for the development of nuclear power based on thermal and fast reactors up to 2100, assuming that in accordance with the scenario 541.7 thous. tons of natural uranium will be used for development of large-scale nuclear power by 2100, and 7.523 thous. tons of RW in the form of long-lived radionuclides will be accumulated.

In case of the *radiation equivalence* in the RW storage period within which the CED from the RW becomes equal to the CED from natural uranium (due to radioactive decay) shall be estimated.

In case of the *radiological equivalence* criterion application to the RW storage period within which the LAR from the RW becomes equal to the LAR from natural uranium (due to radioactive decay) shall be evaluated. The LAR is calculated in the following way:

- Variation of annual equivalent doses to human organs and tissues is determined for different RW storage periods, for each basic radionuclide contained in RW in this period, according to the ICRP dose coefficient database [175];

TABLE 9.2 Standard deviations of ICRP model [207] parameters for estimating risks of all solid cancers in males [207].

STD β_{mult}	STD γ_{mult}	STD $\acute{\omega}_{mult}$	STD β_{add}	STD γ_{add}	STD $\acute{\omega}_{add}$
0.0457	5.4880	0.2744	6.7070	4.8780	0.2744

TABLE 9.3 Standard deviations of the model parameters for estimating the excess absolute risk of leukemia in males after single exposure [135].

Age at the moment of exposure	STD β_{lkm}
0–19 years	0.56
20–39 years	1.31
>40 years	2.63

- The LAR from each of the significant radionuclides contained in RW is calculated for each organ and tissue with the use of the ICRP radiation risk models [78],
- The LAR values for the significant radionuclides contained in RW in the particular time period are added in order to calculate the aggregate LAR from the RW for the given storage period.

The constant LAR from natural uranium is calculated in the same way assuming stability of its radionuclide composition over time. The data on potential biological hazard from different radionuclides contained in RW and natural uranium are given in the Appendix (Tables 9.1–9.3).

The method for calculating LAR from internal exposure is described above [106]. It should be reminded that subject to survival till the current age *e* the LAR shall be calculated by summing of the excess absolute risk (EAR) with the weight of survivorship function (or healthy survival) by attained ages *a* starting from the current age e *to* the age of 100 years.

$$LAR(e) = \frac{1}{DDREF} \times \sum_{a=e}^{100} [EAR(a) \times S(e,a)]$$

DDREF is the dose and dose rate effectiveness factor equal to 2 for solid cancers and 1 for leukemia.

The excess absolute risk is the increment of the background cancer incidence or mortality rates after radiation exposure.

The survivorship function gives the probability to a person of certain sex and age *e* to survive into a certain age *a*. The healthy survival function characterizes the probability for a person of certain sex and age *e* to survive into the given age *a* without suffering from any malignant tumors of the specified site. Attributable

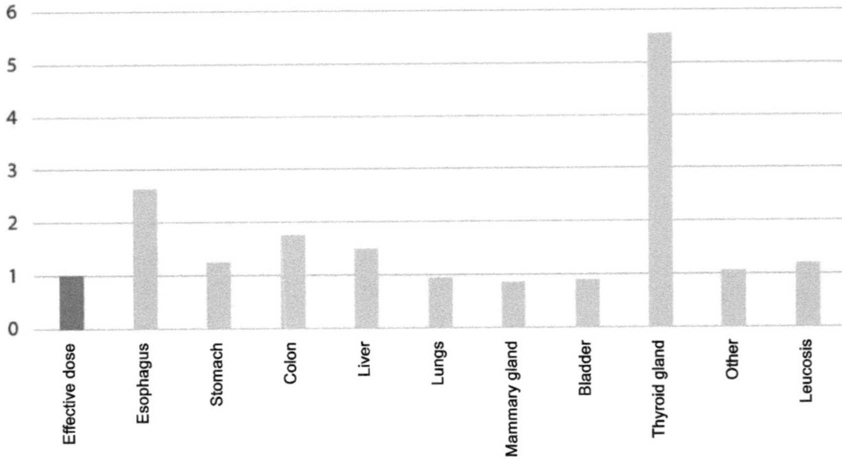

FIGURE 9.2 Ratio of the lifetime attributable risks calculated on the basis of the effective dose equal to1 mSv to the lifetime attributable risks to individual organs and tissues with the equivalent doses corresponding to the committed effective dose of 1 mSv.

risk for possible ages of the population (0–100 years) is used to determine radiological equivalence.

The lifetime attributable risk LAR is calculated on the basis of the equivalent doses derived from the CED. Direct calculation on the basis of the CED is unacceptable as the effective dose prevents from distinguishing between individual organs (Fig. 9.2). As it can be seen from the figure the difference in comparison of the risk calculated on the basis of the effective dose and the risk calculated on the basis of the relevant equivalent dose may exceed 5.5 times.

Calculation of the risks on the basis of the effective dose alone implies that the whole body has been irradiated. It should be also taken into account that in case of internal radiation exposure the CED does not consider time distribution of equivalent doses which can affect the final value of the lifetime attributable risk significantly. Ratios of the lifetime attributable risk calculated only on the basis of the CED equal to 1 mSv to the lifetime attributable risks from different radionuclides with the CED equal to 1 mSv entered the human body are presented in Fig. 9.3. In the first case the dose distribution among the exposed organs and time distribution of the dose are disregarded. As it can be seen on Fig. 9.3 the parameters may differ by 4 times and more.

The ICRP database [175] providing the dynamics of equivalent doses to different organs and radionuclides accumulated over time is used to solve the problem of the CED application. The flowchart to determining f equivalent doses on the basis of the CED was described in detail [106]. The CED (given in Tables 9.1 and 9.2 in the Appendix) normalized to 1 mSv of the aggregate CED of natural uranium (Table 9.3 of the Appendix) is used for calculation. That is the case of using annual radiation risk models quadratic by dose (for malignant

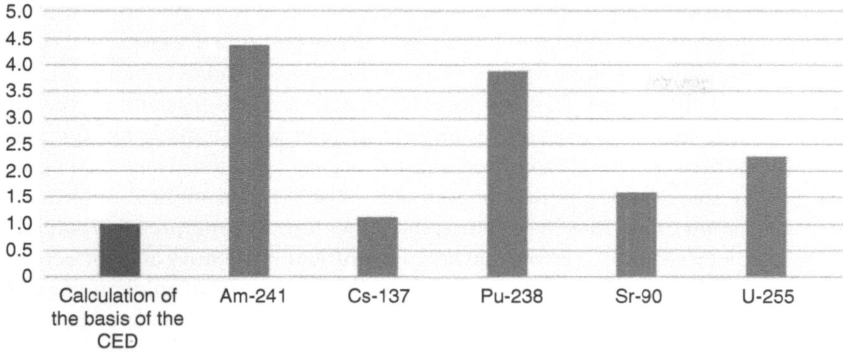

FIGURE 9.3 Ratio of the lifetime attributable risk calculated on the basis of the CED equal to 1 mSv only to the lifetime attributable risk from different radionuclides entered the body with the CED equal to 1 mSv.

bone tumors and leukemia). In evaluating potential biological hazard (PBH) in terms of LAR one must align with very small doses comparative with the doses that a person can receive. That is why it is reasonable to scale potential biological hazard in the LAR terms to 1 mSv.

9.2 Impact of uncertainty in the parameters of the annual radiation risk models on achievement of radiological equivalence in two-component nuclear energy

Uncertainty in calculation of the radiological equivalence achievement time is determined by uncertainties in calculation of the life-long radiation risk (LAR) values.

Uncertainty in the coefficients for the initial annual radiation risk models identified and evaluated on the basis of epidemiological data is the obvious source of uncertainty in the LAR assessment.

Limited amount of data involved into the research is the fundamental source of statistical uncertainty in identification and evaluation of the risk models in the epidemiological studies. Due to this circumstance statistical errors are associated with any risk assessment in any study – the less the available data amount used for the model evaluation the more the statistical dispersion in evaluation of this model parameters becomes.

The statistical uncertainty in any parameter evaluation is usually quantitatively determined with the application of the confidence interval concept corresponding to the probability of the fact that this interval contains the true value of the evaluated parameter. Around 90% or 95% confidence intervals are usually used to characterize uncertainty in evaluation of any parameter. Statistical uncertainty is inevitable in the risk models based on epidemiological studies and may be reduced only subject to availability of greater amount of data for investigation.

The majority of epidemiological studies are observational by virtue of their nature. Observational studies do not represent a controlled experiment, and epidemiology almost completely relies on the data obtained under uncontrollable everyday life conditions. That is why randomization of the research subject in observational studies is impossible, and elimination of systematic background differences between the people exposed to various radiation doses cannot be reached by the means applied for randomized controlled clinical studies. It presents certain difficulties in planning, implementing and interpreting epidemiological studies. Systemic errors that are often difficult to identify and correct are added to statistical errors inevitably presenting in the analysis of epidemiological results thus cause bias in evaluation of the risk model parameters. For the same reason repeatability of the research results presents great difficulties in observational studies.

Systemic errors shall be minimized in the course of any studies in radiation epidemiology by clear and complete establishment of any disease (or death) cases, the relevant definition of the reference groups for comparison, sufficient duration of observations, correct consideration of the interfering factors (distorting the results) as well as by well-organized assessment of exposure doses. Some epidemiological studies solve these problems but others are not sufficiently elaborated in one or several aspects. Consequently, the epidemiological uncertainty degree varies from study to study. Quantitative characterization of various epidemiological uncertainty sources is very important in development of risk models on the basis of epidemiological studies. These uncertainties shall be dispersed through application of the radiation risk models in particular circumstances and considered in the final total uncertainty of the risk estimate.

Statistical radiation risk models may be based on epidemiological data as it was mentioned above. The dose response nature, that is, changes in the cancer development risk depending on the radiation dose is the most important aspect of such modeling. Risk modeling is not usually a simple mathematical procedure as different statistical models may be identified with the use of the same data set.

As a rule, expert judgment is required to select the model that complies the prescribed conditions. In that case the simplest model describing the available radiation and epidemiological data is often selected on the basis of the scientific simplicity principle. However, other considerations may be applied to the model selection, for example, any reasons based on the current (incomplete) understanding of the biological mechanisms of radiation-associated carcinogenesis.

As a result, several different risk models describing changes in the risk of certain cancer in relation to radiation dose to the tissue that may be a potential site for the cancer development. These models may differ in many aspects, for example, if the models are sex-specific, or whether the excess risk changes in relation to various combinations of age at exposure, duration of exposure and the attained age. Incomplete duration of any epidemiological study when not all members of the population under study died is also an uncertainty factor. In this case certain assumptions with regard to the risk evolution outside the time limits of observation are required. A certain pragmatic approach is needed to assess uncertainties associated with selection of models, i.e. the impact of different models on risk assessment.

The feature of research on radiation risk modeling resides in the fact that the current risk assessments have been obtained on the basis of radiation and epidemiological data for medium or high radiation doses (above 100 mSv). These radiation circumstances resulted in the assessment of relatively high excess risks well-defined in terms of quantity. It is still impossible to obtain reliable radiation risk estimates for radiation doses below 100 mSv, even in case of single exposure. The question now arises with regard to the way to apply these risk models in the low radiation doses range in case of long-term radiation exposure, additional assumptions on the human body response to low dose radiation dose response shall be made, although the amount of available epidemiological data on effects of exposure to low dose radiation is not large to make reliable conclusion. In other words, the issue consists in the method of interpolation from the risks at high and medium dose levels (and dose rates) to the zero excess risk at the zero dose. The assumption on the linear no-threshold (LNT) model of dose response is usually made, as it is the simplest model compatible with the data. But the LNT model is not the best for certain cancer types. That is why uncertainty of dose response interpolation in the area of small doses and in case of long-term exposure is inevitably present in forecasting radiation risks.

The dose response curvature decreasing inclination of the LNT model in transition from large doses (and high-dose rates) to small doses (and low-dose rates) is often assumed based on the knowledge of radio-biological carcinogenesis mechanisms. In this case the dose and dose rate effectiveness factor (DDREF) of more than 1.0, that is currently used in the radiological protection system substantiated by the ICRP is applied for interpolation between high and zero radiation risks. For leukemia the quadratic dose – disease relationship of may be demonstrated on the basis of epidemiological observations. The uncertainty in assessment of the DDREF value apparently exists, and some expert groups have made the conclusion for the purposes of radiological protection (i.e., there is no decrease in inclination of the LNT model during transition from large to small doses) that the DDREF magnitude shall be equal to 1.0. In the course of modeling the risk model is usually identified directly on the basis of the available data without involvement of any DDREF. In case of the modeled dose response for the particular type of cancer does not demonstrate any curvature, i.e. the inclination within the framework of the LNT model does not change depending on the dose any further apparent application of the DDREF is not required. But the low doses and low-dose rates range still remains beyond the scope of reliability of statistical epidemiological models.

As a result of the above-mentioned reasons all radiation risk models used for risks prediction have uncertainties reflected in the parameter scattering for these models.

The lifetime attributable risk LAR of solid cancers shall be calculated on the basis of the models presented in the ICRP Publication 103 [78] for annual excess risks: The EAR (excess absolute risk – an additive model) and the ERR (excess relative risk – a multiplicative model).

$$EAR^{ICRP}_{single}(s, c, g, a, d) = d\times_{\beta EAR}(s, c)\times\left(\frac{a}{70}\right)^{\omega' EAR(c)}\times\left(1+\frac{\gamma EAR(c)}{100}\right)^{0,1\cdot(g-30)} \tag{9.1}$$

$$EAR^{ICRP}_{single}(s, c, g, a, d) = d\times_{\beta EAR}(s, c)\times\left(\frac{a}{70}\right)^{\omega' EAR(c)}\times\left(1+\frac{\gamma EAR(c)}{100}\right)^{0,1\cdot(g-30)} \tag{9.2}$$

Where s is sex; c – tumor site; g – age at exposure; a – attained age; d – radiation dose; β_{EAR}, $\acute{\omega}_{\text{EAR}}$, γ_{EAR} – parameters of the additive model; β_{ERR}, $\acute{\omega}_{\text{ERR}}$, γ_{ERR} – parameters of the multiplicative model.

The ICRP proposes to use another radiation risk model [135] for assessment of leukemia risks. The linear-quadratic dose relationship and the use of additive method for risk estimating only are the peculiarities of this model:

$$EAR^{ICRP}_{single}(s, c, g, a, d)=\beta(s, g)\times d\times(1+0,79\times d)\times\exp[\alpha(s, g)\cdot(a-g-25)], \tag{9.3}$$

where β and α are the model parameters.

Standard deviations of the model parameters for solid cancers are presented in Table 9.2 and for leukemia – in Table 9.3.

9.3 Uncertainty in the background morbidity and mortality rates

Cancer risk models may be expressed as either the excess relative risk (ERR) or the excess absolute risk (EAR). The ERR is the proportional increase in risk over the background absolute risk (in the absence of exposure). The EAR is the additional radiation risk above the background absolute risk.

It is of little significance for any irradiated population, used as the basis for the models, whether the risk models are expressed as the ERR or the EAR, if statistical modeling is performed with sufficient care, especially when the excess risks depend on the attained age or time after exposure (so-called consolidated ERR or EAR models).

However, application of identified and evaluated risk models to any groups of people with background risks of cancer differed from the background risks characterizing the cohort which has provided their data for the models presents a problem. This problem arises from the assumptions on the mechanisms of interactions between radiation and background risk factors. Transfer of the EAR model obtained on the basis e data on any population to a different ("foreign" population means that the number of excess of radiation-related cancer cases in the "foreign" population does not depend on its background parameters – radiation-related risk is added to the background risk. This is not the case if the ERR model is applied to the "foreign" cohort as the number of excess radiation-related cancer cases depends on the background parameters – radiation-related risk is multiplied by the background risk and serves as the modifier for the latter.

There is some uncertainty whether the ERR or EAR models, or any combination of these two models is more adequate for risk transfer between the cohorts or populations; in that case the uncertainty varies for different cancer sites. This is mainly due to the absence of the biological knowledge on interaction of radiation and other factors (for example, smoking) associated with the risk of malignant tumors. This source of uncertainty shall be taken into consideration in any assessment of radiation-related risks.

The ICRP [78] recommends to transfer the risk in projecting radiation-related risks of solid cancers in any population by averaging the additive and multiplicative risk models with different weights in order to obtain the average excess absolute risk value:

$$\begin{aligned} EAR^{sol.}_{single}(s, c, g, a, d) &= p(c) \times \lambda^{inc.}_{0.}(s, a, c) \times EAR^{ICRP}_{single}(s, c, g, a, d) \\ &\quad + (1 - p(c)) \times EAR^{ICRP}_{single}(s, c, g, a, d) \end{aligned} \tag{9.4}$$

where $\lambda^{inc.}_{0.}$ is thebackground cancer incidence per a year for the given sex *s*, age *a* and cancer site *c; p(c)* is the weighting coefficient of the multiplicative risk model for site *c*.

The transfer of risk is not applied to project risks of leukemia incidence.

Consequently, the quality of data on cancer incidence and the available data on the death rate in the population subject to the risk model transfer is an important aspect of uncertainty in cancer risk transfer across the populations. The disease incidence and death rate data for the Japanese atomic bomb survivors' cohort are of high quality due to the efforts made to develop the relevant databases. However, the estimated disease incidence and death rate parameters knowingly having uncertainties shall be used for the population subject to application of risk models for prediction purposes.

Age-related relationship of the background lung cancer incidence rate in Russian males and its 95% upper confidence limit defined with due regard for geographical variability of this rate in different regions of Russia are presented in Fig. 9.4 as an example.

Besides, the background parameters of cancer incidence and mortality rates change every calendar year. So, the background incidence of all cancers in the Russian Federation increased approximately 1.2-fold within the period of 2007 to 2016 [32,33].

Thus changing trends in cancer incidence and mortality rates and total death rate affects the final uncertainty of the lifetime attributable risk (LAR) assessment.

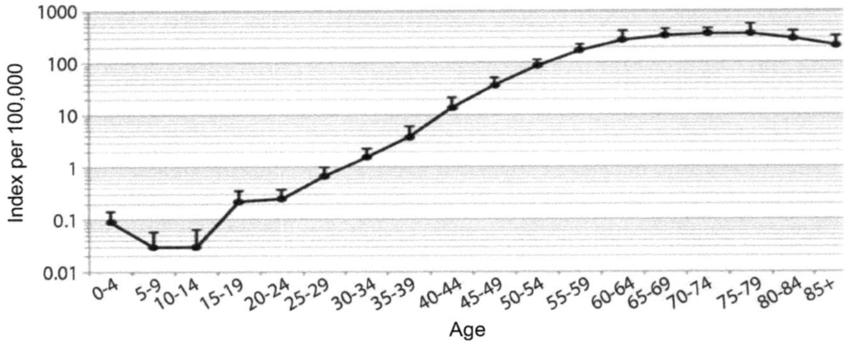

FIGURE 9.4 Background lung cancer incidence rate in males and its 95% upper confidence limit with due regard for variability of this rate in different regions of Russia.

9.4 Effect of uncertainty of radiation doses magnitude on the radiological equivalence achievement

Uncertainty in radiation dose magnitude due to errors in measurement and dose assessment methods significantly impacts on uncertainty in estimates of radiation risks. In case of long-term exposure, the accumulated (cumulative) dose alone cannot provide risk assessment as the risk depends on the exposure duration, nature of dose distribution over time as well as on the age of exposed people – the age at exposure and the attained age. In addition, the minimum latent cancer development period shall be used for risk assessment as the doses within this period do not determine the radiation risk.

Beyond these general considerations uncertainty associated with determination of the individual dose for a person is significant for risk modeling. Even if individual (portable) dosimeters are used for dose monitoring in case of external exposure the doses recorded in this way are unlikely to be absolutely accurate, and this uncertainty varies with time. The situation is even more complicated for assessment of internal exposure doses. Direct dose measurement is usually impossible for radiation with short free path emitted by radionuclides deposited in the tissues, and tissue dose assessment shall be derived from measurement of the biological samples (for example, urine samples). In some cases, such as assessment of the dose accumulated in the lungs from radon decay products, the dose shall be recovered on the basis of the assessed amount of radioactive materials passing through the lungs due to breathing. For other circumstances such as exposure to insoluble plutonium compounds deposited in the lungs, urine as biological specimen are of limited significance because only a small fraction of the deposited radionuclide passes into the blood, potentially it settles down in other tissues and only after that removes from the body with urine.

Assessment of equivalent doses to organs and tissues from internal exposure is based on the modeled generalized concept of a reference man, and knowingly there is uncertainty in actual individual doses k. The uncertainty in estimated doses is most significant in internal exposure doses. The uncertainty should be taken into account in estimating radiation risks from radionuclides entered the human body.

Attention should be paid to the fact that the dose is multiplied by the parameter meaning the specific (per a dose unit) average excess risk factor – the excess absolute (EAR/Sv) or excess relative (ERR/Sv) risk – in additive and multiplicative models of annual radiation risk. These factors are designated by β symbols with different suffixes in the formulae given above (9.1–9.3). Other parameters of the annual radiation risk models mean the modifiers for the average excess risk factor values. Their impact on the averaged risk estimate for the population is insignificant. Besides, in case of the multiplicative radiation risk model application the dose is also multiplied by the background cancer incidence index – in the designation of Formula (9.4).

That is why consideration of the annual radiation risk models parameters spread f and background cancer incidence parameters does not provide any qualitatively new data for assessment of LAR variability as compared with the radiation dose variability consideration. Thus quantitative effect of uncertainty in any non-dose factors on the uncertainty in LAR forecast may be taken into account mainly through increase in radiation dose spread as a result of relevant calculations.

Comparative analysis of estimated time of radiological equivalence (calculated on the LAR basis) and radiation equivalence (calculated on the CED basis) achievement with the use of different test values of uncertainty in CED estimates varying i relatively their average values specified in Tables 9.1–9.3 of the Appendix are provided below in this Chapter.

Time-related intersection of the RW potential biological hazard curve and the constant level for the relevant amount of natural uranium is shown in Fig. 9.5.

Radiation equivalence will be achieved in 287 years after the RW generation in 2100.

Contributions of significant RW radionuclides to the potential biological hazard (Sv) in relation to a calendar year are shown in Fig. 9.6. Only those radionuclides which contribution exceeds 1% at least once per 1000 years are presented. Despite the fact that the mass share of americium in the RW is only 0.23%, The CED of ^{241}Am will begin to dominate in the RW i after 100 years of storage.

Fig. 9.7 demonstrates time-related intersection of the curve of LAR from RW and the permanent LAR rom the relevant amount of natural uranium, that is radiological equivalence will be achieved in 2199, 99 years after RW generation in 2100. This is 188 years earlier than radiation equivalence achievement.

Contributions of significant RW radionuclides to the aggregate average LAR depending on the calendar year are shown in Fig. 9.8. ^{241}Am isotope will begin to provide the maximum contribution over time in the same way as for the CED.

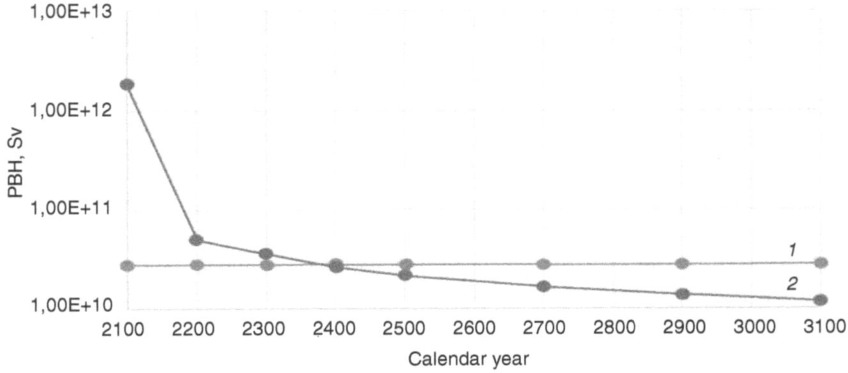

FIGURE 9.5 Potential biological hazard (Sv) of waste from thermal and fast reactors and the relevant amount of natural uranium in different calendar years; the intersection point for the curves is in 2387: 1 – natural uranium; 2 – RW.

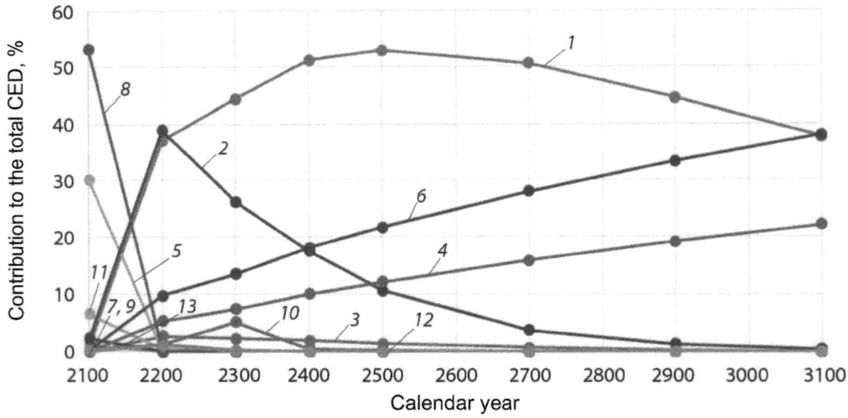

FIGURE 9.6 Contribution of i significant radionuclides to the potential biological hazard of waste from nuclear power based on thermal and fast reactors: 1 – 241Am; 2 – 238Pu; 3 – 242mAm; 4 – 239Pu; 5 – 144Ce; 6 – 240Pu; 7 – 137Cs; 8 – 106Ru; 9 – 154Eu; 10 – 151Sm; 11 – 85Kr; 12 – 90Sr; 13 – 147Pm.

Let us consider time-related changes in radiation and radiological equivalence achievement in case of tenfold increase of potential biological hazard from americium. Potential biological hazard of RW with the americium fraction increased tenfold and also potential biological hazard of natural uranium corresponding to the RW amount are shown in Fig. 9.9. It can be seen from this figure that radiation equivalence will not be achieved even in 1000 years after generation of nuclear power r waste in 2100.

Contributions of significant radionuclides from the RW with 10-fold increased of americium potential biological hazard to the potential biological

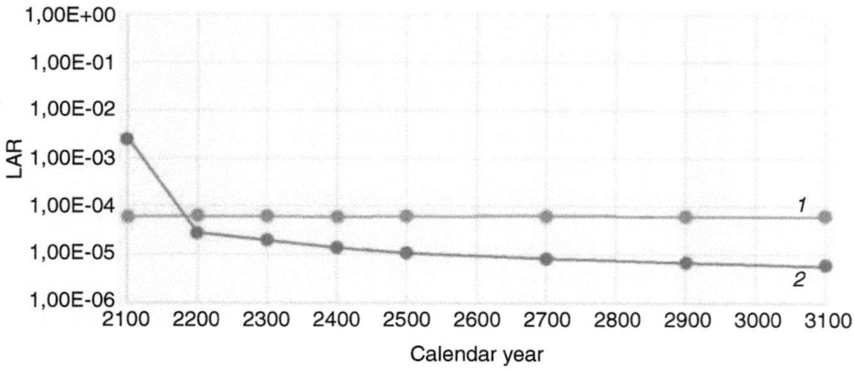

FIGURE 9.7 Lifetime attributable risk from waste generated by thermal and fast reactors and the relevant amount of natural uranium calculated on the basis of the doses scaled to 1 mSv of natural uranium potential biological hazard; the intersection point for the curves is in 2199:1 – natural uranium; 2 – RW.

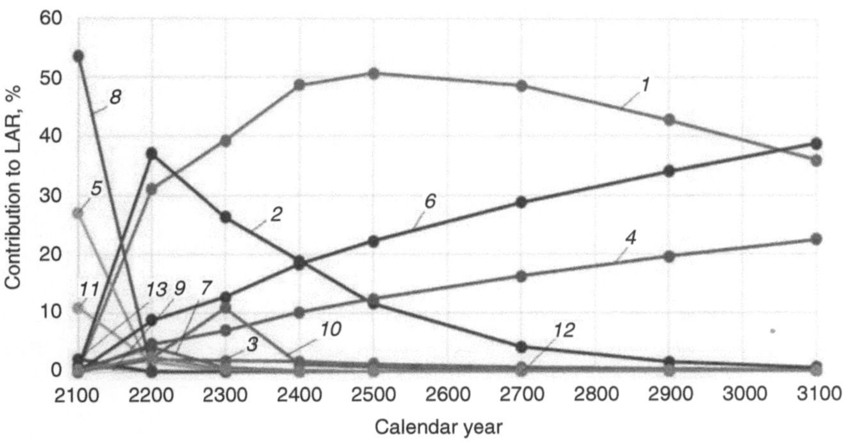

FIGURE 9.8 Contribution of important radionuclides to the lifetime attributable risk from waste from nuclear power nuclear power based on thermal and fast reactors: 1 – 241Am; 2 – 242mAm; 3 – 144Ce; 4 – 137Cs; 5 – 154Eu; 6 – 85Kr; 7 – 147Pm; 8 – 238Pu; 9 – 239Pu; 10 – 240Pu; 11 – 106Ru; 12 – 151Sm; 13 – 90Sr.

hazard (Sv) depending on the calendar year are given in Fig. 9.10. It can be seen from the figure that ^{241}Am makes the predominant contribution into the potential biological hazard.

Despite the fact that radiation equivalence will not be achieved within 1000 years in case of tenfold increase in americium potential biological hazard radiological equivalence will be achieved in 414 years (Fig. 9.11), that is, 315 years later than for the initial RW.

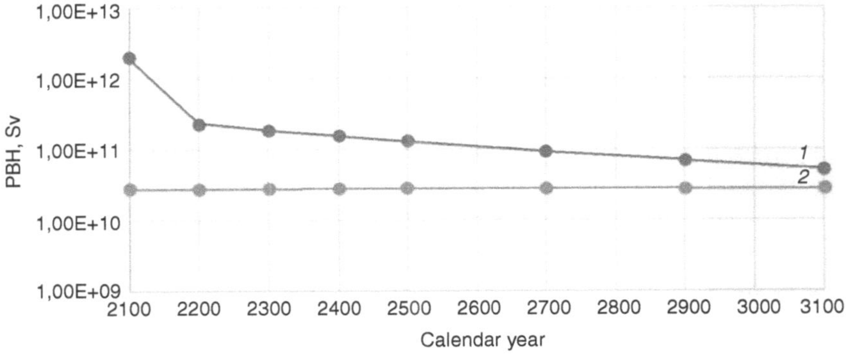

FIGURE 9.9 Potential biological hazard of waste produced by thermal and fast reactors with 10-fold increase in potential biological hazard from americium and the relevant amount of natural uranium in different calendar years: 1 – natural uranium; 2 – RW.

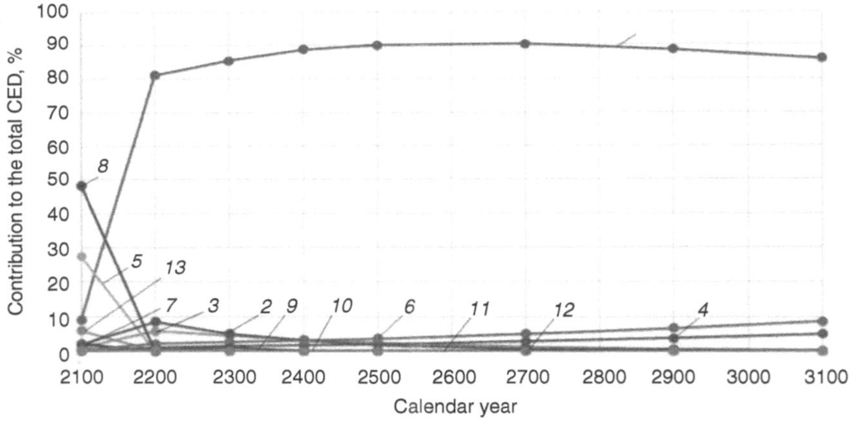

FIGURE 9.10 Contribution of significant radionuclides from nuclear waste generated by thermal and fast reactors to PBH including 10-fold increased PBH from americium 1 – 241Am; 2 – 238Pu; 3 – 242mAm; 4 – 239Pu; 5 – 144Ce; 6 – 240Pu; 7 – 137Cs ; 8- 106Ru; 9 – 154Eu ; 10 – 151Sm ; 11 – 85Kr; 12 – 90Sr; 13 – 147Pm.

Contribution to the aggregate average LAR from RW in case of 10-fold increase in potential biological hazard of americium depending on the calendar year is presented in Fig. 9.12. ^{241}Am isotope also makes the dominant contribution here.

Effect of variations of average CED values (from Tables 9.1 to 9.3 of the Appendix) on the time of achievement of radiation equivalence (calculated on the basis of the CED) and radiological equivalence (calculated on the basis of the LAR) is shown in Fig. 9.13 where the CED from natural uranium is reduced by 30%, and the CED from RW is increased by 30%. In this case radiological

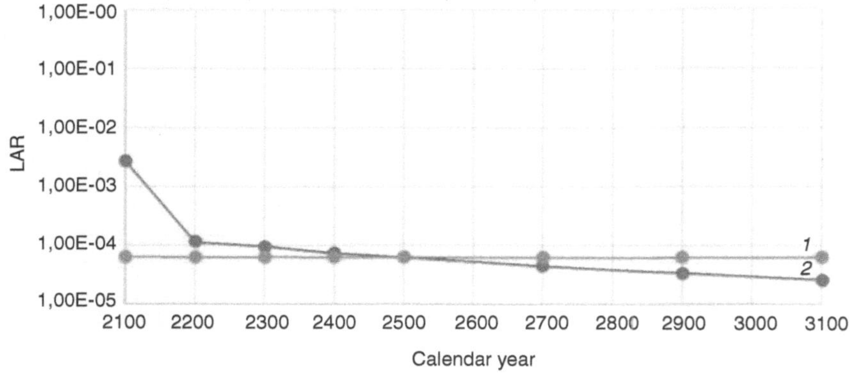

FIGURE 9.11 Lifetime attributable risk from waste generated by thermal and fast reactors with 10-fold increase of americium potential biological hazard and the relevant amount of natural uranium calculated on the basis of the doses scaled to 1 mSv of natural uranium potential biological hazard. The intersection point in case of linear interpolation is in 2514 (315 years later than for the initial RW): 1 – natural uranium; 2 – RW.

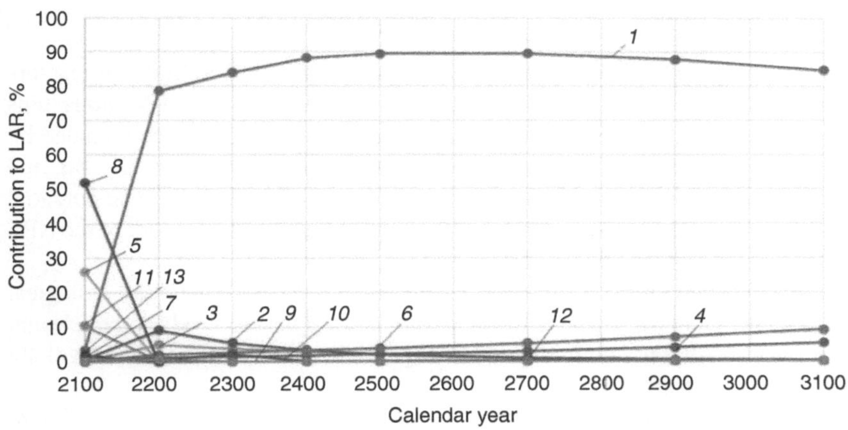

FIGURE 9.12 Contribution of significant radionuclides into the lifetime attributable risk (LAR) from waste generated by thermal and fast reactors with 10-fold increase of americium potential biological hazard: 1 – 241 Am; 2 – 238Pu; 3 – 242mAm; 4 – 239Pu; 5 – 144Ce; 6 – 240Pu; 7 – 137Cs; 8 – 106Ru; 9 – 154Eu; 10 – 151Sm; 11 – 85Kr; 12 – 90Sr; 13 – 147Pm.

and radiation equivalence will be achieved within 100 years and 700 years after RW generation respectively. Similar approximate evaluations demonstrate that in case of double uncertainty in dose assessment radiological equivalence still remains achievable within 270 years after RW generation, while radiation equivalence becomes almost unachievable.

Apart from the development of the scenario for thermal and fast reactors in Russia considered above modeling of the scenario with development of thermal

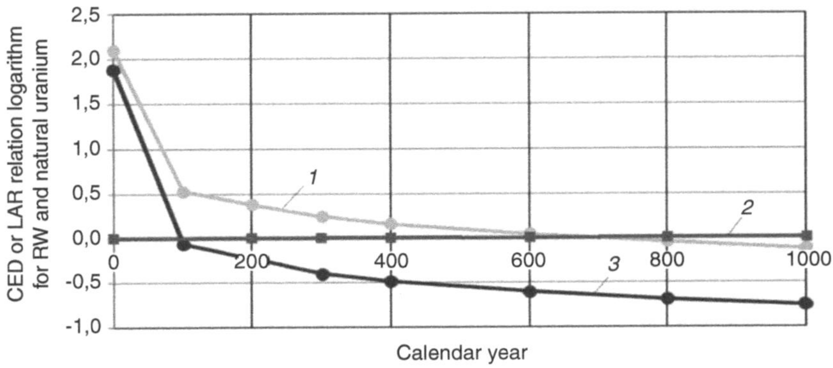

FIGURE 9.13 The time of radiation equivalence (calculated on the basis of the ratio of the CED from RW to CED from natural uranium) achievement and radiological equivalence (calculated on the basis of the ratio of the LAR from RW to LAR from natural uranium); the dose from natural uranium is reduced by 30%, and the doses from RW are increased by 30% in comparison with the values specified in Tables 9.1–9.3 of the Appendix: 1 – ratio between the CED from RW and the CED from natural uranium, dose uncertainty of 30%; 2 – radiation and radiological equivalence level; 3 – ratio between the LAR from RW and the LAR from natural uranium, dose uncertainty of 30%.

reactors alone is also performed. It is assumed that SNF from thermal reactors is not processed but accumulated at the storage facilities. In total 84 thous. tons of VVER and RBMK SNF with different composition will be accumulated by 2100. The aggregate activity of SNF materials amounts to $5.46–10^{19}$ Bq, and the aggregate potential biological hazard amounts to $1.0–10^{13}$ Sv. The potential biological hazard of consumer natural uranium within the framework of the NPE scenario amounts to $2.8–10^{10}$ Sv.

Time-related intersection of the VVER and RBMK SNF potential biological hazard curve and the permanent potential biological hazard level for the relevant amount of natural uranium is shown in Fig. 9.14. Radiation equivalence will not be achieved even in 25,000 years.

Dynamics of the contributions of significant F radionuclides from SNF generated by VVER and RBMK into the potential biological hazard (Sv) is shown in Fig. 9.15. Only the radionuclides with the contribution to PBH that exceeds 1% at least once per 25,000 years are presented. The maximum contribution of ^{241}Amto the PBH is observed at the initial storage stage, but then contribution of ^{240}Pu and ^{239}Pu isotopes takes the first place.

Dynamics of the LARs from SNF generated by VVER and RBMK and constant LAR level from the relevant amount of natural uranium is shown in Fig. 9.16. Radiological equivalence will be achieved in 20,513 years after SNF storage commencement.

Contributions of significant radionuclides to the aggregate average LAR from VVER and RBMK generated SNF depending on the calendar year are shown in Fig. 9.17. As in the case of potential biological hazard ^{241}Am isotope makes the

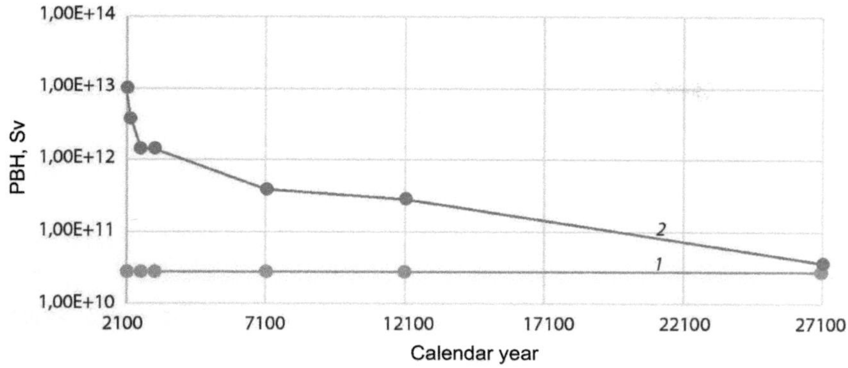

FIGURE 9.14 Dynamics of potential biological hazard from SNF of thermal reactors and potential biological hazard from the relevant amount of natural uranium: 1 – natural uranium; 2 – VVER and RBMK SNF.

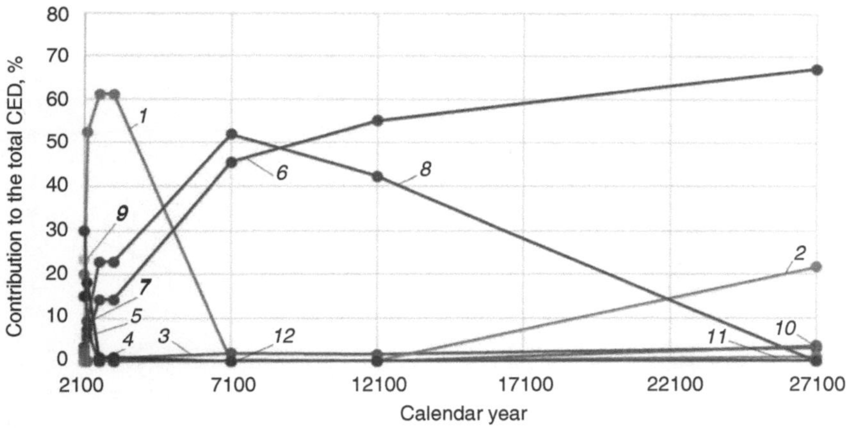

FIGURE 9.15 Dynamics of contribution (%) of significant radionuclides from SNF generated by thermal reactors: to pPBH of 1 – 241 Am; 2 – 210Pb; 3 – 243Am; 4 – 238Pu; 5 – 244Cm; 6 – 239Pu; 7 – 134Cs; 8- 240Pu ; 9 – 137Cs; 10 – 241Pu; 11 – 237Nr; 12 – 90 Sr.

maximum contribution to the risk at the first storage period, contribution of ^{240}Pu and ^{239}Pu isotopes PBH begin to dominate later.

In accordance with the adopted development scenario for large-scale nuclear power in Russia 541.7 thous. tons of natural uranium will be consumed by thermal and fast reactors, and 7.523 thous. tons of long-lived RW will be accumulated.

Taking into account these mass ratios equity of the committed effective dose CED to the public from RW and natural uranium (radiation equivalence) will be achieved in 287 years of RW storage, and equity of radiation risks (radiological equivalence) – in 99 years of storage. Radiological equivalence will be achieved 188 years earlier than radiation equivalence.

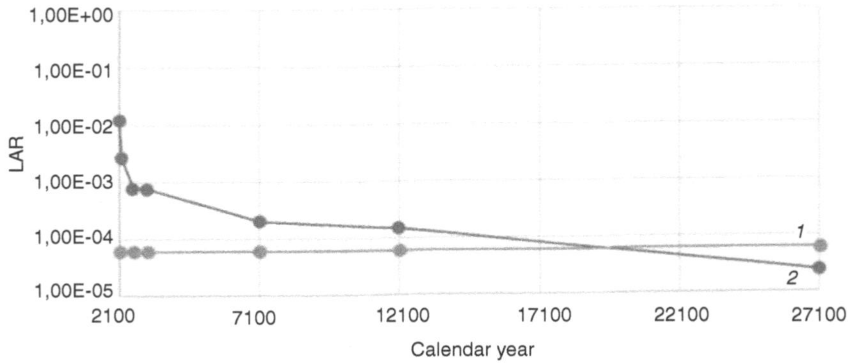

FIGURE 9.16 Dynamics of the LAR from SNF generated by thermal reactors and the LAR value from the relevant amount of natural uranium calculated on the basis of the doses scaled to 1 mSv of PBH of natural uranium; the intersection point for the curves in case of linear interpolation is in 22613: 1 – natural uranium; 2 – VVER and RBMK generated SNF.

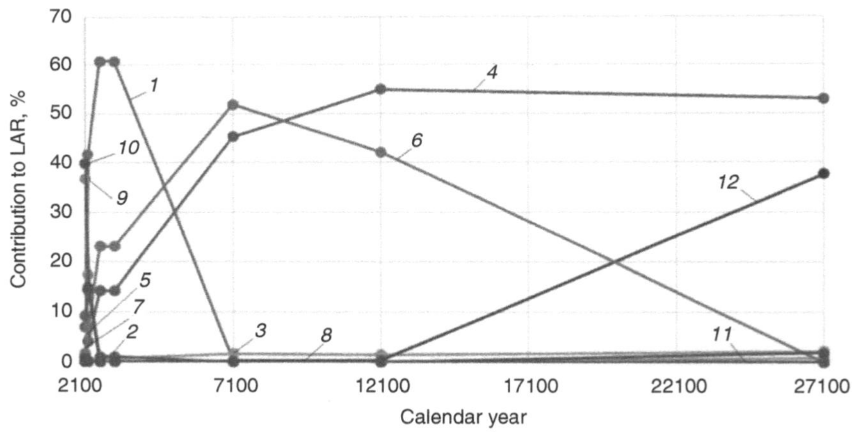

FIGURE 9.17 Dynamics of contributions (%) of significant radionuclides into the LAR from SNF generated by thermal reactors. 1 – 241Am; 2 – 238Pu; 3 – 243Am; 4 – 239Pu; 5 – 244Cm; 6 – 240Pu; 7 – 134Cs; 8 – 241Pu; 9 – 137Cs; 10 – 90Sr; 11 – 237Nr; 12 – 210Pb.

In case of double uncertainty in dose assessment radiological equivalence remains achievable within 270 years after RW generation, while radiation equivalence becomes almost unachievable. In case of the dose assessment uncertainty of approximately 30% radiological and radiation equivalence will be achieved within the time limits of 100 years and 700 years after RW generation respectively.

^{241}Am isotope makes the maximum contribution both to radiation dose and radiation risk during RW storage periods exceeding 100–150 years.

Thus extraction of americium from RW considerably reduces the period for achievement of both radiation and radiological equivalence, and application of the radiological equivalence principle enables to substantiate significant reduction of the RW storage time.

Consideration of the development scenario for nuclear power based only on thermal reactors demonstrates that in this case radiation equivalence achievement is impossible. The storage time of more than 20,000 years is necessary to achieve radiological equivalence.

Chapter 10

Technology support of the nonproliferation regime and conditions for export of the CNFC and FNR technologies

The Treaty on the Non-Proliferation of Nuclear Weapons effective from 1970 and prolongated for an indefinite term in 1995 was not able to prevent emergence of new states having nuclear weapon. At the same time the nonproliferation considerations in the USA hampered development of fast reactors and the closed nuclear fuel cycle: President G. Ford prohibited processing of NPP SNF in 1976 due to plutonium proliferation hazard, and later J. Carter confirmed this prohibition. But plutonium had not disappeared because of that, and the expression "deferred decision'' was closely associated with SNF. The USA searched for the solution in the area of direct SNF disposal according to Yucca Mountain project. After 22 years of work and expenditures amounting to about 9 bln. dollars the project was closed in 2009.

Strategy-2000 set the task for simultaneous solution of the SNF problem and technology enhancement of the nonproliferation regime. The possibility to give-up short doubling time and to eliminate uranium blanket, transition to equilibrium fuel in fast reactors with BR ~1 and gradual departure from uranium enrichment create the required preconditions for technology support of the nonproliferation regime. The fuel processing technology shall eliminate any possibility to use it for Pu, ^{235}U or ^{233}U extraction. In this case it is mainly reduced to fuel purification from fission products with the residue of 1–10%.

The nonproliferation problems are not urgent for nuclear power in Russia, or rather reduced to prevention of any theft of nuclear materials, but nonproliferation assurance is one of the most important requirements for export of fast reactors and the relevant technologies.

Development of fast reactors with inherent safety seems the most feasible technique for disposal of plutonium accumulated in irradiated fuel of modern power reactors with its transfer from spent fuel pools to the most protected conditions of reactors and fuel cycle facilities. If this technology turns out to

Closed Nuclear Fuel Cycle with Fast Reactors: Handbook of Russian Nuclear Power.
DOI: https://doi.org/10.1016/B978-0-323-99308-1.00003-9

be attractive from the viewpoint of economy, safety and waste management the concerned countries will be able to arrange gradual replacement of light-water reactors operated on enriched uranium with this technology, thus closing the legal channel for proliferation of weapon-grade materials.

Of course, no new NFC technology will be able to prevent illegal use of the existing plutonium extraction technologies, particularly from the fuel of thermal reactors, as well as uranium enrichment for production of weapon-grade materials. But the above-mentioned set of arrangements provides "technology barriers'' for proliferation separating the nuclear power fuel cycle and weapon production and thus enhancing the efficiency of political measures and inspections of the non-proliferation regime implementation in combination with satellite supervision over configuration maintenance for the NFC buildings and structures.

Chapter 11

Economic competitiveness of innovative nuclear power

11.1 Requirements for competitiveness of FNRs with the CNFC

At present development of nuclear power is based on NPPs with pressurized water reactors with subsequent transition to more innovative NPP design with RP VVER-TOI similar in terms of technology. Arrangement of the sound structure for two-component nuclear power providing for transition to the closed cycle based on fast neutron reactors (FNRs) is planned after 2030. Arrangement of the sound structure for nuclear generation is associated with both development of traditional technologies for NPPs with TNRs subject to improvement of their technical and economic indices (TEI) and development of high-capacity fast reactors with sodium coolant and subsequent transition to NPPs with sodium-cooled fast reactors subject to their proven competitiveness.

The estimated TEI of NPPs under design and planned for construction in Russia are presented in Table 11.1. In order to assess competitiveness of thermal neutron NPPs their parameters are considered with regard to the requirements for the enhanced design of NPPs with VVER-TOI+ and VVER-S operated within the open NFC and NPPs with RBN-1200 operated within the closed NFC. For the specified calculation conditions, the parameters of VVER-TOI+, VVER-S and RBN-1200 are target ones taking into account confirmation of the requirements for further extension up to 2030.

Expenditures of the closed NFC for NPPs with RBN-1200 include SNF processing with subsequent fabrication of fresh nuclear fuel from the regenerated product with due regard for RW disposal costs. Expenditures of the open NFC for NPPs with VVER-TOI+ and VVER-S include the expenses for fresh fuel and SNF processing with subsequent disposal of the generated RW amounts.

The possibility to achieve average values of MNIT fuel burnup of at least 12% ha in high-capacity FNRs in accordance with the planned R&D programs is taken into account in competitiveness analysis.

Closed Nuclear Fuel Cycle with Fast Reactors: Handbook of Russian Nuclear Power.
DOI: https://doi.org/10.1016/B978-0-323-99308-1.00030-1

TABLE 11.1 TEI of advanced NPP designs by 2035 (cost parameters at 2018 values, VAT exclusive).

Parameter	VVER-TOI+	VVER-S	RBN-1200 (requirements)
Installed capacity, MW (e)	2 x 1255	2 x 1255	2 x 1220
Relative capital investments (for the Center conditions), thous. RUB/kW (e)	96	90	84–90**
1-st power unit construction period, years	6	6	6
Type of fuel	UO_2	MOX (single recycling)	MNIT/MOX
Fuel factor, kop./kW·h	21	26*	20
Auxiliary power consumption, %	6.4	6.4	5.0
Service life, years	60	60	60
Capacity utilization factor, %	~ 90	~ 90	~90

Expenses for MOX fuel fabrication and handling of MOX SNF; expenses for U-SNF processing used as the source of regenerated product for MOX fuel fabrication are accounted within the traditional U-NFC costs (FF of VVER-TOI+).
***The minimum value – the competitiveness requirements within the framework of Proryv Project Framework, the maximum value – at the level of NPPs with RP VVER-S (~20% lower than the capital investments to the basic NPP design with RP VVER-TOI).*

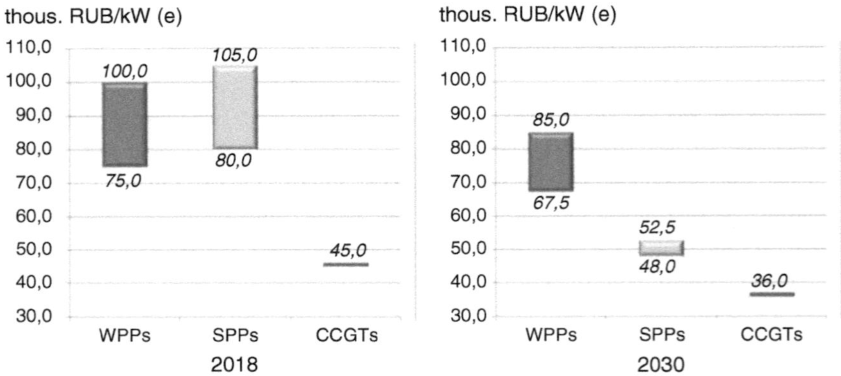

FIGURE 11.1 Relative capital investments into alternative generation at 2018 values, VAT exclusive.

Comparison of alternative power technologies competing with nuclear generation shall be also performed with due regard for potential improvement of their economic parameters. Estimated relative capital investments into generation alternative to nuclear one in Russia are presented in Fig. 11.1. Cost reduction for wind and solar power plants is planned in accordance with the global forecast.

Reduction of the capital investment level (at the fixed level of prices) in combined cycle turbine gas units (CCGTs) implies arrangement of import substitution for the main equipment by 2030.

Capital investments into CCGTs are considerably lower than capital investments into WPP and SPP construction. As distinct from NPPs, changes of relative capital investments and other TEI for large-scale thermal power plants competing with NPPs in the long run are only marginally related to further improvement of technologies. Dependence on import of the main CCGT equipment and its successful localization or replacement with domestic designs will remain the most crucial uncertainty factor at present and for further extension at least up to 2025–2030. The pricing policy at the market of materials and equipment is also significant. Almost twofold decrease of the currency exchange rate as compared with 2014 did not result in any similar benefits for material and equipment costs. Other TEI of the generation alternative to nuclear power including variations in the natural gas prices are assumed to be at the level specified in Chapter 1.

The LCOE parameter calculations for various generation types under the conditions of the Central European part of Russia, Ural and Siberia are presented in Table 11.2. The calculations are carried out with the nominal discount rate of 10, 7, and 5%. The calculations for regions are different in the LCOE parameters for CCGTs (the capital investments and natural gas prices vary) and NPPs (the capital investments vary).

The following main conclusions may be made subsequent to analysis of the calculation results:

- Significant reduction of the SPP LCOE will take place by 2035. The WPP LCOE will also decrease but at the lower rate;
- The LCOE of power units with CCGTs is decreased due to construction cost reduction because of import substitution for the main equipment. CCGTs will remain the main competitive technology for NPPs till 2035;
  - optimized NPPs with TNRs (VVER-TOI+ and VVER-S) will be more efficient power generation sources than SPPs and WPPs in the long run up to 2035;
  - NPPs with TNRs will be more competitive in comparison with CCGTs subject to achievement of the target parameters with low discount rates for all location areas under consideration. In this case their equal efficiency with CCGTs will be maintained even with high (market) discount rates for the conditions of location in the Center and Ural;
  - NPPs with FNRs will be more efficient than CCGTs subject to fulfillment of the competitiveness requirements with all discount rates under consideration and in all location areas.

It should be noted that the option with redundant gas turbine power units is considered in the modern economic analysis of RES that are not capable to ensure stable power supply for reasons beyond control. The necessity for construction of the relevant units together with RES is provided. In accordance

TABLE 11.2 LCOE of power plants in the Central European part of the RF, Ural and Siberia with different discount rates for TEI of 2035, kop./kW • h.

Generation type	Center		Ural		Siberia	
	Min.	Max.	Min.	Max.	Min.	Max.
	Nominal discount of 10%					
WPPs	434	546	434	546	434	546
SPPs	473	518	473	518	473	518
CCGTs	223	249	212	237	213	238
NPPs with TNRs	226	231	232	236	241	247
NPPs with FNRs	212	223	216	226	225	235
	Nominal discount of 7%					
WPPs	352	444	352	444	352	444
SPPs	384	420	384	420	384	420
CCGTs	204	229	192	215	192	215
NPPs with TNRs	181	182	184	186	191	193
NPPs with FNRs	162	169	164	172	170	177
	Nominal discount of 5%					
WPPs	292	368	292	368	292	368
SPPs	319	348	319	348	319	348
CCGTs	191	215	178	200	177	199
NPPs with TNRs	150	152	152	154	157	158
NPPs with FNRs	128	133	130	135	134	138

with the estimates of Energy Research Institute of the Russian Academy of Sciences such redundancy will result in the WPP and SPP LCOE increase by ~50%, and consequently – in the RES competitiveness reduction.

In this case development of RES technologies resulting in reduction of their relative capital investments and LCOE should not escape our attention. Development of nuclear generation technologies shall be ensured in order to improve their TEI and provide sustainable competitiveness. The main areas of such development are defined and associated with closing of the NFC and implementation of the inherent safety reactor technologies.

Fulfillment of the established requirements for TEI of innovative NPP designs with high-capacity FNRs and the requirements for the CNFC economy become the key objectives for the nuclear industry.

The established competitiveness requirements will ensure economic efficiency of the industrial power supply complexes (IPC) with high-capacity FNRs in comparison with alternative generation (including CCGTs) with all discount rates under consideration taking into account potential improvement of their TEI.

11.2 Effect of load following on the NPP economy

Under the conditions of the expected increase of the nuclear generation share in the national power balance structure the contribution of NPPs into the total electricity generation may exceed the basic load limits. This explains the requirements for the possibility to adjust the capacity in accordance with changing load in the grid. On the other side, maintenance of the maximum possible capacity utilization factor is feasible for NPPs with their high relative capital investments, and any deviations from this regime negatively affect their competitiveness in comparison with other generation technologies. Production of hydrogen as storable energy source material for subsequent use within the peak load periods or sale as a commercial product may be one of the techniques for the maximum capacity utilization factor maintenance at NPPs under the lower power demand conditions.

The issue related to the effect of the load following operation mode for RPs with FNRs is considered in the course of design from the viewpoint of transient process dynamics, but remains open with regard to determination of the quantitative changes in technical and economic characteristics.

The main areas requiring further investigation:

- Limitations with regard to fatigue strength of fuel rods in case of cyclic power variations;
- Limitations with regard to corrosion resistance of fuel rods and fuel assemblies upon the fuel campaign duration increase in order to achieve the specified burnout depth taking into account the average capacity utilization factor reduction;
- Effect of the fuel campaign duration increase of the CNFC fuel balance at the system level.

In accordance with the data of the IAEA report «Hydrogen Production from Nuclear Power» [169] fossil hydrocarbon fuel will remain the basic raw material for hydrogen production on the immediate and mid-term horizon. Usage of natural gas at the initial stage of hydrogen power engineering is beneficial from the viewpoint of relatively low environmental impact and economy. But in the long run it is expected that hydrogen will be produced mainly from water with the use of even more environmentally friendly technologies where the final cost of the produced energy source material will not depend on the prices of gas or any other fuel types. The opportunities and scope of nuclear capacity utilization for hydrogen production will depend on the following factors:

- Economy of large-scale hydrogen production and its subsequent storage and transportation;
- The public opinion and the adopted policy line of the global community with regard to climate changes and CO2 emissions;

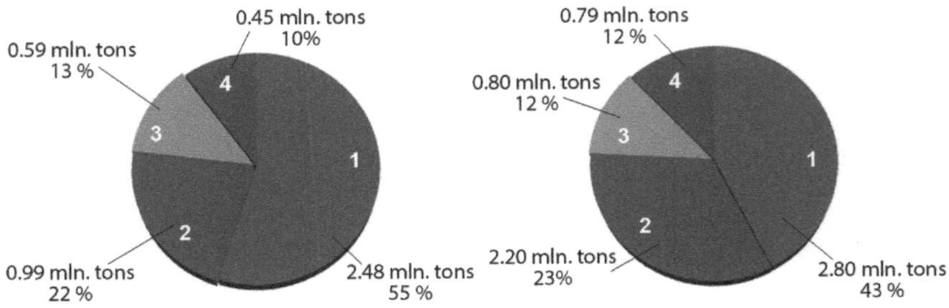

FIGURE 11.2 Hydrogen consumption structure in 2013 (a) and 2020 (b):1 - ammonia production; 2 - oil refinery; 3 - methanol production;4 - metal reduction (glass, food and power industry).

- Saving of hydrocarbon resources for more feasible and environmentally friendly subsequent usage;
  - Amounts of oil and natural gas extraction;
  - commitment to enhancement of energy security and independence from any external supplies of energy source materials.

In the beginning of the XXI century the majority of hydrogen produced in Russia is used for ammonia production. Taking into account decreasing resources of high-quality sweet light crude oil hydrogen will be more and more sought after for petroleum product refining in the nearest future. The current and estimated hydrogen consumption parameters determined on the basis of the data presented at the second CREON ENERGY International Conference "Hydrogen 2014" are given in Fig. 11.2. According to these data, hydrogen consumption in Russia will increase from 4.5 mln. tons in 2013 to 6.6 mln. tons in 2020.

Taking into account decreasing resources of high-quality sweet light crude oil hydrogen will be more and more sought after for processing enhancement. The share of hydrogen consumption in metallurgy, glass, food and power industry will also increase at the moderate rate. The opportunities for arrangement of the commercial hydrogen market in Russia strongly depend on involvement of the government and the fuel and energy industry enterprises into the programs for reduction of carbon emissions and environmental protection. The demand for environmentally friendly means of transport and decentralized power and heat generation techniques may considerably facilitate emergence of fuel rods as a new hydrogen consumer class. However, it should be noted that even upon formation of the conditions (infrastructure, grants) favorable enough for development of hydrogen-powered transport there is no guarantee that vehicles with fuel cells will rapidly start to replace the existing car park. Besides, the market of vehicles equipped with storage batteries is already developing at present which can eventually prevent from widespread use of hydrogen-powered transport.

Wider application of fuel cells is expected in the distributed generation area for combined production of electric power and heat. Fuel cells are interesting for Russia with regard to development of the remote regions in the North and the

Far East where they could be used as independent power sources for residential houses and machinery [100]. Solid oxide fuel cells that are more resistant to fuel contamination in comparison with other types are the most promising in this aspect. In case of pure hydrogen usage as fuel harmless water steam is the only exhaust of the system as distinct from natural gas or methane.

The following areas and objectives of the fuel and energy industry development were considered in the development scenarios for industrial power supply complexes based on NPPs with FNRs for comparison of generation options in the load following mode and with the constant capacity utilization factor on the basis of hydrogen production technologies:

- Increase in the consumption share of more environmentally friendly energy products with higher quality within the entire production cycle as well as the power generation share in the fuel and energy balance (FEB) will enlarge the share of nuclear power in the energy consumption structure;
- Increase in the refinery yield for extracted heavy oil due to deterioration of its physical and chemical characteristics;
- Development of the petrochemical industry with formation of hydrocarbon deep processing centers represented by small and medium-sized enterprises.

The specific discounted profit of the system of NPPs with FNRs (with the discount rate of 5%) including additional equipment for hydrogen production was assessed as the main economic parameter. The integral parameters were determined on the basis of the time interval up to 2050 including the period for transition to large-scale distribution of industrial power supply complexes.

Continuation of NPP operation with the rated capacity utilization factor during the grid load reduction periods not only helps to eliminate any negative impact on the reactor plant equipment but also enables to generate excessive power for production of hydrogen as storable energy source material.

Excessive electric power may be used both in low-temperature electrolysis (LTE) of water and high-temperature electrolysis (HTE) of steam from the reactor plant. In the latter case, the specific energy consumption for production of 1 kg of hydrogen is reduced, and the hydrogen yield increases with the same excessive power generation, but the expenditures for more expensive electrolysis equipment also increase. Eventually the option with HTE turns out to be more preferable from the economic viewpoint with the assumed initial data.

Hydrogen and oxygen accumulated within the reduced load periods may be used for additional electricity generation during the peak load periods. For this purpose, the industrial power supply complex shall include an after-burning steam generator in addition to the electrolysis unit; it may be used either for steam temperature increase upstream of the main NPP turbine or for supply of an additional turbine generator. The first option requires special-purpose analysis of the possibility to force the turbine parameters set in the IPC design including its dynamic characteristics, but no such analysis has been performed as of the time of this book writing.

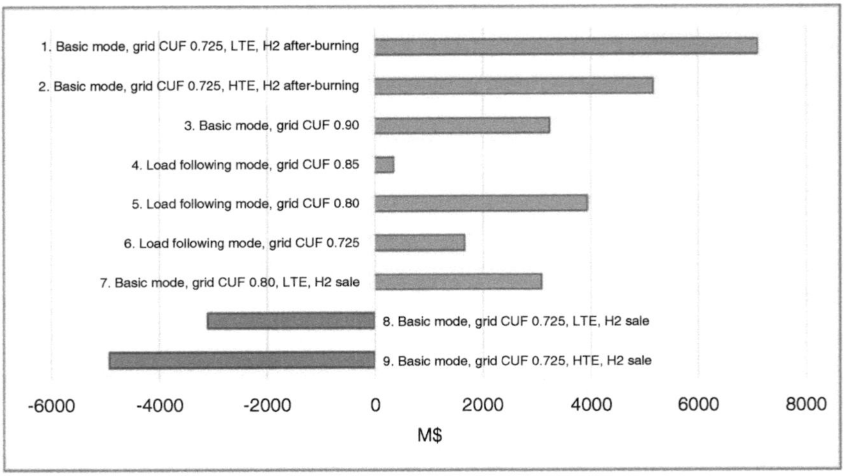

FIGURE 11.3 Specific integrated profit.

Arguments in favor of the scenarios with hydrogen after-burning and peak energy generation include absence of any need for hydrogen transportation to remote consumers, independence from the market price of hydrogen and the possibility to gain additional profit from peak energy sale at the premium price. Disadvantages of these scenarios are due to additional expenditures for the after-burning steam generators and especially for the additional turbine generator.

In case of absolute necessity for adjustment of the power output to the grid the options with sale of hydrogen produced through the use of excessive electric power become more beneficial than simple reduction of the capacity utilization factor (compare Scenarios 3 and 5, 4 and 6 in Fig. 11.3). In this case HTE of steam from the reactor plant enables to increase hydrogen yield and to gain higher profits (compare Scenarios 6 and 7).

The options of near-plant electrolysis with after-burning of the produced hydrogen and usage of an additional turbine generator for power generation during the peak grid load periods seem unfeasible from the economic viewpoint with the expected TEI of high-temperature electrolysis units.

The presented evaluations are indicative of the necessity to coordinate the development process for the IPC system based on FNRs with the industrial environment not only with regard to satisfaction of its needs (including, in particular, adjustment of power generation capacities) but also through shaping of these needs towards arrangement of opportunities to achieve the maximum efficiency of the nuclear power generation technologies being developed. These may include combination of large-scale production of hydrogen (or any other energy source materials) based on NPPs with petrochemical and metallurgical facilities, road transport fuel supply networks, etc.

Part III

Nuclear fuel and closing of the nuclear fuel cycle

Chapter 12

Uranium and uranium-plutonium nuclear fuel

12.1 Uranium fuel

Uranium is the main fuel for modern nuclear reactors. Many uranium compounds (metal, alloy, oxide, nitride, or carbide) may be used, but uranium dioxide (UO_2) enriched to 3–5% of ^{235}U is the main nuclear fuel at present.

Uranium minerals (with the oxides from UO_2 to U_3O_8) are the main natural uranium-containing raw materials although development of sandstone deposits is also feasible subject to uranium content of at least 0.1%. Besides, extraction of uranium from the deposits with lower uranium content (up to 0.004%) as a by-product in the course of phosphate processing for fertilizers or gold mining may be also feasible.

After mining of uranium ore (i.e., uranium minerals in barren rock) uranium minerals are extracted from it (mechanically and chemically) at leach plants.

The produced uranium concentrate (U_3O_8) is purified from any remaining impurities at refinery plants and transformed into uranium hexafluoride (UF_6) which is supplied to uranium enrichment plants where uranium is enriched with ^{235}U isotope and transformed into uranium dioxide (UO_2) – the basic fuel for modern nuclear reactors.

12.2 Uranium-plutonium nuclear fuel

Unlike ^{235}U contained in natural uranium (although in small quantities [0.72%]) ^{239}Pu can be produced only in nuclear reactors. Apart from fission products trans-uranium nuclides with plutonium atop are also accumulated in irradiated nuclear fuel (SNF) in the course of the nuclear reactor operation with initial uranium nuclear fuel. Apart from fission products SNF unloaded from a thermal reactor with burnup of 33,000 MW·day/t contains 0.83% of initial fissionable ^{235}U isotope and 95% of fertile ^{238}U isotope, as well as 0.68% of secondary fissionable plutonium isotopes.

Closed Nuclear Fuel Cycle with Fast Reactors: Handbook of Russian Nuclear Power.
DOI: https://doi.org/10.1016/B978-0-323-99308-1.00034-9

Extraction of fissionable and fertile nuclides from SNF for their use in production of new nuclear fuel is the necessary precondition for efficient functioning of large-scale nuclear power.

The following fuel was considered and developed for fast reactors at different times:

- Metal (UZr and UPuZr) – in the USA, United Kingdom (UK), Russia, India, and the Republic of Korea;
- Oxide (UO and $UPuO_2$, or MOX fuel) – in Russia, France, Japan, UK, the USA, Germany, India, and China;
- Carbide (UC, UPuC) – in Russia, France, India and the USA;
- Nitride (UN and UPuN,) – in Russia, the USA, France, and Japan.

At the initial stage of the fast reactor concept development and in the first experimental plants priority was given to dense fuel types (metal, nitride, carbide). However, oxide fuel was specifically selected for the first prototype plants BN-350 and BN-600 (UO_2) as well as for BN-800 ($UPuO_2$). Oxide fuel was also developed in France for Phenix and SuperPhenix reactors and in the UK for PFR reactor. After accidents at EBR-1 (1956) and E. Fermi (1966) reactors the projects of FFTF and Clinch River reactors in the USA were also reoriented from metallic fuel to oxide one.

It was a reasonable solution for the above-mentioned initial period in development and acquisition of the technology for sodium-cooled fast reactors due to maturity of oxide fuel technology for thermal reactors.

Advantages of dense fuel types (they have higher density, higher thermal conductivity and higher compatibility with coolant) for fast reactors enable to achieve the BRC of > 1 and to reduce the amount of heat accumulated in the fuel significantly thus enhancing the reactor plant safety. That is why transition from oxide fuel to dense fuel types is considered in all countries developing advanced fast neutron reactors (FRs). The problem of fuel selection (nitride, carbide or metal) is still subject to discussion. Analysis of the available data on the properties of dense fuels, their advantages and drawbacks and the arguments for selection of nitride dense fuel for implementation in Proryv Project are provided in the study [96].

At the initial development stages of the physical basis and the first FR designs the focus was on the possibility to achieve the highest fuel breeding ratio values. It was due to poor fuel base for FRs in the form of plutonium absent in nature and the necessity for its rapid production that is most clearly demonstrated in the concept of "minimum doubling time T_2" proposed by V.V. Orlov [63]. Metallic fuel is the most suitable for implementation of this strategy.

By 2019 the list of priorities defining the place of fast reactors in the nuclear power system has changed considerably. The safety problems (particularly environmental safety), competitiveness, SNF and RW accumulation, nonproliferation, optimal utilization of natural resources came to the fore. The main requirements for the new generation nuclear power based on the inherent

TABLE 12.1 Comparison of the main properties of different fuels.

Parameter	UPuO$_2$ U$_{0.8}$ • Pu $_{0.2}$ O$_2$	UPuC U$_{0.8}$ • Pu $_{0.2}$ C	UPuN U $_{0.8}$ • Pu $_{0.2}$ N	U-Pu-Zr U-19Pu-10Zr
Theoretical density, g/cm^3	11.04	13.6	14.32	15.73
Heavy metal density, g/cm^3	9.3	12.4	13.1	14.1
Recommended smeared fuel density, % of theoretical density (g/cm^3)	~85 (9.3)		80 (11.5)	~75 (11.8)
Thermal conductivity, W/(m-K)				
T = 1000 K	2.6	19.7	15.8	25
T = 2000 K	2.4	24	20.1	
Melting temperature, K	3023	2750	3050	1400
Permissible fuel temperature limit, °C	~2500	~2200	Not more than 1800* (dissoc.)	Not more than 800

**Insufficient data, additional investigations are required.*

TABLE 12.2 Comparison of MOX and MN fuel.

Parameter	MOX	MN	Comparative assessment	
			MOX	MN
Safety				
BRC > 1	With the breeding layer in the core	Provided	More difficult	Simpler
Quasi-equilibrium mode (stable reactivity margin)	With the breeding layer in the core	Possible	More difficult	Simpler
Void reactivity effect	About 0	About 0	=	=
Minimum possible reactivity margin at the rated power depending on the core design, % Δk/k	BN (SFR) 1.0–2.5 (3.0 β_{ef} – 7.5 β_{ef})	BN (SFR) 0.5–1.0 (3.0 β_{ef} – 7.5 β_{ef})	Prompt neutron runaway is possible	
Minimum possible reactivity margin at the rated power depending on the core temperature, β_{ef}	BREST – not considered	BREST β_{ef}(hot) – 1.5 β_{ef}(cold)	Prompt neutron runaway is possible	Prompt neutron runaway can be prevented
Consequences of severe accidents without prompt neutron runaway	UTOP and ULOF		Greater fuel damage	Less fuel damage
Thermal conductivity, W/(m-K)	2.6	15.8	Worse	Better
T = 1000 K				
T = 2000 K	2.4	20.1		
Maximum fuel temperature, °C	-2200	1100 liquid metal sublayer	Worse	Better
		1400 gas sublayer		
Melting temperature, °C	3023	3070	=	=

(*continued on next page*)

TABLE 12.2 Comparison of MOX and MN fuel—cont'd

Parameter	MOX	MN	Comparative assessment MOX	Comparative assessment MN
Compatibility with cladding	Normal	Normal	=	=
coolant	Normal	Normal		
Swelling	Moderate	Moderate	=	=
Heat capacity, kJ/mol·K	95	70	Worse	Better
Energy accumulated in NF	High	Medium	Worse	Better
Pyrophoricity of NF powders and the possibility of the fuel kernel ignition in case of any contact with air (high temperatures)	None	Yes	Better	Worse
Ecology				
SNF accumulation			=	=
Long-lived radioisotope production	None	C-14	Better	Worse
Radiation equivalence achievement	Possible for h.n. subject to complete U, Pu, MA recycling and partial recycling of fission products	Possible for h.n. subject to complete U, Pu, MA recycling and partial recycling of fission products	=	=
Breeding				
Theoretical density, g/cm^3, for h.n.	$U_{0.8}{\cdot}Pu_{0.2}\ O_2$	$U_{0.8}{\cdot}Pu_{0.2}\ N$	Worse	Better
	11.04	14.32		
	9.3	13.1		
BR	<1.2	<1.35	Worse	Better

Note: h.n., heavy nuclei.

safety principles were specified in a gist in the study [43] for the first time. The comparative analysis results for two main types of dense fuel – nitride and metal – are provided below in order to define the optimum option for implementation of this concept (Table 12.1).

It should be noted that nitride fuel has high density and thermal conductivity (1.4 and 10 times higher in comparison with oxide). Even higher values of density and thermal conductivity are typical for metallic fuel.

Despite the fact that metal has the maximum theoretical density nitride fuel (UN and UPuN) is almost equal to alloy metallic fuel where Zr and increased porosity necessary for swelling reduction and creep growth decrease its density.

Phase changes of metallic fuel and especially its interaction with steel claddings resulting in formation of low-melting eutectics (510°C for Pu is below the coolant temperature at the core outlet; 710°C for U and 750°C for (U + 15%Pu + 10%Zr) determine low margins before breakage in case of any accidents with temperature increase or require the coolant temperature reduction.

In order to use higher thermal conductivity of dense fuel efficiently it is reasonable to use fuel rods with liquid metal sublayer.

In comparison with carbide nitride has an advantage due to its better manufacturability and higher continuity of the process cycle to the cycle of its predecessor – oxide fuel.

A relative drawback of nitride resides in neutron absorption during $^{14}N(n,p)^{14}C$ reaction resulting in certain deterioration of neutron balance and generation of carbon (^{14}C) with long half-life.

The maximum experience in operation of fast sodium-cooled reactors (more than 147 reactor-years) is accumulated in Russia (the USSR): ~ 45 % of the entire global experience, including operation of BR-5/10 (1959— 2002, Obninsk), BOR-60 (1969, Dimitrovgrad), BN-350 (1972-1990, Aktau, Mangyshlak), BN-600, and BN-800 (1980 and 2015, Zarechny) reactors.

Different fuels (PuO_2, UC, UCN, UN, UPnN, UO_2, $UPuO_2$, alloy and non-alloy metallic fuel, inert matrices fuels) were irradiated and studied in domestic fast reactors. Comparison of MOX and mixed nitride (MN) fuels according to the main parameters is presented in Table 12.2.

Chapter 13

Dense nuclear fuel for fast reactors

13.1 Metallic fuel

The most extensive program for metallic fuel investigation was implemented in the USA.

Metallic fuel had been studied in EBR-II reactor for 25 years. Almost 9000 kg of U-5%Fs were used as regular and experimental fuel. The alloy composition is the following: 0.2%Zr, <0.01%Nb, 1.8%Mo, 1%Tc, 1.5%Ru, 0.3%Rh, 0.2%Pd, 95%U. The main drawback preventing from the alloy use in case of high burnup resides in its poor compatibility with the cladding. From the beginning of 1989 the core was completely loaded with U-Pu-10%Zr and U-10%Zr fuel. Alloying with zirconium was considered as promising for the solidus temperature increase and reduction of diffusive interaction with the cladding.

About 13,000 fuel rods with U-10%Zr and 600 fuel rods with U-Pu-10%Zr were irradiated in EBR-II in total, including about 100 fuel rods – to the burnup of at. 18%. The maximum burnup achieved in fuel rods with U-10%Zr is at. 20.6% and with U-Pu-10%Zr – at. 18.5% [166].

A total of 1050 fuel rods with U-Zr (maximum burnup – at. 14%) and 37 fuel rods with UPuZr (maximum burnup – at. 9%) were irradiated in FFTF reactor. The main parameters of irradiated fuel rods are presented below.

Irradiation conditions and geometrical parameters of fuel rods with doped metal alloy irradiated in EBR-II and FFTF reactors [166].

Smeared density, %	75
Cladding diameter, mm	5.84–7.37
Cladding thickness, mm	0.38–0.56
Ratio of the gas plenum height to the core height	1.0–1.5
Diameter of the fuel column, mm	4.32–5.66
Linear power, kW/m	36–59
Maximum cladding temperature, °C	522–651
Maximum fuel temperature, °C	640–745
Sublayer type	Na
Cladding material	Austenitic steel 316, ferrite steel HT9

Closed Nuclear Fuel Cycle with Fast Reactors: Handbook of Russian Nuclear Power.
DOI: https://doi.org/10.1016/B978-0-323-99308-1.00005-2

Behavior of fuel in emergency situations was studied in FBTA and WPF off-pile units, TREAT, and EBR-II reactors: experiments with power ramp in the reactors, investigation of fuel compatibility with steels under increased temperature in the exvessel units. Seven fuel rods with artificial defects were irradiated in EBR-II reactor.

The studies enabled to determine:

- Dependence of fission gas release and fuel swelling on burnup;
- The scale of fuel axial elongation with different burnup;
- The scale of corrosive cladding damage on the fuel side depending on the temperature;
- The main causes of fuel cladding damage under steady-state and emergency operating conditions.

The accumulated database on the results of studies enables the American specialists to hope that burnup of about 200 GW/t_{hm} can be achieved in fast reactors with metallic fuel, and the simple production and processing technology for U-Pu-Zr fuel facilitates transition to its widespread use, although probably additional verification of this technology is required. But statistical representativeness of the experiments is insufficient, particularly for fuel rods with UPuZr under the conditions of FFTF reactor (effect of the scaling factor, as the fuel rod length is twice larger than for EBR-II), for fuel rods with low-swelling ferrite claddings as well as with high burnup. Additional studies of fuel behavior under emergency conditions are also necessary.

But there are opinions postulating that the existing database is sufficient for use in demonstration reactors subject to introduction of some uncertainty factors that can be excluded subsequently in the course of experience gaining. The assumed parameters of reactors with metallic fuel are presented below.

Parameters of a fuel rod with metallic fuel recommended on the basis of the USA experience [166].

Nominal composition	U-Pu(20%)-Zr(10% wt.)
Pu/(U+Pu) ratio, %	17–28
Theoretical density, % g/cm^3	100/(15.6–15.9)
Smeared density	75% of theoretical density
Fuel and gas plenum volume ratio	1.4
Fuel slug length, cm	91*
External fuel slug diameter after manufacturing, cm	0.5*
Sublayer type	Na
Cladding material	HT9 or 316cw
External/ internal cladding diameter, cm	0.69/0.57*
Maximum linear power, kW/m	49–52
Maximum cladding temperature, °C	620

** Dimensions of the fuel slug and cladding are specified on the basis of experience of UPuZr use in FFTF reactor. The data from testing in EBR-II reactors show that usage of fuel rods with larger diameter is also possible. Thus, the dimensions of fuel rods may be different but the smeared density and the fuel gas plenum volume ratio shall be ensured.*

At present the activities with metallic fuel are going on in the USA within the framework of AFC (Advanced Fuel Campaign) program with regard to investigation and development of transmutation fuel. It is noted that adjustment of the process for commercial production of fuel is necessary. All experiments on fuel with minor actinides are carried out in ATR research reactor. Improved designs of fuel rods with doped metallic fuel are studied for significant enhancement of its burnup as it potentially results in lower losses of actinides in the course of recycling. The experiments include the following studies:

1. Reduced smeared fuel density for swelling accommodation in case of higher burnup;
2. Annular pellets for enhanced geometrical stability at low smeared density as well as to avoid liquid metal sublayer;
3. Microalloying of fuel for immobilization of fission products (lanthanides) inside the fuel matrix and prevention of its movement towards the cladding and subsequent chemical interaction with it;
4. Advanced metal fuel for transmutation of minor actinides in fast reactors capable to ensure reduced corrosive interaction with the cladding and higher burnup. MTZ (5Mo-4.3Ti-0.7Zr) with (2–4)% wt. of palladium is introduced instead of Zr as well as the chromium coating of the cladding.

Investigations of metallic fuel were also carried out in the UK for DFR reactor (1960). Studies of zirconium-doped metallic fuel are going on in the Republic of Korea (KAERI) for the sodium-cooled fast reactor PGSFR as well as in India (declared as the opportunity for the future reactor of $N = 1000$ MW).

In Russia the first experiments with doped metallic fuel for fast reactors on the basis of U-Pu were started in VNIINM in 1950-1960s [6.5], and then active development in this area was suspended.

As it was mentioned above, zirconium alloying provides dimensional stability under the irradiation and also enhances corrosion resistance and ensures the required solidus and liquidus temperature values. But the negative consequences of alloying (fissionable material density reduction, thermal conductivity decrease, etc.) and filling of elements with sodium (increase of the gas plenum dimensions, restrictions for the storage and processing techniques, etc.) are also obvious.

Studies of nondoped metallic fuel in helium-bonded fuel rods have been carried out in RIAR for many years in order to minimize the above-mentioned drawbacks [61].

The following dependencies were investigated:

- Radiation swelling of uranium and its alloys (particularly with plutonium) on such parameters as the content of alloying additives (0–40% wt.), the irradiation temperature (200–800°C), the burnup rate ($1\cdot10^{13}$–$9\cdot10^{13}$ fis/cm^3·s), the rod sublayer (sodium, helium),

- Corrosive damage of steel claddings on the type and thickness of antidiffusion layers (nonmetal, metal, combined layers) in the area of contact between metal slug and the cladding, the fuel and cladding temperature under design-basis and some emergency conditions.

Besides, the degree of mechanical limitation (free swelling, partial or full limitation) was investigated.

The established dependencies were used in development, manufacturing and testing of full-scale fuel assemblies with experimental elements of various purpose filled with helium and having high smeared density (γ_{ef}=12.5–18 g h.a./cm^3) in BOR-60 and BN-350 reactors.

The studies carried out in RIAR demonstrated that the initial porosity value and shape of metal fuel were the main factors affecting the fuel rod reliability under the preset conditions (cladding dimensions and its material, thermal loads and cladding temperature, fuel burnup, damage dose).

The following parameters depend on selection of the initial fuel porosity and shape:

- The rate of fuel-cladding mechanical interaction (FCMI) achievement and, consequently, the kinetics of fuel-cladding gap thermal conductivity increase as well as the temperature difference in the contact area;
- Porosity of the swelled slug and consequently its thermal conductivity and temperature difference across its section;
- The temperature in the fuel slug center which should not exceed the critical value;
- Gas permeability of the slug and consequently efficient use of gas plenums;
- Gas release and consequently effect of the fission gases in fuel matrix and released under the cladding on the cladding stress-strain state.

The fuel slug design and the initial porosity value also determine the technologies for slug manufacturing and elements assembly, the smeared density (mass, fraction) of uranium in an element (in a fuel assembly, in the core) and consequently nuclear-physical, technical- economical and other parameters of the core.

The set of scientific and technology research works including development and testing of experimental sodium-bonded rods with doped U-Zr and U-Pu-Zr fuel was performed in SSC RIAR in 1988–2002. Irradiation of four full-scale fuel assemblies up to the maximum burnup of ~10atwas carried out in BOR-60 (Table 13.1).

It was impossible to perform all these studies to the full extent in BOR-60 reactor due to the core peculiarities: the gas plenum of regular fuel rods was located in the bottom part, while for the tested fuel rods with three-component alloy it had to be located in the upper part that was difficult to implement. Besides, the cladding material with insufficient irradiation stability (EI847 steel)

TABLE 13.1 Irradiation characteristics for alloyed metallic fuel in BOR-60.

Parameter		Fuel assembly number			
Fuel		1	2	3	4
		U-11% Zr	U-10% Zr	U-18% Pu-10% Zr	U-18% Pu- 10% Zr
Irradiation time	Start	11.07.91	13.11.92 13.07.95	25.11.94	14.01.1999
	completion	16.11.93	10.10.94 12.11.95	10.07.97	26.01.2001
Burnup, at. %		7.1	9.7	9.6	9.7
Neutron fluence (E>0.1 MeV), cm^2		7.4×10^{22}	1.09×10^{23}	1.05×10^{23}	$1.28 \times \cdot 10^{23}$
Maximum power density, W/cm		314	420	350	400
Temperature of cladding inner surface, °C		545	618	570	600

was used. Nevertheless, good experience in manufacturing and irradiation of doped metallic fuel was obtained [61,92].

Nondoped metallic fuel is also studied in India together with doped fuel.

Foreign and national experience showed that smeared density of U-Pu-Zr fuel equal to 75% shall be ensured in order to accommodate the fuel swelling and to prevent FCMI.

13.2 Carbide fuel

Experience of India. The most extensive foreign experience in irradiation of mixed carbide used as regular fuel for the experimental FBTR reactor has been accumulated in India [156]. Fuel composition: 70%PuC+30%UC. At the first stage of irradiation in the experimental FBTR reactor the maximum linear power was 250 W/cm and later was increased to 320 W/cm. For a short period of time the fuel was operated at 400 W/cm. All fuel rods are leak-tight.

Special attention was paid to investigation of two problems related to carbide fuel: carbonization of the cladding and formation of low-temperature eutectic at burnup increase. It was established that carbonization of the cladding was not the factor restricting high burnup. This phenomenon was not registered in fuel elements with high burnup.

At the first stage the maximum linear power and burnup values of 250 W/cm and 25 GW·day/t were taken due to the high plutonium content in carbide and the relevant melting temperature and thermal conductivity decrease. Carbide is initially manufactured as hyper-stoichiometric in order to prevent formation of the metallic phase. The C/Me ratio is decreased with burnup growth to the

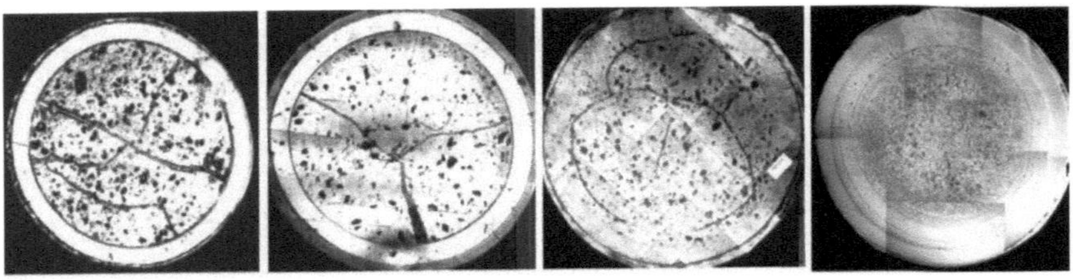

25 GW*day/t 50 GW*day/t 100 GW*day/t 155 GW*day/t

FIGURE 13.1 Microstructure of 70%PuC–20%UC at different burnup in FBTR reactor.

value below the one resulting in formation of (U, Pu) alloy with low melting temperature. This alloy in its turn can interact with the cladding resulting in formation of low-temperature eutectic. These reactions can break the cladding, and this is one of the restricting factors. High carbide swelling is also one of the problems. It should be noted that insignificant carbide swelling was observed in FBTR reactor even at high burnup due to low carbide temperature.

At lower burnup level when free carbide swelling was observed its rate was of 1–1.2% per 1at% Positive results of irradiation and postirradiation examinations allowed gradual burnup increase to the maximum value of 160 GW·day/t, damage dose of 80 dpa. The fuel microstructure in the core midplane for fuel rods with different burnup is shown in Fig. 13.1.

At burnup more than 50 GW·day/t with closure of fuel-cladding gaps radial cracks turned into circular ones. Gradual decrease of fuel porosity with burnup increase is observed. Metallographical fuel examination at burnup of 155 GW·day/t showed a clear zone without any porosity at the fuel pellet periphery. Complete closure of fuel-cladding gap and circular cracks in the fuel both in the core midplane and at the ends of the fuel column are indicative of free fuel swelling absence along the entire core height.

Fuel with 70% Pu content has rather low melting temperature (2150 K) and demonstrates high plasticity thus enabling to accommodate its high swelling due to creep without any high stresses in the cladding. Maximum gas release is 16% at the burnup of 115 GW·day/t.

Despite the positive experience in irradiation of low-temperature carbide fuel up to high burnup values MOX fuel was selected for the PBFR reactor under construction, and metal fuel (UPuZr) in the long run.

Experience of Europe. Very limited experience in carbide (UPuC) irradiation in reactors is available in Europe (1970s – early 1980s):

- HFR (Petten) (2 fuel rods, maximum burnup of at. 0.5%);
- KNK-II (21 fuel rods, maximum burnup of 70,000 MW-day/t);
- DFR (several fuel rods with the burnup of about 6.5 at. %);
- Phenix – NIMPHE2 experiment (2 helium-bonded fuel rods with UPuC, maximum burnup of 6%, helium intermediate layer);
- Rapsodie (maximum burnup of 12.5%, Na intermediate layer).

Carbide irradiation experiments in Rapsodie reactor demonstrated that sodium –bonded carbide rods with fuel density of 91 %th. could be irradiated at the linear power of 800–900 W/cm up to the burnup of about 12 at. % without any coating.

Analysis of the irradiation results enables to make the following conclusions [208]:

- Performance of helium-bonded UPuC rods is determined by the design parameters to a great extent, particularly by the pellet and smeared densities as well as the cladding diameter;
- Cladding failures are caused by fuel swelling and loss of the cladding ductility due to its carbonization.

It was established that cladding carbonization in hyperstoichiometric $UPuC_1$+x fuel was one of the most serious problems, in fuel rods with the helium sublayer – to a lesser extent and in fuel rods with the sodium sublayer – to a greater extent. Carbonization is mainly due to the formation of $M_{23}C_6$ carbide because of carbon transfer from fuel to the cladding.

Experience of USA. The most extensive experience in the studies of experimental fuel elements with mixed carbide and the sodium sublayer both under normal operation conditions and in case of operational occurrences was accumulated in the USA [166]. Irradiation of mixed carbide in the following reactors was carried out: EBR-II – about 500 fuel rods – 10 helium-bonded rods reached the burnup of 20 at. % without any breakage; FFTF – more than 200 fuel rods, particularly with microspheric fuel; TREAT – 10 experiments (including 5 fuel rods subjected to power ramp of 15% after burnup of 12 at.%).

Testing parameters for UPuC in EBR-II, FFTF, TREAT reactors

Sublayer	Helium, sodium
Smeared density, %	73–86
Pellet density, %	80–98
Type of fuel	Pellet, microspheres
Cladding material	316SS, PE16, D9, D21
Cladding temperature, °C	555–695
Linear power, kW/m	50–150
Burnup, at. %	up to 20

Destructive postirradiation examinations of 90 fuel rods were made. The following was determined [166]:

- Although carbonization of the claddings made of 316SS steel and irradiated in EBR-II was observed no fuel rods failures due to this cause were recorded;
- In order to ensure the performance of helium–bonded rods to high burnup values the smeared density shall be 78–79%, and the pellet density – 80–82%;
- The carbide composition shall be MC+10 vol%M2C3 in order to reduce fuel swelling and cladding carbonization.

Thus USA program demonstrated operability of fuel elements with mixed carbide at high linear powers and high burnups.

Domestic experience. The first testing of carbide fuel was started within the 1-st core of BR-10 reactor with PuO_2 (since 1963) – 7 fuel assemblies with UC were irradiated. In 1965–1971, the reactor was operated with the UC full core loading, the maximum burnup of 6.1 at%. All fuel rods had helium sublayer. 127 fuel assemblies (more than 1000 fuel rods) were irradiated in total. The experiment could not be called successful, as 58 fuel assemblies contained defective fuel rods [38].

Regular fuel assemblies of BR-10 reactor contain 7 He-bonded fuel rods enclosed in a hexagonal wrapper made of stainless steel with the wall thickness of 0.3 mm. The assembly height is 833 mm. The stainless cladding diameter is of 8.7 mm, the wall thickness of 0.4 mm and the rod length of 500 mm. The UC pellets diameter is of 7.65 mm, core height of 380 mm. The gas plenum with the height of about 80 mm is located in the upper part of the rod.

In 1965–1971, the reactor operated with the sodium temperature of 430°C at the core inlet, and the sodium temperature at the core outlet was 500°C under the rated power conditions. The maximum cladding temperature was 570°C, maximum fuel temperature ~1000°C. The average linear power of fuel rod was 230 W/cm.

Fuel claddings made of two steel grades were used: 8Cr9NiTi and 16Cr15Ni3MoNb; in this case 52 fuel assemblies out of 63 inspected fuel assemblies with the claddings made of 8Cr9NiTi contained defective fuel elements, and 6 fuel assemblies out of 60 inspected fuel assemblies with the claddings made of 16Cr15Ni3MoNb contained defective fuel elements.

18 fuel assemblies with uranium monocarbide with the burnup of 0.8 to 5.2at% were examined in the IPPE hot laboratory.

Four fuel assemblies contained failed fuel rods, and in two of them the hexagonal wrappers were also failed. Breakage of the wrappers was located along the ribs and corresponded to the midplane of the core. Analysis of fuel rod performance showed that nonuniform swelling along the cladding perimeter deteriorated its performance. The damage factor for the claddings made of 8Cr9NiTi steel is close to 1, and their failure becomes possible. The claddings made of 16Cr15Ni3MoNb steel have significant operability reserve (due to lower swelling) accounting for their leak-tightness.

A diffusive interaction rim caused by carbonization with the depth of up to 50 microns is observed on the external surface of the claddings. Different diffusion rims with the thickness of up to 30 micron and pittings with the depth of up to 50 microns are observed on the claddings on the fuel side.

According to the hydrostatic weighing results the average monocarbide swelling value is $(1.9 \pm 0.4)\%$ per 1at% of burnup.

The bad experience is primarily due to the usage of high-swelling steels in the BR-10 fuel assembly design (the fuel assembly contains only 7 fuel rods, high temperature gradients and consequently cladding swelling gradients are present).

TABLE 13.2 Comparison of the characteristics for carbide irradiated in BR-10 and BOR-60 reactors.

Parameter	BR-10	BOR-60
Pellet manufacturing method	Uranium dioxide carbidization	Initial metal carbidization
Carbon content	4.8–5.4	4.7–5.1
Cladding diameter x wall thickness, mm	8.7 x 0.4	6.9 x 0.4
Fuel cladding gap, mm	0.08–0.45 He	0.12–0.4 He 0.2–0.6 Na-K
Maximum fuel temperature, °C	1000	1250–1550 He 800–900 Na-K

Besides, sodium used at that time was contaminated with up to 0.017% C and up to 0.003% N, and consequently carbonization and nitridation of the claddings took place.

Irradiation of six experimental fuel assemblies (19 fuel rods in each) with UC, UPuC, UC-UN, including 8 fuel rods with UPuC was carried out in BOR-60 reactor. The maximum burnup was 3.3–10.4 at%, the fuel rods had Na-K and He sublayers. All fuel rods were leak-tight [61].

The postirradiation examination results showed that fission gas release did not exceed 10–18%, and the fuel swelling rate was 1.2–2.5% per 1 at% The performed tests enabled to develop the recommendations on the certain parameters of mixed carbide fuel with the helium sublayer (Table 13.2) required to ensure fuel performance up to the burnup of 10 at%:

- Smeared fuel density – 80%
- Carbon content – 4.7–5%
- Maximum fuel temperature – 1500°C.

13.3 Nitride fuel international experience

Experience of the USA. The most extensive foreign experience in nitride irradiation (prior to Proryv Project launch in Russia) was accumulated in USA where about 100 fuel elements of various geometry with UPuN were manufactured and irradiated in EBR-II reactor within the framework of the first program on development of advanced FR fuel types [121]. Fuel rods with two sublayer types were used: helium and sodium (shrouded and nonshrouded fuel); with two cladding steel types (304 and 316); smeared density of 75–86%; maximum linear power of 67–107 kW/m, maximum burnup of 9at% . The majority of fuel rods had the sodium sublayer and high pellet density (94–97%). Besides, the works for manufacturing of several dozens of fuel elements with UN and their irradiation in FFTF and EBR-II reactors were started in the USA in 1984 for substantiation of the SP-100 space nuclear installation. Fuel rods had helium sublayers, the cladding material was heat-resistant Nb-1Zr alloy. Irradiation conditions: maximum cladding temperature – up to 1500 K, maximum fuel temperature – up to 1950 K, maximum burnup – 0.7% to 6at%, pellet density – 87 and 94.5%. Several fuel rods with UN were irradiated in the research reactors with thermal spectrum, pellet density 93, ~~and~~ 95%, maximum fuel temperature 2050, 1750 K, maximum burnup 2 and 4.5at% [159].

Experience of Europe. The European experience is limited to irradiation of several rods with UPuN in DFR reactor and 16 rods with UPuN in Phenix reactor. All fuel rods had helium sublayers.

Mixed nitride was irradiated in DFR at linear power of 130 kW/m, maximum fuel temperature 2300 K, maximum burnup – 1.3–5.0 at%, pellet density – 85%, smeared density 71% [163].

Two fuel assemblies were irradiated in Phenix reactor within the framework of Nimphe experiment: in the first one – 11 MN rods up to maximum burnup of 6.9at% at maximum linear power of 42 kW/m, and in the second one – 5 MN rods up to burnup of 5.8at% at linear power of 73 kW/m. Pu was detected on the cladding inner border in the fuel rod irradiated at the maximum linear power of 73 kW/m that was indicative of nitride dissociation at the temperature below 2000 K [129]. It should be noted that the MN pellets in the experiment had a high open porosity fraction (about 90%) reached through the pellet sintering time reduction. Domestic studies of mixed nitride fuel irradiated in BOR-60 reactor even at a higher linear power level (up to 100 kW/m) did not reveal any signs of dissociation [61].

Experience of Japan. 4 fuel rods with mixed nitride were manufactured in Japan and irradiated in the experimental reactor JMTR up to the burnup of 4.1 and 5.5 at.% at the maximum linear power of 65 and 73 kW/m, and 2 fuel rods were irradiated in JOYO reactor up to the burnup of 4.5 at.% at the maximum linear power of 80 kW/m. The pellet density was 83–86%, the smeared fuel density 78–83% [200].

13.4 Domestic experience in nitride fuel development prior to Proryv Project launching

BR-10 reactor [188] Two loadings of BR-10 reactor with UN (660 fuel rods and 590 fuel rods). All fuel rods had helium sublayers, maximum burnup about 9at%, maximum linear power 45 kW/m, maximum fuel temperature 1175 K, pellet density – 85–94%.

The fuel rod length was 615 mm, the cladding was made of EI-847 steel (16Cr15Ni3MoB) with the diameter of 8.4 mm and the wall thickness of 0.4 mm. The fuel column height was 400 mm, the fuel pellet diameter 7.4 mm. A molybdenum gasket with the thickness of 0.5–1.0 mm and a nickel or steel reflector with the height of 80 mm and the diameter of 7.3 mm were installed above the fuel column. The gas plenum with the height of 100 mm with a spring aimed to prevent displacement of the fuel pellets was located in the upper part.

TABLE 13.3 Operation conditions and design of BR-10 reactor fuel rods with uranium nitride rods.

Parameter	Loading IV	Loading V
Irradiation duration, eff. days	882	970–1130
Fuel pellet density, g/cm^3	12–13.4	
Maximum linear power, W/cm	450	
Maximum burnup, at. %	8.0	8.8
Sodium temperature at the core inlet, °C	325–350	
Maximum coolant heating in the fuel assembly, °C	165	
Cladding diameter x wall thickness, mm	8.4 x 0.4	
Fuel pellet diameter, mm	7.4	
Core height, mm	400	

Fuel assemblies consist of three parts: head, hexagonal part and tail. The fuel assembly length is 833 mm. All assembly parts are made of 18Cr10NiTi steel. The fuel assembly contains 7 fuel rods (5 rods in the central truncated assemblies). The fuel bundle is enclosed in a hexagonal wrapper with the flat-to-flat size of 26.1 mm, the length of 643 mm and the wall thickness of 0.5 mm.

The initial core loading – 83 fuel assemblies. During the first stage (IV reactor loading) with uranium nitride BR-10 reactor was operated from May 1983 to August 1989. During the second stage with uranium nitride (V reactor loading) BR-10 reactor was operated from 1990 to 2002.

The data on operation and design parameters of nitride rods of IV and V loadings are presented in Table 13.3. Operation of BR-10 reactor with uranium nitride for eighteen years demonstrated its good performance. The maximum fuel burnup in the first loading was 8at%, and in the second one – 8.8 at%. More than 99% fuel rods were operated up to the design burnup value (8.0 at%) without any loss of cladding integrity. For the second core the number of cases with rods failures increased significantly, mainly at the burnup level exceeding 8 at%. 24 cases of fuel rod damages were registered in the second core. The analysis of calculation data on nitride performance showed that the most probable cause of fuel rod failures was fuel – cladding mechanical interaction.

BOR-60 reactor. One fuel assembly with UN, maximum burnup more, than 8 at%, maximum fuel temperature – 1775 K.

Several He-bonded rods with UPuN, maximum burnup of 4 and 8.95 at%, maximum fuel temperature of 2475 and 1750 K respectively, pellet density of 85–86%.

Four He-bonded rods with mixed nitride with increased Pu content – 45 and 60w%.The works were performed within the framework of the Russian –

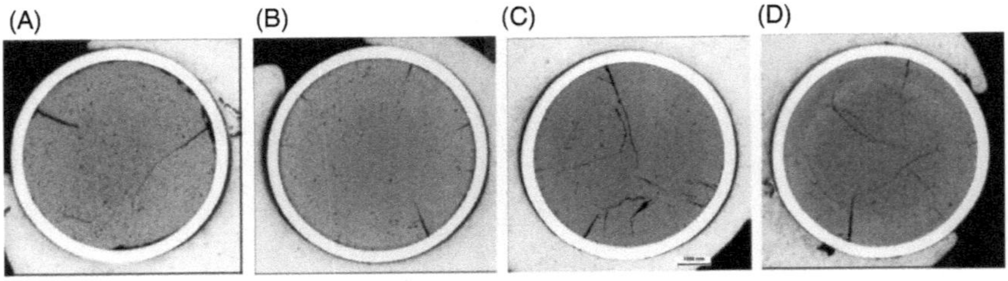

FIGURE 13.2 Cross-section of BORA-BORA fuel rods after the second stage of irradiation in BOR-60: (A) UPu0.45N – top, (B) midplane; (C) $UPu_{0.6}N$ – top, (D) midplane.

TABLE 13.4 Irradiation parameters for UPuN rods in BOR-60 reactor (BORA-BORA).

Parameter	$UPu_{0.6}N$	$UPu_{0.45}N$
Irradiation duration, eff. days	900/514*	900/514
Maximum burnup, at. %	12.1/7.0	9.4/5.4
Maximum dose, dpa	43/23	43/23
Maximum cladding temperature, °C	604/587	567/588
Maximum linear power, kW/m	54.5/35	41.9/26
Maximum fuel temperature, °C (calculated through the use of DRAKON code)	1760/1110	1490/930

**Irradiation parameters in the end of the 2-nd/1-st stage.*

French BORA-BORA program including three phases: 1) fuel manufacturing; 2) irradiation in BOR-60 reactor; 3) postirradiation examinations [210].

The following fuel compositions with claddings made of austenitic steel ChS 68-ID c.w. were studied:

- UPu0.45 O2 oxides (4 fuel rods with pelletized fuel and 4 fuel rods with vibro-compacted fuel);
- Nitrides (2 fuel rods with UPu0.45N, 2 fuel rods with UPu0.6N);
- Fuel compositions based on inert matrices (2 fuel rods with PuN+ZrN, 2 fuel rods with PuO2+MgO).

In August 2000, two irradiation devices with experimental fuel rods were loaded into the core of BOR-60. The first device included 4 fuel rods with pelletized MOX fuel and 4 fuel rods with vibro-compacted MOX fuel, and the second device – 4 fuel rods with mixed nitride, 2 fuel rods with PuN+ZrN, 2 fuel rods with PuO_2+MgO. Irradiation was completed in 2005.

In 2009, the final postreactor studies of the fuel rods with $UPu_{0.45}$ N and $UPu_{0.6}N$ were completed (Fig. 13.2). The main irradiation parameters for these fuel rods are presented in Table 13.4 [86].

Nondestructive and destructive examination of fuel rods with $UPu_{0.45}N$ and $UPu_{0.6}N$ demonstrated the following.

The fuel-cladding gap was closed in all cross-sections except for the top and bottom one, as early as the end of the first irradiation stage. Fuel density measurements in the cross-sections with maximum linear power showed that the average restrained swelling rate of $UPu_{0.45}N$ was (0.48–0.68) ±0.04% /1at% of burnup, and for $UPu_{0.6}N$ – (0.64–1.11) ±0.04%/1at%.

Presence of gas fission products under the fuel claddings confirms their leak-tightness and enables to evaluate krypton and xenon release from the fuel: 19 and 19.3% of the generated amounts.

The structure of both fuel types after the second irradiation stage is characterized with intra- and intergranular porosity; in this case a higher fraction of open porosity in $UPu_{0.45}N$ fuel facilitates better release of gas fission products from the fuel and provides for lower gas swelling rate.

The studies demonstrated almost complete absence of any corrosive interaction of fuel and fission products with the cladding made of ChS68-IDcw austenitic steel. The maximum corrosion of the internal cladding surface in the upper section of $UPu_{0.6}N$ rod was about 15 μm.

No nitride dissociation was detected after two stages of irradiation with regard to the microstructure state, any changes of nitrogen content in the gaseous phase under the fuel cladding or any other signs. This is probably due to the so-called autoclaving effect, when partial pressure of nitrogen in the gaseous phase is equal to partial pressure of nitrogen in the course of nitride dissociation because of closed pores formation under irradiation at high temperatures.

Positive results of the irradiation testing of high-purity mixed mononitride fuel in BOR-60 reactor up to the burnup of 12.1at% at maximum linear power of up to 545 W/cm should be attributed to high initial plutonium distribution homogeneity, low oxygen and carbon content (less than 0.15 and 0.1% wt. respectively), uniform porosity distribution and combination of pores on the borders and inside the graines.

Irradiation tests of fuel rods of BREST-OD-300 type with pelletized uranium – plutonium nitride and lead sublayer were also carried out in BOR-60 reactor within two dismountable irradiation devices [66]. The main characteristics of fuel rods are provided below.

Main characteristics of experimental fuel rods of BREST type with lead sublayer in BOR-60 reactor

Fuel	(72%U+18%Pu)N
Fuel pellet density, % th	91–92
Pellet diameter, mm	7.9
Cladding diameter/ wall thickness, mm	9.4/0.5
Fuel rod/ fuel column stack length, mm	625/390
Maximum linear power, W/cm	420
Maximum cladding temperature, °C	630
Maximum fuel temperature, °C	810

The PIE results after two stages of irradiation (with the maximum burnup of 2.8 and 4at%) demonstrated progressing interaction of the cladding with the lead sublayer characterized with its dissolution in some areas and formation of metallic deposits in other areas. The cladding thickness in some points reduced by approximately 40% due to its dissolution. For this reason, it was decided to stop the reactor testing upon achievement of the maximum fuel burnup equal to 5.5at%. But the obtained results directly related to the fuel characterize its good state after irradiation.

As distinct from helium-bonded fuel rods no migration of fission products was present in lead-bonded fuel rods due to lower fuel temperature, for example axial distribution of cesium repeated the burnup distribution to the full extent, no cesium transportation to the less heated areas of the fuel rod was present, the fuel microstructure was characterized with absence of gas pores and second phase separation from the fission products typical for fuel with helium sublayer.

Chapter 14

Development of nitride fuel within the framework of Proryv Project

14.1 Requirements for the design of nitride fuel rod

Nitride was defined as promising fuel for fast sodium-cooled reactors due to its attractive physical and chemical properties: high density, thermal conductivity, high melting temperature (> 2700°C), good compatibility with stainless steels, sodium, water (at T < 60°C), air (at T < 25°C), as well as with water-based reprocessing methods.

According to the authors of [130] the global and domestic experience in development of dense fuels (nitride, carbide) enable to determine the main requirements for the design and operating conditions of the optimized fuel element with nitride from the viewpoint of the modern criteria for a safe and cost-efficient fast reactor:

- Helium sublayer;
- Smeared density < 80 %;
- Maximum burnup > 15–20at%

At high burnup of dense fuel the main problem in gas-bonded fuel rods is fuel-cladding mechanical interaction which causes in the worst case cladding integrity loss. However, according to the expert opinion [209] as well as on the basis of the positive irradiation results in BOR-60 it can be solved by reduction of fuel smeared density and arrangement of open porosity for fission gas release. The required smeared density can be provided by the initial pellet porosity and also by the selection of the relevant fuel-cladding gaps. But for helium-bonded fuel rod the gap increase is restricted by the relevant fuel temperature increase due to reduction of the gap thermal conductivity resulting in the fuel swelling rate growth. Increase of the pellet porosity results not only in the swelling rate reduction but also in the pellet thermal creep rate increase after beginning of fuel – cladding mechanical interaction and the relevant decrease of stresses in the cladding.

Closed Nuclear Fuel Cycle with Fast Reactors: Handbook of Russian Nuclear Power.
DOI: https://doi.org/10.1016/B978-0-323-99308-1.00011-8

Selection of the sublayer type for fuel rod is a separate issue requiring additional consideration. On the one side, the helium sublayer is more preferable from the viewpoint of technology as well as the existing experience on fuel behavior under normal and emergency conditions. But for high burnup cores the problems can arise at proving of helium-bonded rods performance as higher fuel temperature is typical for them. As it is commonly known, the fuel temperature considerably depends on conductivity of fuel-cladding gap which is in its turn determined by thermal conductivity of its medium (gas or liquid metal). Thermal conductivity of liquid metals (Na, Pb-Bi, and Pb) is higher by several orders than helium thermal conductivity. Besides, gap conductivity is deteriorated under irradiation due to release of fission gas Xe, Kr with even lower thermal conductivity under the cladding. Filling of a fuel rod with liquid metal results in significant reduction of the fuel temperature thus decreasing the swelling level for dense fuel and providing the possibility to operate without fuel – cladding mechanical interaction.

It is important to provide the maximum possible gap between fuel and cladding and to ensure its nonclosure up to the end-of-life. As the development experience shows, the tendency towards increase of the initial gap is primarily restricted with the relevant increase of the cracking degree in fuel pellets and relocation of fuel fragments as it is confirmed by practice. However, no loss of integrity was registered in the six fuel rods with the maximum gap (0.76 mm) irradiated in EBR-II reactor up to the maximum burnup of 9at% [186]. The authors could not explain the obtained results to the full extent. They suppose that damage of the cladding in fuel rod with larger gap is decreased even in case of considerable fuel cracking either due to smaller pellet fragments in such fuel rod, or due to the fact that a larger gap results in accommodation of the cladding ovality changes with less severe local deformations.

On the basis of the fuel irradiation results the authors recommend to increase the initial fuel-cladding gap to at least 0.51–0.76 mm. Irradiation of fuel rods with the Na-K sublayer in BOR-60 reactor confirmed good performance of the fuel rods with a gap of 0.6 mm (the maximum burnup 8.9at%).

The American experience in irradiation of UPuN rods with sodium sublayer showed that considerable number of integrity losses (about 1/3 of the total number of fuel rods tested) occurred for these fuel elements. Fuel cracking, displacement of its fragments and local straining of the cladding were named among the most probable causes. The claddings maintained integrity only subject to application of internal shrouds preventing from any displacement of fuel fragments.

So, additional experimental data are required to confirm performance of both helium and sodium- bonded rods.

Experience on advanced fuel irradiation in FRs is significantly less than the experience gained for oxide fuel. Nevertheless, selection of fuel for innovative FRs within the framework of large-scale nuclear power engineering shall be primarily determined by the safety and cost-effectiveness requirements even given importance of the previous experience and familiarization with the oxide technologies.

For high-capacity power reactors nitride fuel enables to implement the advantages of the cores with BRC of ~1 and the corresponding fuel cycle: low reactivity margin for fuel burnup, self-production of fuel, absence of any need for uranium and plutonium separation as well as the required feedback parameters determining the reactor safety (reactivity coefficients and effects).

As it was mentioned above, the most extensive work program for metallic fuel investigation was implemented in USA where the main experience in its operation was accumulated. Development of the reactor concepts with metallic fuel (PRISM, KALIMER) as well as the experience in operation of EBR-II demonstrates the potential of metallic fuel as applied to low-capacity reactors. But it is probably hardly applicable to high-capacity reactors.

The priority of safety issues enabled to decide on nitride fuel in development of reactors with both sodium and lead coolant within the framework of Proryv Project.

The experience with nitride in Russia is much more extensive than for other dense fuels, and it has received more study (18 years of use in BR-10, BORA-BORA experiments in BOR-60 reactors with the burnup of 12.1 at%). Nevertheless, the existing experience in nitride investigation is not sufficient for reliable prediction of fuel performance with the preset characteristics at the operating parameters of BN-1200 and BREST-OD-300 reactors. The integrated computational and experimental substantiation program is developed within the framework of Proryv Project for fuel elements with mixed nitride fuel for BN-1200 and BREST reactors up to 2020. The results of this program will become the basis for verification and validation of the design fuel codes and operability substantiation for fuel elements in the course of fuel licensing for BREST and BN-1200 reactors [68].

The objectives of the integrated computational and experimental substantiation program are the following:

- Improvement of the manufacturing technology;
- Optimization of the mixed nitride composition and structure for enhancement of the manufacturing technology quality and operability;
- Pre-reactor investigation of the properties;
- Irradiation of nitride in the MIR and BOR-60 research reactors (fuel rods with helium and lead sublayers) and in the BN-600 commercial reactor (fuel rods with helium sublayers);
- Postirradiation examination;
- Development of the methods, codes and criteria for the fuel performance judging, improvement and verification of the fuel codes subsequent to the postirradiation examination results.

14.2 Nitride manufacturing technologies

Two technologies are used for manufacture of nitrides:

1. Carbothermal reduction from initial oxides;
2. Mixed nitride synthesis from initial metals.

The technology for carbothermal synthesis of nitride powders is considered the most extensively studied worldwide from the viewpoint of commercial development. Its main advantage resides in usage of oxides – products of the existing spent nuclear fuel radiochemical processing plants – as the initial materials, and the main problem is production of fuel with the required (low) oxygen and carbon content.

In the USSR this process was developed mainly for production of uranium mononitride with different enrichment degrees. Mononitride uranium fuel for the two cores of BR-10 reactor was produced through the use of this method.

Carbothermal mixed mononitride production process is based on the reaction described in the form of a generalized equation:

$$2MeO_2 + 4C + N_2 = 2MeN + 4CO \tag{14.1}$$

Oxygen removal and production of mononitride takes place in the course of gradual heating. The mononitride generation process is finished at 1700–1800 °C. Uranium and plutonium oxides, carbon, argon, nitrogen and nitrogen-hydrogen mixture ($N_2 + 6\%\ H_2$) are used as initial materials. The produced batch is pressed at the pressure of 1.5–2.0 t/cm^2 into the briquettes with (45 ± 5) % of theoretical density (t.d.); the briquettes are loaded into a vacuum-compression furnace with a graphite heater and heated gradually. Hydrogen (N2+6-8 vol. % H_2) is introduced at the final stage of the carbothermal process at 1600–1800°C for binding and removal of residual carbon from mixed mononitride in the form of gaseous hydrocarbons thus providing for the carbon concentration reduction to the approximate allowable level of 5–0.2% wt. Solid single-phase solution of UPuN with the specified composition is produced as a result of synthesis.

The laboratory technology of mixed uranium-plutonium mononitride synthesis from the initial oxides and manufacturing of fuel pellets from it was developed in VNIINM within the period of 2008–2011. In 2011, 2.5 kg of fuel pellets were manufactured on the laboratory equipment representing the prototype of industrial equipment to a certain extent, and experimental helium-bonded fuel rods were made of them for testing in an experimental fuel assembly of BOR-60 reactor.

Application of the electromagnetic vortex mixer enabling to prepare highly homogeneous powdered mixtures within a short period of time was the peculiarity of this technology. Homogeneity is especially important for efficient solid-phase reactions with carbon where close contact of the components is necessary. The flow diagram of carbothermal mixed mononitride synthesis is presented in Fig. 14.1.

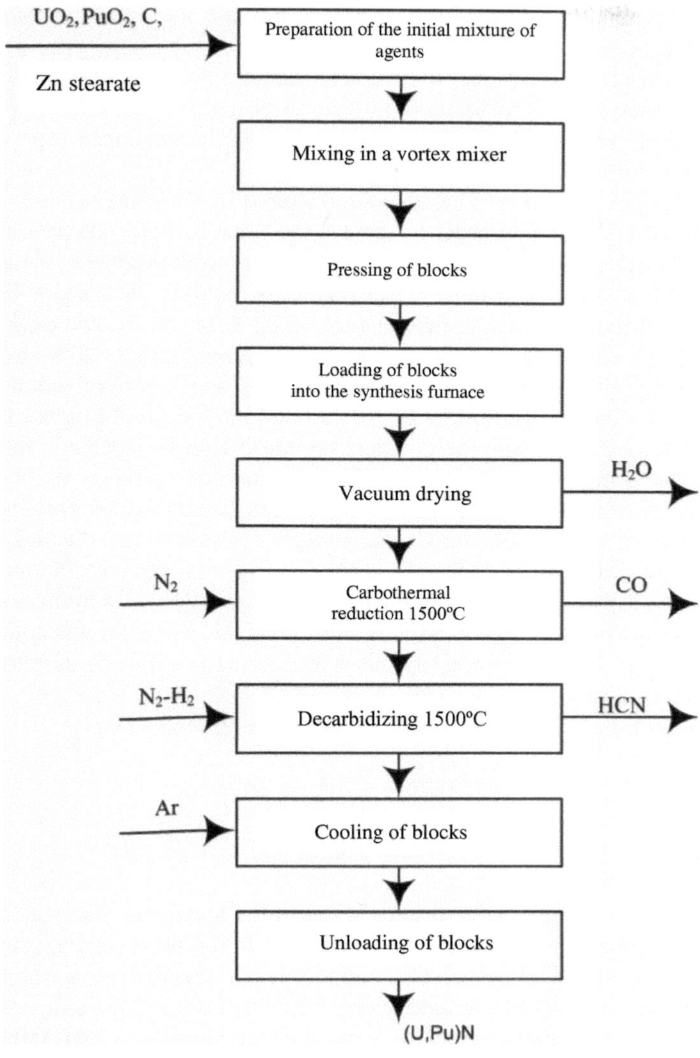

FIGURE 14.1 Flow diagram of carbothermal mixed mononitride synthesis.

Analysis of the developed technology parameters shows that the process for removal of residual carbon from nitride at the final stages of synthesis is crucial (the process duration can reach 36 hours due to residual carbon removal).

In the experimental technology carbothermal synthesis is performed in thin disks pressed together in order to ensure release of gaseous reaction products. The disk thickness determines the decarbidizing reaction duration. Preliminary pressing of thin disks made of mixed components is also applied in the global practice for manufacturing of mixed nitride through the use of carbothermal synthesis process.

It is known from experience that the scale factor is very important in this case. Thus it is necessary to focus on optimization of the processes directly on the prototypes or models of industrial equipment taking the developed experimental technology as the basis.

The oxygen content in (U,Pu)N manufactured in VNIINM is surely within the limits of 0.15% wt. In order to achieve the same carbon content in the fuel additional operations for its removal at the initial powder production stage were necessary. For example, in Japan it was possible to achieve the acceptable purity of mixed nitride in the experimental fuel: (0.02–0.18) % wt. for oxygen and (0.11–0.3) % wt. for carbon with the process duration of about 20 hours [160]. The French published data (CEA) on experimental samples of mixed nitrides – 0.05% wt. for oxygen and 0.1% wt. for carbon [185]. It should be noted that a ball mill is used to prepare the powdered mixture both in Japan and in France for manufacturing of mixed nitride by the carbothermal synthesis method.

Mixed nitride synthesis from initial metals. As U and Pu alloy is supposed to be produced after electrochemical regeneration of irradiated fuel the initial metal U and Pu are alloyed in vacuum at the temperature of ~ 1200°C for 30 minutes in order to develop the process under the laboratory conditions. Uranium content in the alloy obtained after regeneration is adjusted at this stage through addition of waste uranium. In case of necessity this synthesis method may be used to utilize plutonium of weapon origin.

Mixed nitride synthesis is based on the following reactions:

$$2\text{Me (alloy)} + 3\text{H}_2 = 2\text{MeH}_3 \qquad (14.2)$$

$$\text{MeH}_3 + 1/2(1 + \text{x}) * \text{N}_2 = \text{MeN}_1 + \text{x}1.5\text{H}_2 \qquad (14.3)$$

Synthesis of mixed nitride is carried out in the apparatus made of stainless steel and heated with a sliding electric furnace. Initial uranium and plutonium ingots are placed into a steel pan (or a firepot) which is loaded into the apparatus. The resulting alloy is hydrotreated with purified hydrogen at the temperature of up to 220°C and then nitrated with chemically pure nitrogen at 220–550°C. The selected mode provides the nitrogen content of 6.4–6.7% in mixed nitride. The resulting mixed nitride has the form of powder with the size of dendritic particles of up to 30–40 μm suitable for further manufacturing of fuel kernels. It should be noted that hydrotreating and nitrating operations are carried out sequentially in the same apparatus without any reloading of intermediate products.

The block diagram of mixed nitride production from individual metals is presented in Fig. 14.2.

Multiple studies of the technology for production of nitride powders from individual metals have been carried out in JSC "VNIINM." Two nitride power production plants were developed for manufacturing of experimental nitride

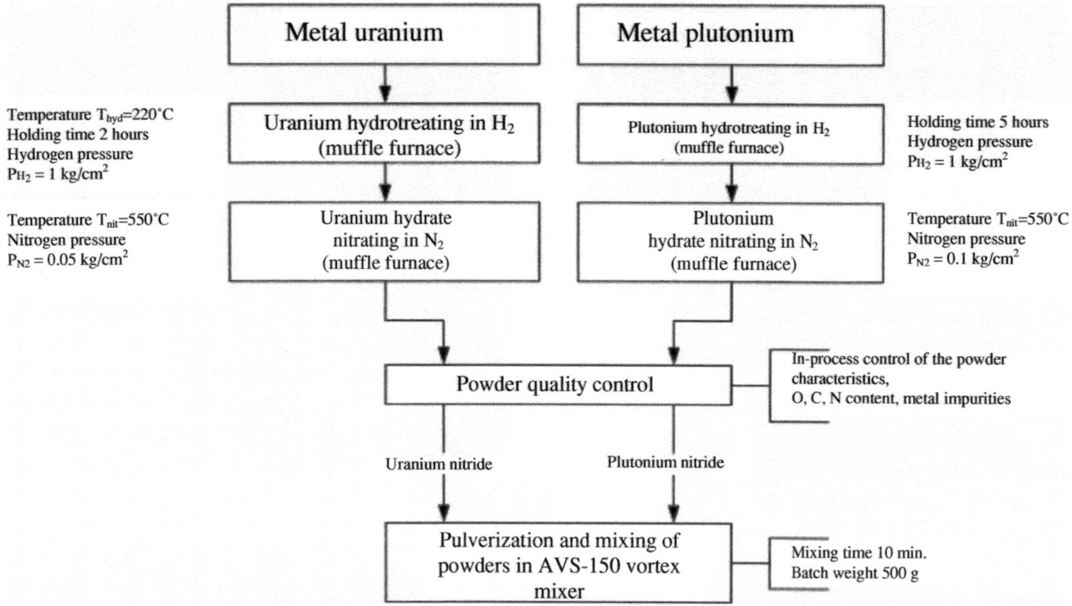

FIGURE 14.2 Flow chart for production of uranium and plutonium nitrides mixture from individual metals.

fuel: uranium nitride and plutonium nitride. Engineering studies of pilot nitride synthesis plants were carried out.

All fuel rods tested in BOR-60 reactor prior to Proryv Project launch were manufactured particularly with the use of the hydrotreating and nitrating technology. Powders were mixed in ball and vibration mills providing for powder activation in order to enhance efficiency of uranium-plutonium nitride sintering.

Before 2001 mononitride produced from initial individual metals synthesized into nitride by the hydrotreating and nitrating method with subsequent mechanical mixing of the resulting powders in the ball mill and sintering with simultaneous removal of excessive nitrogen was developed and studies in VNIINM laboratories [86,198]. Within the period of 2002–2004 the laboratory technology for manufacturing of mixed nitride pellets through the use of a vortex mixer (AVS-150) was developed in VNIINM. 4 kg of fuel meeting to the specification requirements were manufactured according to this technology for 15 experimental fuel rods with the lead sublayer of BOR-60 experimental fuel assembly.

At present the hydrotreating and nitrating technology is applied at Siberian Chemical Combine for large-scale manufacturing of (UN) for the BFS test facility of JSC "SSC RF IPPE". But the requirements for oxygen content in uranium nitride are considerably lower.

It is known from the international experience that in the late 1970s several kilograms of ($U_{0.8}Pu_{0.2}$) powder were produced in the USA by hydrotreating and dehydration of U-Pu alloy.

Certain risk is inherent for production of mononitrides from metals as hydrogen and plutonium are present in the process at the same time. Feasibility of this technology selection shall be determined by the fuel cycle concept.

14.3 Nitride fuel studies

Laboratory facilities for mixed nitride (MN) fuel manufacturing and research were arranged in the following organizations:

VNIINM – samples, fuel produced by the carbothermal method and BN-600, MIR and BOR-60 fuel rods;
VNIITF – samples, fuel produced by the hydrometallurgical method;

RIAR – samples with MAs, fuel, BOR-60 fuel rods and experimental fuel assemblies.

Experimental plants for manufacturing of fuel rods and experimental fuel assemblies for BN-600 (up to 12 fuel assemblies per a year) are arranged at Siberian Chemical Plant:

KEU-1 – experimental mixed nitride fuel technology (up to 200 kg/year);
KEU-2 – acquisition of industrial mixed nitride technology since 2016.

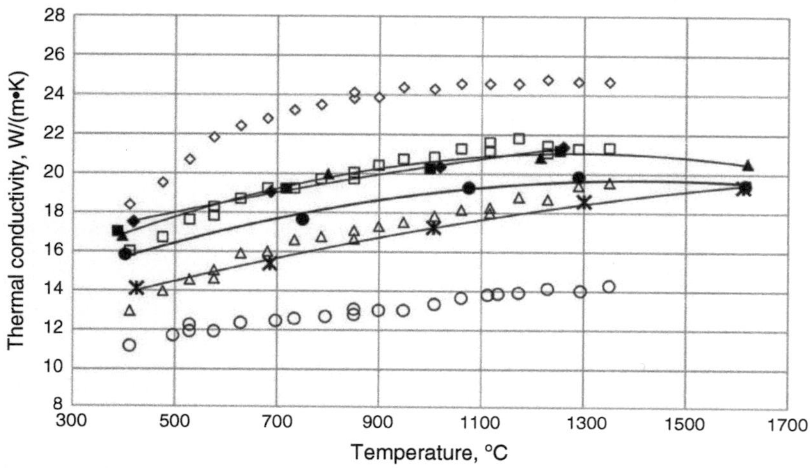

FIGURE 14.3 Temperature dependencies of (U1-xPux) N thermal conductivity scaled to theoretical density obtained in VNIINM (the cumulative error confidence limits $\delta = \pm 5$ %) and in [147]: ◇ – UN [147], Δ – $U_{0,65}$.$Pu_{0,35}$N [147], Δ – $U_{0,95}Pu_{0,05}$N [VNIINM], ▲ – $U_{0,85}Pu_{0,15}$N [VNIINM], Ж – $U_{0,60}$ Pu $_{0,40}$N [147], □ – $U_{0,80}Pu_{0,20}$N [147], O – PuN [147], ■ - $U_{0,90}Pu_{0,10}$N [VNIINM], ● – $U_{0,75}Pu_{0,25}$N [VNIINM].

Twenty-three full-scale experimental fuel assemblies for BN-600 (more than 1000 fuel rods), 11 experimental fuel assemblies for BOR-60 and 1 experimental fuel assembly for MIR research reactor were manufactured at laboratory, experimental and pilot facilities.

Pre-reactor studies. Thermal conductivity of uranium – plutonium nitride depending on the temperature within the range of 400–1600°C, plutonium content of 5–40% wt. and density of 83.2–93.7th. was measured by the laser burst method in vacuum. The obtained domestic data were compared with the equation recommended for the calculations (Fig. 14.3) [77]. It was shown that the recommended equation described the domestic data adequately. The results were mainly within the above-mentioned equation error. The difference resides in the fact that the domestic data show continuous increase of MN thermal conductivity versus temperature (experimental data up to 1880 K). Thus the recommended dependence provides conservative assessment at the temperature exceeding 1600 K. Foreign data are absent for any temperature above 1600 K. Thermal conductivity of nitride increases with density growth.

In 2018, the studies of heat capacity and thermal conductivity of MN for densities (80–95%th.), Pu content (0–50%), temperature (200–1600°C) were almost completed for the fuels manufactured by the carbothermal reduction and direct synthesis methods. The measurement methodology provides for determination of the temperature dependence for thermal conductivity of ceramic materials within the operating temperature range of 400–1600°C and the thermal

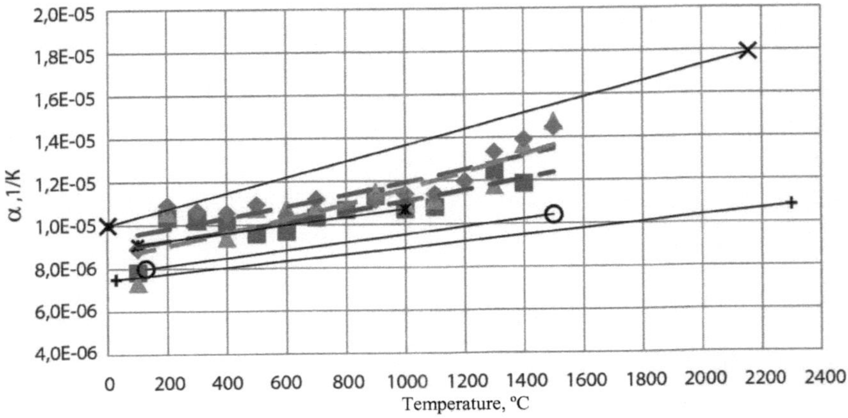

FIGURE 14.4 Temperature dependence of the linear thermal expansion coefficient of samples (the permissible relative error limit for measurement of linear increments 3%) as compared with the published data [216]: Ж – $(U_{0.95}$- $Pu_{0.05})$ N, □ - $(U_{0.90}$- $Pu_{0.10})$ N, ♦ – $(U_{0.85}$- $Pu_{0.15})$ N.

conductivity range of 1.30 W/m·K with the maximum relative error of 8% and the confidence coefficient of $P = .95$.

The studies of the Young's and Poisson's modules as well as the linear thermal expansion coefficient of nitride in case of variations in density (80–89% th.), Pu content (0–15%), temperature (100–1600°C) are completed for the fuel manufactured by the carbothermal and direct synthesis methods (Fig. 14.4).

The nitride dissociation process at high temperatures was studied for non-irradiated nitride (Pu content of 0–50%) including the samples with 0.25% Np. Thermogravimetric studies were carried out in the dynamic helium atmosphere (purity class of 7.0–99.99999%) with the use of a thermal analyzer in a glove box with inert atmosphere and a gas treatment system combined with a quadrupole mass spectrometer. Mass loss of mixed nitride and uranium nitride during isothermal exposure for 30 minutes within the range of 1800–2100°C takes place with the constant rate in a single stage (Fig. 14.5). Mixed nitride begins to lose mass at lower temperature than uranium nitride. In this case the mixed nitride mass loss rate at 2000°C is higher than for uranium nitride at 2100°C.

The intensive evaporation onset temperature of uranium nitride and mixed nitride as well as kinetic dependencies of the decomposition process rate on temperature and plutonium content in MN fuel was determined. It was demonstrated that the mass loss rate increased with increase of plutonium content in nitride. In combination with the experimental and published data it is indicative of predominant plutonium nitride evaporation from the sample surface.

The performed research confirmed that the manufacturing method (from initial oxides or from initial metals) did not affect the pre-reactor properties of MN fuel meeting the technical specifications.

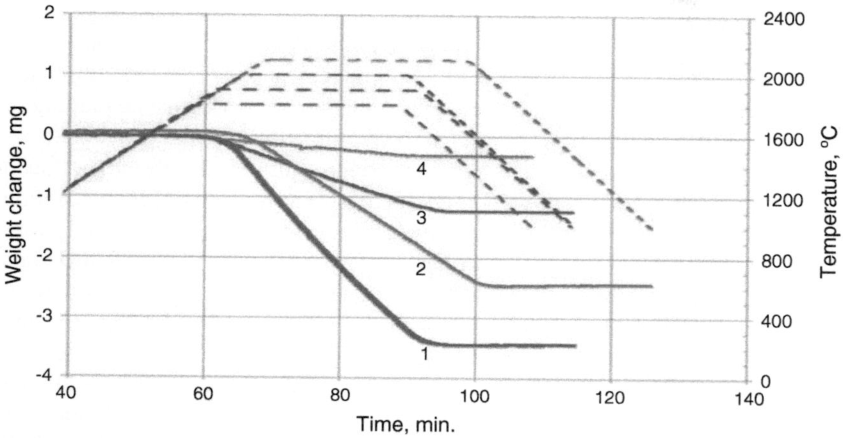

FIGURE 14.5 Thermogravimetric study results for uranium nitride at 2100°C and mixed nitride (U0,5Pu0,5)N at the temperatures of 1800, 1900, and 2000°C: 1 – (U, Pu)N 2000°C; 2 – UN 2100°C; 3 – (U, Pu)N 1900°C; 4 – (U, Pu)N 1800°C; dashed line – temperature.

14.3.1 Reactor testing

MIR reactor: Research was performed with the use of the irradiating device including 7 fuel elements, 6 of them were equipped with sensors for measurement of pressure of fission gas under the cladding, the fuel column elongation and the fuel temperature. The maximum fuel temperature was 1050°C.

BOR-60 reactor: Testing is performed in dismountable irradiation facilities with the possibility for intermediate unloading of individual fuel rods for postirradiation examination. Experimental fuel rods of different types with various cladding materials, different dimensions of fuel – cladding gap and (U,Pu)N pellets of different density are irradiated. As of the end of 2018 five dismountable experimental fuel assemblies (7 fuel rods in each) are being irradiated in BOR-60 reactor; one experimental assembly contains lead-bonded fuel rods, and one assembly contains fuel with Np, Am. All fuel rods are leak-tight. Postirradiation examinations of two fuel rods from OU-1, one fuel rod from OU-2 and one fuel rod with the lead sublayer from OU-4 are completed.

BN-600 reactor: In total 18 experimental fuel assemblies out of 21 manufactured ones are loaded into the core of BN-600 reactor as of the end of 2018.

In BN-600 reactor the most representative fuel testing conditions with regard to irradiation duration under the constant operating parameters and with the minimum impact of transient modes and power variations are provided. In order to ensure testing representativeness for the fuel elements of BREST-OD-300 and BN-1200 type with regard to compliance of burnup, burnup rate and dose they are located in the core periphery and in the breeding zone of BN-600. It results in significant radial irregularity of burnup and damage dose across the experimental fuel assembly section. The irradiation parameters for regular fuel rods in real cores of BREST-OD-300 and BN-1200 reactors have significant ranges depending on the fuel assembly location. Thus the irradiation results for experimental fuel assemblies in BN-600 under the conditions of significant gradients across the section enable to obtain the data for performance judging of fuel rods with different linear powers at the required burnup rate.

Due to limited experience in nitride irradiation in FR and absence of such experience for BN-600 reactor it was decided to perform the first MN irradiation in BN-600 within a combined experimental fuel assembly containing only four fuel rods with MN so that to ensure tests safety. The remaining 123 fuel rods contained uranium dioxide. The design of the first fuel rods with mixed nitride for testing in BN-600 reactor maintained the regular fuel rod geometry and the regular cladding material (ChS68-IDcw). It was made for more credible substantiation of fuel testing safety in BN-600 power reactor. These tests are not fully representative for the fuel rods of BREST-OD-300 and BN-1200 reactors with regard to both fuel rod geometry and the irradiation parameters, first of all – with regard to the mixed nitride burnup rate. But they enable to obtain the data on swelling of mixed nitride and gas release depending on the fuel temperature at different burnups as well as on corrosive fuel-cladding compatibility within the shorter time. In total three combined experimental fuel assemblies with fuel rods of regular geometry (CEFA-1, CEFA-6 and CEFA-7) were manufactured for irradiation in BN-600 reactor core.

CEFA-2 with the fuel rods of BN-1200 geometry and CEFA-3 with the fuel rods of BREST-OD-300 geometry were also manufactured. At present irradiation and postirradiation examination of all CEFAs are completed.

All other EFAs with the fuel rods of BN-1200 and BREST-OD-300 type (so-called full-scale EFAs) contain 61 fuel rods with MN fuel each. Five EFAs are unloaded; irradiation of eight EFAs is in progress. Postirradiation examination of three full-scale EFAs is completed.

The maximum MN fuel burnup and the maximum damage dose in the fuel rods with cladding made of austenitic steel (ChS68-ID hd) are accomplished in CEFA-7 – 7.5 at%, 74 dpa. The maximum MN burnup and the maximum damage dose in the fuel rods with cladding made of EP823 ferritic-martensitic steel were accomplished in EFA-11 – 6.4 at%., 77 dpa (as of 15.03.2019).

Integrity of fuel rods in all examined fuel assemblies was confirmed.

Subsequent to PIE results the experimental data necessary for the technical specification up-dating, verification of fuel codes as well as for substantiation of the fuel design and performance for the first loadings of BREST-OD-300 and BN-1200 with MN fuel were obtained.

14.4 Development of methods, codes, and criteria for substantiation of fuel performance

Engineering codes DRAKON (JSC «SSC RF IPPE»), KORAT (JSC «VNI-INM») are applied in Proryv Project, and mechanistic code BERKUT (Nuclear

Safety Institute of the Russian Academy of Sciences (IBRAE RAN) is under development. All these codes are developed and being elaborated for calculations of container-type cylindrical fuel rods with ceramic pellet fuel.

Pretesting calculations of experimental fuel assemblies are carried out by DRAKON and KORAT codes at the engineering design stage in order to judge MN fuel performance and to obtain the permission for its irradiation. Additional stress-strain state and temperature calculations are carried out prior to completion of postirradiation examinations with the application of DRAKON and BERKUT codes; in this case the actual irradiation history is taken into account. Additional verification of the codes subsequent to the results of post irradiation examinations is performed for each experimental fuel assembly.

Subsequent to the calculation studies of CEFA-1 and CEFA-6 fuel rods performed for the actual irradiation history by DRAKON and BERKUT codes the conclusion was made that in general the expected results were obtained on MN fuel swelling, gas release, cladding deformation that was important from the viewpoint of adequate fuel operating life prediction.

A validation certificate is obtained for the engineering version of BERKUT code for the MN fuel burnup up to 5.5at%. The validation procedure for DRAKON code has been in progress since 2018, and the validation procedure for the mechanistic version of BERKUT code – since 2019.

Prior to implementation of the relevant regulatory documents "Temporary Guidelines on Application of the Data on the Properties of Structural Materials and MN Fuel to Judge Performance of Experimental Fuel Rods for Experimental Fuel Assemblies of BN-600 Reactor " (TG) were developed: the 1-st revision – in 2013, the 2-nd revision – in 2014, and the 3-rd revision – in 2018.

The TG include recommendations on the design criteria for judging fuel performance of sodium-cooled fast reactors, recommendations on the properties of steels and mixed nitride that may be used for experimental fuel rods of BN-1200 and BREST-OD-300 type in the course of their irradiation within experimental fuel assemblies in BN-600 reactor. The design criteria for fuel performance judging were updated in the 3-rd revision subsequent to the joint meeting of R&D Board 1 (Nuclear Power Installations and Nuclear Power Plants), R&D Board 2 (Nuclear Materials and Nuclear Fuel Technologies) of Rosatom and the R&D Board of JSC «Rosenergoatom» on the item «Damage limits for the fuel rods of fast power reactors» dated 06.12.2017.

The TG will be upgraded to the relevant standard subsequent to approval of the consolidated list of the applicable standardization system documents for Proryv Project.

In order to achieve the objectives of Proryv Project the R&D Program for Development of Mixed Nitride (MN) Fuel with Average Burnup of up to 12 at% was developed. The strategic aim of the Program is to achieve the average burnup of 12 at% in the fuel rods of BN-1200 and BR-1200 cores.

This R&D Program includes the following works:

- Improvement of the composition and structure of MN fuel and fuel rod claddings in order to achieve the average burnup of 12 at% and the damage dose of more than 150 dpa (reduction of the swelling rate, increase of the creep rate and fuel crack resistance; increase of the irradiation stability and high-temperature strength of the claddings);
- The set of in-depth out-of-pile studies of experimental MN fuel samples in order to get the dependencies of the improved MN fuel properties;
- Optimization of the irradiation parameters and the design of fuel rods for BR-1200 and BN-1200 cores in order to get the conditions for good fuel performance with increased fuel burnup (liquid-metal sublayer, axial heterogeneity, etc.);
- MN fuel investigation and testing in research reactors (MIR, IVV-2M, BOR-60, MBIR) and power reactors (BN-600, BN-800, BREST-OD-300);
- Experimental studies in BOR-60, MBIR reactors in order to determine the permissible loading limits for MN fuel rods.

Basic designs of fuel elements for BR-1200 and BN-1200 reactors designed for the MN fuel burnup depth limit with the average target value of 12at% proved through experimental judging represent the main ultimate results of the works according to the program. Some experts think that further increase of burnup can cause certain difficulties in assurance of the reactivity margin not resulting in the increased probability of a reactivity-related accident. It seems that additional expenditures for burnup increase will not result in significant LCOE reduction as it has not happened in evaluation of burnup increase from 8 to 12 at%.

Chapter 15

Mixed oxide fuel for fast reactors

Extensive experience on manufacturing of experimental fuel assemblies from mixed oxide (MOX) uranium-plutonium fuel and their in-reactor testing in BOR-60, BN-350 and BN-600 reactors has been accumulated in Russia. MOX fuel was manufactured with the use of two technologies – pellet production and vibro-compaction.

15.1 Pellet technology

The technology for production of pelletized uranium-plutonium oxide fuel for BN reactors has been developed in VNIINM since 1960s. From the very beginning the works were aimed to produce homogeneous MOX fuel with uniform distribution of the fissionable component (plutonium) through the pellet volume.

Correct selection of the homogeneous fuel concept at the early stages was confirmed by numerous Russian and foreign studies. Besides, this concept provides consistent and smooth transition from the uranium fuel cycle of BN-600 reactor to the closed fuel cycle with minimum risks due to similar structures of uranium and homogeneous MOX fuel.

The main focus in development of the pelletized MOX fuel production technology was on the mold powder preparation methods. These methods in particular distinguish the existing and ever developed global technologies from each other. As a rule, the main know-hows of the developing companies are concentrated in the powder part of the technology. Both mechanical mixing of oxides and chemical methods for production of homogeneous mixed powder were considered in the course of the technology development and enhancement: co-deposition of uranium and plutonium with ammonia (Granate Plant), carbonate codeposition, plasmochemical conversion, and sol-gel process [83].

Granulated powders are the most acceptable for development of the processes for plutonium-containing fuel manufacturing. Usage of the granulated material produced through the use of sol-gel and Granate methods guarantees low level of dust formation at the powder stages of the fuel manufacturing process, but

Closed Nuclear Fuel Cycle with Fast Reactors: White Book of Russian Nuclear Power.
DOI: https://doi.org/10.1016/B978-0-323-99308-1.00029-5

application of the chemical granulation methods does not provide high stability of granule properties required for manufacturing of ceramic fuel with stable quality. Unstable powder characteristics were the reason for termination of any further domestic works with the application of the sol-gel method.

The Granate method involves deposition of uranium and plutonium hydroxide in presence of surface-active agents. It was demonstrated that the granulated material produced through the use of this method enabled to manufacture fuel pellets with the required structure and composition. About 700 kg of mixed oxides were produced at «Granat» facility since 1988. Manufacturing of codeposited powder batches on the pilot line of PA «Mayak» was terminated due to the plant operating life expiry. Investigations of the carbonate co-deposition and plasmochemical conversion methods finished with production of small experimental batches and were terminated for various reasons.

Further works for manufacturing of experimental fuel assemblies with MOX fuel were performed only with the application of the mechanical mixing of individual uranium and plutonium oxides. The method involves single-stage pulverization and mixing of powdered individual oxides with subsequent granulation, pressing and sintering of the produced mixture. The main problem of the mechanical mixing method resides in the mixture homogeneity accomplishment so that to fulfill the requirements for the fuel with regard to uniform plutonium content within the fuel pellet volume also providing for smooth SNF processing.

A ball mill is widely used in the global practice as mixing equipment ensuring compliance with the requirements for the mixture homogeneity. A vortex mixture is used for these purposes in the Russian technology for production of pelletized MOX fuel. Introduction of vortex mixing carried out in the rotating electromagnetic field into the process enabled to improve fuel homogeneity significantly. Uniformity of plutonium distribution in a fuel pellet manufactured through the use of this method is comparable with the similar parameter for chemically codeposited fuel. In addition, due to mechanical processing in the mixer powder acquires peculiar properties facilitating more active sintering of the pressed pellet in the course of subsequent thermal treatment [37].

Pulverization and mixing of powders takes place due to intensive rotation of ferromagnetic needles located inside the container made of nonmagnetic material in the electromagnetic field. High intensity of the processes taking place in the vortex mixer accounts for short time required for preparation of finely dispersed homogeneous mixture (not more than 10 minutes). The particles of initial PuO_2 powders with the size of up to 20 μm are reduced to the micron size and uniformly distributed in submicron UO_2 powder. The entire process was called "vortex pulverization process" (VP-process) according to the name of the applied apparatus.

At the stages following mixing and granulation the VP-process is slightly different from the common global technologies accumulating the most successful experience in production of pelletized nuclear fuel. The vortex pulverization technology is notable for its flexibility and enables to manufacture homogeneous mixed oxide and nitride fuel of the required quality.

15.2 Vibration compaction technology

The Research Institute of Atomic Reactors (RIAR) in Dimitrovgrad has been carrying out the studies on development of the advanced fuel cycle with the use of pyro-electrochemical processes for irradiated fuel processing and production of the granulated material directly suitable for manufacturing of fuel rods by vibro-compaction for more than 40 years [20,101]. Pyro-electrochemical processing and vibro-compaction technology enable to implement the process for production of granulated fuel and manufacturing of fuel rods in the remote automated mode.

Optimization of the fuel rod design enabled to achieve burnup of more than 30 at% for individual experimental fuel rods in BOR-60 reactor. 30 experimental fuel assemblies have been irradiated in BN-600 reactor by 2010, the maximum burnup is 10.5 at %.

The pellet technology is adopted as the basic technology for the future BN reactors.

15.3 Experience of MOX fuel use in fast reactors

Works related to manufacturing and testing of experimental fuel assemblies with pelletized MOX fuel aimed to judge the performance of fuel rods and fuel assemblies of BN-800 reactor have been started in 1980s. Testing of 53 experimental fuel assemblies in total with pelletized MOX fuel in BN-350 (14 experimental fuel assemblies) and BN-600 (39 experimental fuel assemblies) was completed by the date of BN-800 start-up. In addition, three experimental fuel assemblies of BN-800 design were manufactured and loaded into BN-600 in summer of 2010. All fuel rods were manufactured at PA «Mayak».

All experimental fuel assemblies maintained their leak-tightness after testing in BN-350 reactor. Four experimental fuel assemblies were subjected to postirradiation examinations.

MOX irradiation parameters in BN-350	
Maximum linear power, kW/m	48
Maximum cladding temperature, °C	690
Maximum burnup, at. %	10.2
Maximum damage dose, dpa	53

The results of BN-350 fuel rods investigations enabled to make the conclusion that the operability resource of fuel rods with pelletized MOX fuel was not exhausted after the maximum burnup of 10 at%, and the fuel rods maintained their performance.

In 1988–2003, the irradiation in BN-600 reactor and postirradiation examinations of experimental fuel assemblies with uranium-plutonium oxide pelletized

TABLE 15.1 Maximum irradiation parameters of experimental fuel assemblies in BN-600 reactor.

Parameter	MOX	
	Pellet	Vibro
Operating time, eff. days	560	572
Maximum burnup, at%	11.8	10.6
Maximum damage dose, dpa	78	80.9
Maximum linear power, kW/m	49.2	47
Total number of experimental fuel assemblies/ with failed rods, pcs	39/1	30/6 (including 1 with fuel contact)

fuel was carried out. Pelletized MOX fuel on the basis of low-background plutonium was used in all experimental fuel assemblies.

MOX irradiation parameters in BN-600	
Maximum linear power, kW/m	49.2
Maximum cladding temperature, °C	710
Maximum burnup, at. %	11.8
Maximum damage dose, dpa	78

The maximum parameters achieved during the testing of experimental fuel assemblies with pelletized and (for comparison) vibro-compacted MOX fuel in BN-600 are given in Table 15.1.

Nondestructive examinations of five experimental fuel assemblies from the 25-th to 38-th cycles were performed in the "hot" cell of Beloyarsk NPP. During visual examination of the fuel rods no signs of overheating and cladding defects were detected. According to the results of visual examination and the outer diameter control some fuel rods from four fuel assemblies (BN-16, BN-22, BN-25 and BN-31) were taken for destructive examinations in "hot" cells at the Institute of Reactor Materials. At the same time comparative studies of fuel rods from regular fuel assemblies irradiated simultaneously with experimental fuel assemblies were carried out at the Institute of Reactor Materials.

The measurement results of cladding diameters of irradiated fuel rods and their mechanical properties did not reveal any fundamental differences from the results of similar investigations for the regular fuel rods claddings under the corresponding irradiation parameters. The examination results of cladding corrosion demonstrated that pelletized MOX fuel was almost identical to uranium dioxide in its compatibility with the claddings made of austenitic steel under similar irradiation conditions. The nature and degree of uranium dioxide and MOX restructuring are similar under identical operating conditions.

The results of studies for the fuel rods of BN-600 reactor with pelletized MOX fuel enable to make the conclusion that the fuel performance is not exhausted up to the maximum burnup of ~12 at%.

In 2009, three experimental fuel assemblies with pelletized MOX fuel of BN-800 design with the sodium intermediate layer between the fuel bundle and the top blanket were manufactured and supplied to Beloyarsk NPP. The top blanket includes steel shielding rods and absorbing elements with boron carbide. The claddings made of ChS68-ID cw new generation steel were manufactured at PJSC «Mechanical Engineering Plant» according to the improved technology. Reactor tests of the experimental fuel assemblies were carried out for four cycles. The operating parameters of the fuel assemblies are given in Table 15.2.

The experimental fuel assembly No.02.09 with the maximum burnup was selected for postirradiation examination.

Subsequent to experimental research for the irradiation behavior of BN-800 design fuel assembly the conclusion was made that in general the design was operable under the operating conditions of BN-600 reactor. The first three experimental fuel assemblies with pelletized MOX fuel were irradiated in the BN-600 core of the third modification with the life time extended to 585 eff. days with fuel rods integrity. Both pelletized and vibro-compacted MOX fuel is used in BN-800 reactor plant within the hybrid core.

Pelletized MOX fuel was used as regular fuel for the following fast reactors: PFR (UK), Rapsodie, Phenix, Super-Phenix (France), KNK-2 (Germany), Joyo, Monju (Japan), and FFTF (USA). The consolidated information on the characteristics and operating conditions of fuel assemblies with MOX fuel is provided in Table 15.3. All reactors were shut down by 2018.

Fuel burnup increase is one of the most efficient ways for improvement of the technical and economic parameters of NPPs with fast reactors. The objective to achieve the maximum fuel burnup of ~20 at% corresponding to damage dose of more than 180–200 dpa is set for further .extension. But in 2019 the maximum fuel burnup in fast power reactors does not exceed ~10 at% for regular fuel. This is primarily due to the unstable behavior of claddings materials and high cladding deformations under irradiation.

Subsequent to long-term works aimed to enhance the irradiation stability of cladding austenitic stainless steels the improved ChS68-ID cw steel is developed in our country in order to provide the irradiation dose of 90–95 dpa, and studies of EK164-ID cw austenitic steel planned for use up to 110–115 dpa are near completion. Achievement of higher doses is associated with application of ferritic-martensitic steels with almost no swelling.

15.4 Industrial production of MOX fuel

Industrial production of MOX fuel is arranged at MCP FSUE within the framework of the Federal Special Program "New Generation Nuclear Power Technologies for the Period of 2010–2015 and for Further Extension to 2020" approved by Resolution of the Russian Federation Government No. 50 dated 03.02.2010 and intended to supply fuel to Unit 4 of Beloyarsk NPP with BN-800 reactor plant with the design capacity of 400 FA/year. The facility was

TABLE 15.2 Operating parameters of experimental fuel assemblies with pelletized MOX fuel of BN-800 type.

Experimental FA number	Cell number	Cycle number	Date of loading	Operating time, eff. days	Max. dose, dpa	Max. burnup, at. %
03.09	15–22 LEZ	59–62	08.04.10	584.63	82.1	11.3
	23–06 in-reactor storage	63	03.05.12			
02.09	24–21 LEZ	59–62	08.04.10	584.63	84.7	11.6
	18–03 in-reactor storage	63	02.05.12			
01.09	15–10 LEZ	59–62	08.04.10	584.63	83.0	11.4
	06.23 in-reactor storage	63	02.05.12			

TABLE 15.3 Characteristics and parameters of MOX fuel bundles operation in fast reactors.

Reactor	Operation period	Thermal/electric capacity, MW	Number of fuel bundles in the core	Fuel bundle case size, mm	Number of fuel rods in a bundle/Core	External fuel rods diamter, mm	Max linear capacity, KW/m	Max case temperature, °C	Max burnup, MW ◦ day/t (% h.a.)
PFR	1975–1994	650/250	72	142	325/23 400	5,8/6,6/8,5	48	670	200 000 (23%h.a.) (Exp. Fuel bundles)
Rapsodie	1967–1983	40/0	66	49,8	61/4 026	5,1	43	635 ном.	102 000 (11,8% h.a.) (radiated ~30 000 rods with standard fuel, among them: ~10 000 fuel rods with burnup more then 10% h.a.; max burnup of exp. fuel bundles 27% h.a. (173 dpa)
Phenix	1974–2009	563/255	103	124	217/22 350	6,6	45	650 ном.	150 000 (17,4% h.a.) (ETVS)

(*continued on next page*)

TABLE 15.3 Characteristics and parameters of MOX fuel bundles operation in fast reactors—cont'd

Reactor	Operation period	Thermal/ electric capacity, MW	Number of fuel bundles in the core	Fuel bundle case size, mm	Number of fuel rods in a bundle/ Core	External fuel rods diamter, mm	Max linear capacity, KW/m	Max case temperature, °C	Max burnup, MW ◦ day/t (% h.a.)
Super-phenix	1986–1996	2990/1242	364	173	271/98 700	8,5	48	620 ном.	90 000 (10,4% h.a.) (design)
KNK-2	1978–1991	58/20	14	108	169/2 366	8,2	45	600 ном.	172 000 (20% h.a.) (exp. fuel rods)
Joyo	1978–2007	140/0	85	78,5	127/10 795	5,5	42	675 (midwall)	86 900 (10,1% h.a.)
Monju	1994–2009	714/280	198	111	169/33 462	6,5	36	675 (midwall)	94 000 (10,9% h.a.) (design)
FFTF	1980–1994	400/0	73	118	217/15 841	5,84	41,3	680	155 000 (18% h.a.) (more then 63 500 fuel rods with burnup 80 000 MW ◦ day/t (9,3% h.a.), more then 4300 fuel rods with burnup 100 000 MW ◦ day/t (11,6% h.a.)

commissioned in December 2014. The arranged production facility is one of the nuclear fuel cycle closing infrastructure components at MCP FSUE Federal Nuclear Organization including the process capacities for nuclear fuel storage, processing and fabrication. The production facility is located in the existing mine openings of the Radiochemical Plant in the underground section of the facility. The rock acts as a natural containment protecting against any external natural and human-induced impacts.

Production of MOX fuel consists of the following basic facilities and sections: the plutonium re-purification facility, the pellet manufacturing section, the fuel rod manufacturing section, the FA assembly section and the analytical control systems.

The plutonium repurification facility has a flexible flow chart enabling not only to repurify plutonium dioxide but also to process substandard MOX fuel pellets (scrap) with plutonium extraction and purification as well as to extract americium concentrate from waste solutions for further storage or disposal. The possibility to use both extraction and sorption technologies are provided at the plutonium re-purification plant enabling to purify plutonium and to process liquid RW generated in the course of repurification with extraction of valuable components (plutonium) and its return to the MOX fuel production, depending on the objective. In this case generation of liquid HLW is eliminated, and the amounts of discharged waste are minimized.

Subsequent to implementation of the primary refurbishment and adjustment manufacturing of the finished product was started at all MOX fuel production sections.

The product is manufactured in the automated mode in a chain of localizing radiation protection chambers and boxes through the use of some unique equipment. The production facility provides for the possibility to manufacture MOX fuel assemblies with the use of both low-background ("weapon-grade") and high-background ("energy-grade") plutonium dioxide. Depleted or natural uranium dioxide is used as uranium raw material. 45 analytical, process and final inspection methodologies were developed and validated for the product quality control and confirmation.

Chapter 16

Remix fuel

Reprocessing of spent fuel is considered in order to use the energy potential of the remaining uranium and produced plutonium, to eliminate storage of spent nuclear fuel (SNF) and accumulated reprocessing products and to save natural uranium resources [36].

Accumulation of ^{232}U radionuclide generating the chain of short-lived intensive gamma emitters is one of the main problems in reuse of uranium from SNF [26]. The recovered plutonium requires deep re-purification prior to refabrication into MOX or REMIX fuel for the same reason (decay of ^{236}Pu into ^{232}U).

Another reason restricting reuse of uranium resides in accumulation of ^{236}U being a neutron absorber. Presence of even isotopes in the fuel composition requires additional enrichment thus reducing the fuel cycle efficiency [53].

Approximately 24,000 tons of SNF have been accumulated totally in the Russian Federation by 2019. About 650–700 tons of spent fuel is annually unloaded from the reactors of Russian NPPs, while only ~15% of this amount is processed [109]. Regenerated nuclear materials are traditionally used separately in Russia. Regenerated uranium is enriched and used mainly in RBMK reactors. Plutonium is accumulated for the fast reactor program launching, and production of MOX fuel for BN-800 has already started.

The concept of two-component nuclear power system including both reactor types (VVER - with the thermal neutron spectrum and FR - with the fast neutron spectrum) is adopted in the Russian Federation [52]. The transition period may include reuse of regenerated nuclear materials as mixed fuel for thermal reactors (for example, VVER) possibly more efficient than MOX fuel with partial loading of the core.

16.1 Modeling of nuclear fuel cycles

Khlopin Radium Institute in cooperation with Kurchatov Institute has been engaged in development of REMIX fuel (REMIX - Regenerated Mixture of U-Pu oxides) involving joint reuse of regenerated uranium and plutonium for the entire core loading in the existing VVER 1000/1200 since 2005 [36,53].

Closed Nuclear Fuel Cycle with Fast Reactors: Handbook of Russian Nuclear Power.
DOI: https://doi.org/10.1016/B978-0-323-99308-1.00014-3

Numerous calculations of the nuclear fuel cycle have been performed with the use of software tools to model nuclear fuel burnup according to the Monte Carlo method. SCALE 6.2 software was the main tool for calculations [203]. SCALE is the set of tools enabling to carry out calculations of criticality, fuel burnup, activation of materials, characteristics of radiation sources and radiation protection. The calculations were performed with the use of the library of continuous neutron interaction cross-section dependencies on energy based on the qualified nuclear data files ENDF/B VII. 1. A three-dimensional model of the infinite reactor core with TVS-2M fuel assemblies were used for modeling [104]. The neutron multiplication factor in the course of fuel burnup was maintained at the level of $k_{eff} = 1$ by boron concentration control in the coolant.

Different patterns of nuclear fuel cycles with the use of mixed uranium-plutonium fuel were modeled. A group of thermal reactors of the Russian design (VVER-1000/1200 types) within the entire period of their operation was considered in the analysis. Subsequent to fuel burnup and storage the isotope composition of SNF was assessed for the possibility and feasibility of its further use. In case the residual energy potential of SNF is higher than for natural uranium nuclear materials from this fuel are used in the next recycle modeling.

Several potential variants of nuclear fuel cycles based on mixed uranium-plutonium fuel are considered in this Chapter.

REMIX A - Basic variant. In the course of spent fuel reprocessing regenerated U and Pu are separated (not necessarily to the full extent) and then remixed in the solution at the initial ratio, and the required energy potential is ensured by 19% ^{235}U make-up within the standards for non-proliferation of nuclear materials. Multiple recycling of such fuel is possible, but ~15% of excessive regenerated uranium is generated at each recycle. In case the make-up of highly enriched uranium is used 55–65% recycling takes place without generation of any excessive regenerated uranium.

REMIX A recycling scheme is presented in Fig. 16.1. Seven cycles are considered, Cycle 1 (highlighted with green color) is the starting loading of fuel made of enriched natural uranium.

Approximately 4-fold reduction of the final accumulated SNF amount in comparison with the open fuel cycle is possible in this variant due to reprocessing and multiple reuses of regenerated materials. Saving of natural uranium at the level of 25–30% is achieved. The mixed fuel fraction in REMIX NFC is 86%.

REMIX C. This variant provides for supply of regenerated uranium from SNF reprocessing for re-enrichment up to ~4–5% ^{235}U and its subsequent mixing with the heat-treated plutonium-uranium master mixture remaining at the production facility. Make-up of natural enriched uranium is used. Excess of nuclear material is eliminated, but transportation of regenerated uranium to the enrichment plant and back takes place.

REMIX C recycling scheme is presented in Fig. 16.2. Subject to implementation of 7 cycles in this variant the amount of SNF in storage is sevenfold reduced.

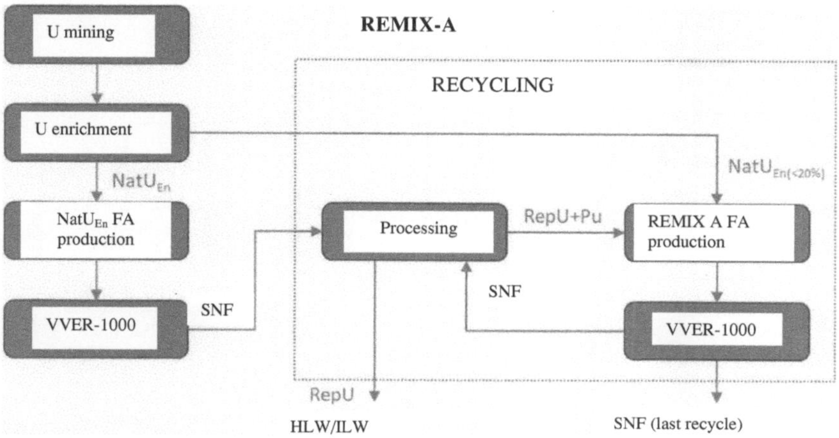

FIGURE 16.1 REMIX-A recycling with the make-up enrichment of up to 20% of ^{235}U.

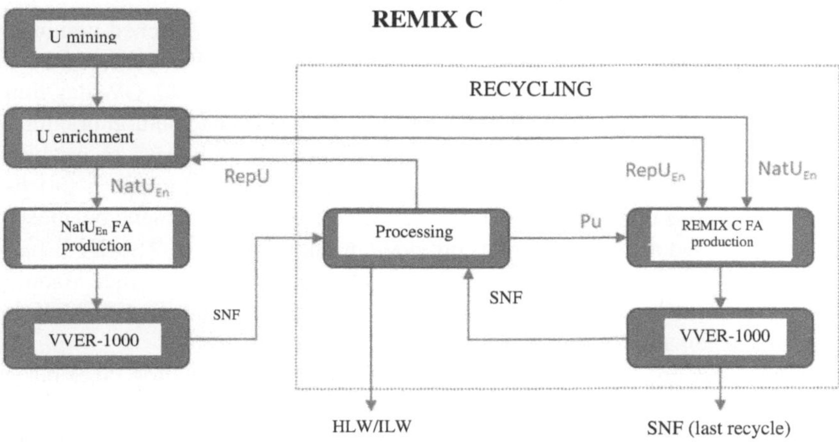

FIGURE 16.2 REMIX-C recycling.

Saving of natural uranium is 30%. The mixed fuel fraction in REMIX NFC is 86%.

In the context of fuel supply for the reactors of Russian design constructed abroad within the entire lifecycle period the concept of REMIX C seems optimal as the Russian Federation plans to provide the services for SNF reprocessing and return of the customer's nuclear materials but in the form of mixed uranium-plutonium fuel and radioactive waste. The number of such fuel recycles (up to 7) is sufficient for the entire operating life of the reactor plant. In case of such approach the customer avoids long-term storage of SNF as it is not accumulated in the course of operation.

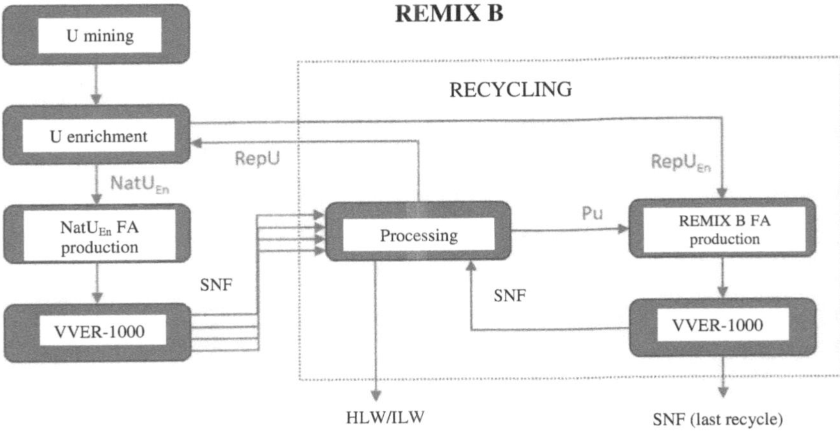

FIGURE 16.3 REMIX-B recycling.

REMIX B is the NFC option without any make-up with enriched uranium, with reduction of the total amount of nuclear materials; the content of Pu in the fuel can exceed 4%. Subsequent to fuel burnup of ~47 GW-day/thm approximately 1.25% of plutonium is produced from enriched natural uranium. Consequently, 3–4-fold larger amount of SNF shall be processed to produce mixed fuel with the plutonium content of 4%. The regenerated uranium enrichment method is used to provide the required effective fuel enrichment. No make-up with enriched natural uranium is provided. Manufacturing of REMIX-B fuel requires higher component mixing accuracy, but the secondary fuel amount is several times less than initial SNF from natural uranium. The REMIX B recycling scheme is presented in Fig. 16.3.

"Compression" of fuel and increase of plutonium concentration takes place at each recycle due to make-up absence.

The number of recycles for this option is limited because of high plutonium content. At the first recycle the plutonium content starts from 4%, at the second - from 8%. 4-fold reduction of SNF is achieved in case of one uranium fueling with a single REMIX B recycle as in the case of seven REMIX A recycles. In case of two recycles the SNF amount is reduced more than 10 times as compared with the open NFC. The saving of natural uranium is about 25–30%. The mixed fuel fraction in the REMIX NFC is 20–30%. This fuel composition is patented in Russia [103].

REMIX E. The economic evaluation of the NFC based on the prices taken from published sources showed [119] that in case of the NFC with recycling of nuclear materials the main extra charge was due to the mixed fuel fabrication price. Consequently, the NFC option with the lower percentage of mixed fuel use at the comparable saving of natural uranium and SNF accumulation reduction seems the most attractive from the economic viewpoint.

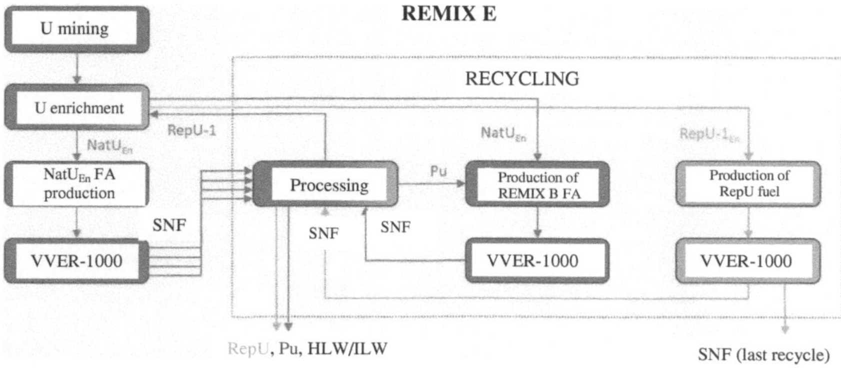

FIGURE 16.4 REMIX-E recycling.

The new approach - REMIX E - was proposed involving simplification of mixed fuel fabrication through the use of increased plutonium concentration together with enriched natural uranium; in this case regenerated uranium is planned to be enriched and used separately. This NFC option seems optimal with regard to fast start and enables to reduce SNF accumulation at the level of other options. This fuel composition is being patented in Russia and abroad.

Three fuel types are used in the nuclear fuel cycle with REMIX E: standard enriched natural uranium fuel (56%), fuel made of enriched regenerated uranium (ERU 18%) and mixed U_{nat} -Pu fuel (25%). Low fraction of mixed fuel is due to increased initial plutonium concentration (5%); fuel "compression" takes place. It is important to note that enriched natural uranium and not regenerated product is used in this U_{nat}-Pu fuel. Such production facilities can be arranged even now on the basis of the existing MOX fuel production capacities of MCP FSUE. Single-regenerated uranium is planned to be used at the existing production facilities as ERU fuel. Standard enriched natural uranium fuel is the majority in this NFC.

The REMIX E recycling scheme is presented in Fig. 16.4. The scheme of nuclear material use is given in Fig. 16.5. The fuel used is marked with rectangles: from natural uranium, mixed uranium and plutonium and ERU. The material extracted after reprocessing and supplied for the new fuel production is shown with arrows. Regenerated nuclear material not used after reprocessing and supplied for storage is marked with an oval. Seven cycles were considered. The last cycle is finished with fuel made of ERU (marked with dashed line) which is not supplied for reprocessing and is considered as SNF. The amount of generated SNF is directly related to duration of the cross-cycling and fuel "compression" with regard to plutonium and ^{232}U limitation in ERU.

There is no limitation for duration of cross-cycling in the presented NFC, but in case of the NFC consideration at a small group of reactors the amount of SNF becomes so insignificant at some point due to continuous reprocessing

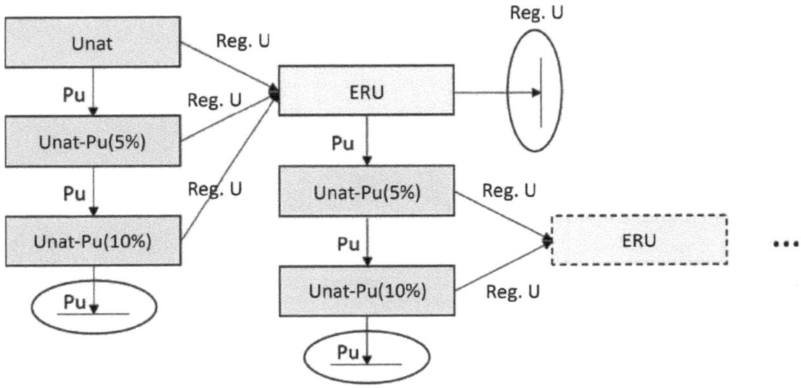

FIGURE 16.5 Cross-use of nuclear materials.

and fuel "compression" that it is more feasible to store or dispose it instead of reprocessing. That is why 7 cycles are considered as an example. Two cycles of mixed U_{nat}-Pu fuel with "compression" are used in the cross-cycling variant presented in Fig. 16.5. "Compression" to 5% Pu takes place at the first cycle, and to 10% Pu - at the second cycle. The second (optional) Pu-cycle is used only for plutonium after-burning, and the number of odd plutonium isotopes after it is almost equal to the number of even plutonium isotopes. The peculiarity of this variant due to cross-cycling resides in generation of burnt regenerated uranium (instead of SNF) with the energy potential comparable to natural uranium which is not acceptable for further recycling in thermal reactors.

REMIX concept. Fabrication of pilot fuel rods with REMIX fuel has been completed, and their irradiation has started by 2019 within the framework of REMIX concept promotion. The program for irradiation of REMIX composition consisting of 10 fuel rods is implemented in MIR research reactor. At present 4 fuel rods are irradiated, and their post-irradiation examinations are completed. The block plan of REMIX concept implementation is given in Fig. 16.6.

The second campaign with the use of REMIX fuel is in progress at the third power unit of Balakovo NPP simultaneously with irradiation in the research reactor. No deviations in the operation of fuel assemblies are detected, and the supervisory body has issued the license for refueling.

The key prerequisite for the REMIX NFC product resides in availability of domestic experience in its application (reference) in Russia, including the required infrastructure for SNF reprocessing and fabrication of REMIX fuel assemblies. Development of large-scale production facilities for fabrication of uranium-plutonium fuel combined with SNF reprocessing is necessary to arrange the appropriate REMIX NFC in Russia. Rosatom approved the REMIX Fuel Reviewing Program. The Program includes the projects on safety analysis for pilot operation of VVER-1000 reactor plant with TVS-2M fuel assemblies

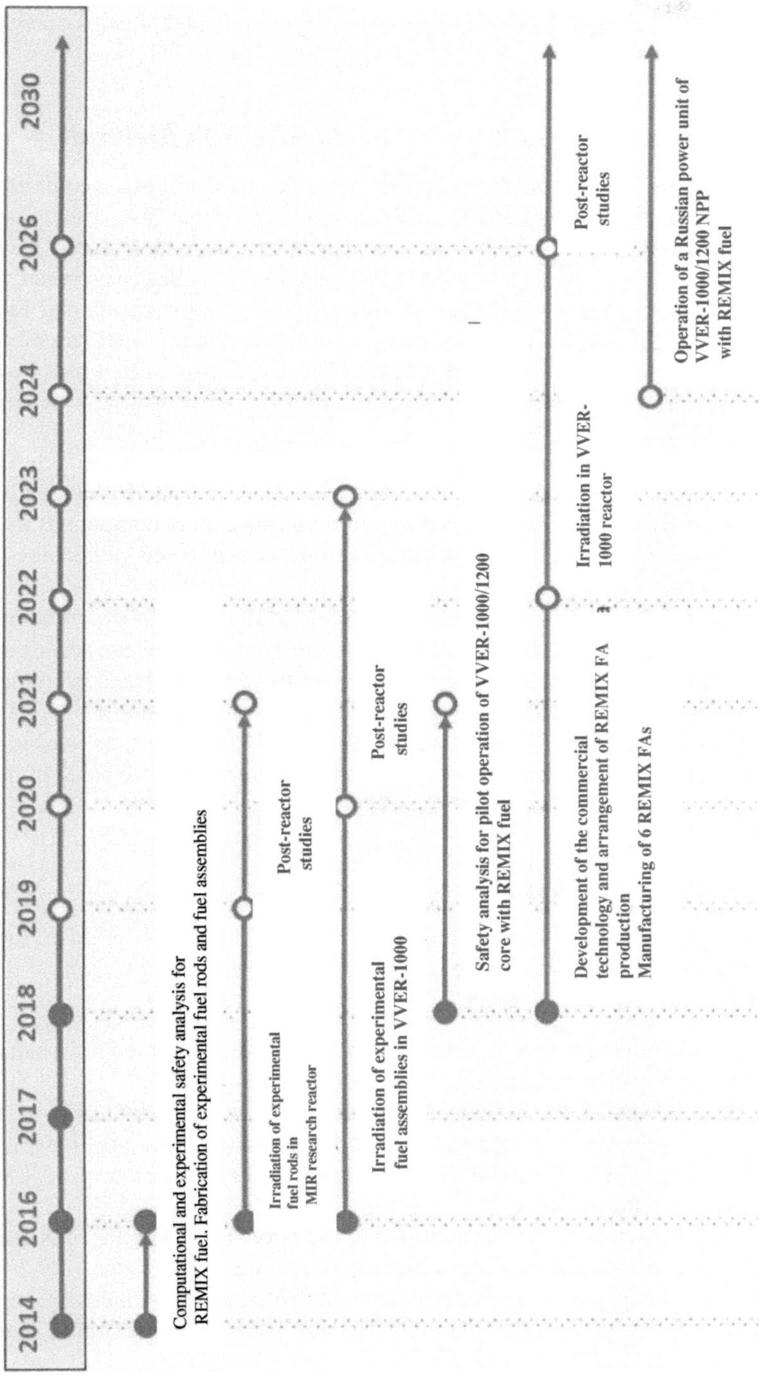

FIGURE 16.6 Implementation of REMIX concept.

with REMIX fuel as well as development of the technology and pilot production facilities for REMIX fuel.

16.2 Manufacturing of the pilot batch of REMIX fuel rods

Khlopin Radium Institute produced a pilot batch of powdered master mixture of uranium and 20% of plutonium with the weight of about 3 kg [34]. Fuel pellets with the plutonium mass fraction of ~1% were manufactured from this master mixture in JSC «VNIINM». The majority of pellets were used to manufacture fuel rods(JSC «Siberian Chemical Plant») and experimental fuel assemblies (NCCP «PJSC») for irradiation of the experimental fuel assembly TVS-2M in VVER-1000 reactor at Balakovo NPP. Experimental fuel rods with REMIX fuel for irradiation in the pilot facility of MIR research reactor were also manufactured in JSC «VNIINM».

The master mixture of uranium and plutonium oxides was manufactured according to two different deposition technologies (ammonia and formate) with subsequent conversion of the produced uranium and plutonium compounds into the oxide form. Deposition process for ammonia uranates and plutonates is presented in Fig. 16.7.

The results of the studies performed in JSC «VNIINM» with experimental pellets showed that the pellets made from the master mixture produced according to the ammonia deposition technology complied with all required characteristics to a greater extent. For this reason, this particular ammonia uranate and plutonate deposition technology with subsequent filtration of the sediment, its washing, drying and heat treatment was adopted for manufacturing of a large batch of the powdered master mixture for REMIX fuel.

16.3 Tests of REMIX fuel in MIR reactor

Irradiation and post-irradiation examination of 4 fuel rods with REMIX fuel were performed in MIR research reactor [71]. Experimental fuel rods have the following peculiarities:

- Unification with the fuel rod on the basis of TVS-2M is provided for irradiation in VVER-1000 reactor core;
- Shortened design of the fuel rod ~600 mm enables its irradiation in MIR reactor core with the maximum power density uniformity along the height;
- Pelletized REMIX fuel similar to the fuel placed for irradiation in VVER-1000 reactor core is used;
- Availability of a gas plenum enables to study gas releases from REMIX fuel and perform re-fabrication without any damage of the fuel pellets.
- Testing of 10 experimental fuel elements with REMIX fuel includes:
- Basic irradiation of fuel elements in MIR reactor up to the burnup of ~50 MW-day/kg_{hm};

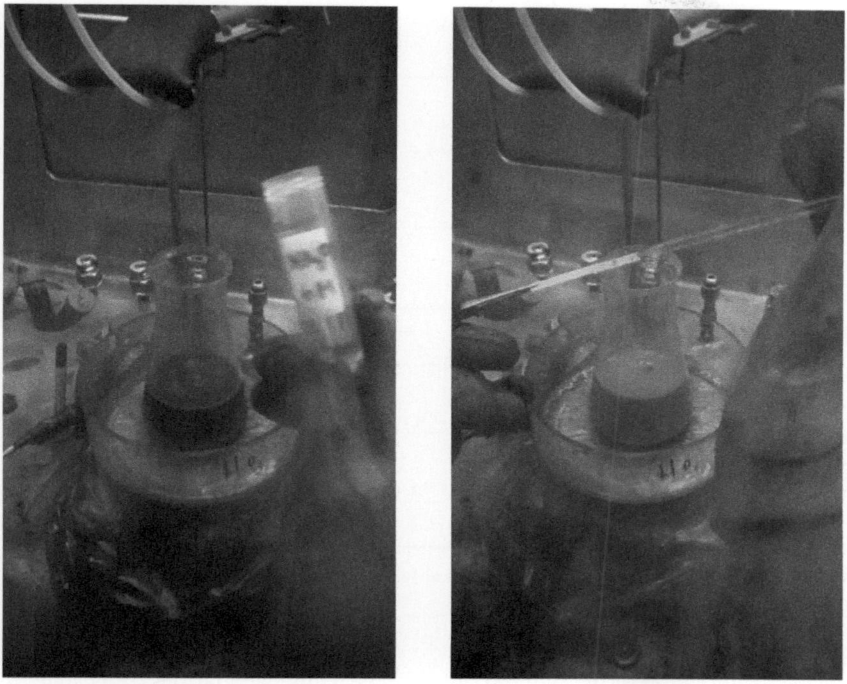

(A) (B)

FIGURE 16.7 Reaction mixture in the course of (A) and after (B) deposition.

- Special experiments for fuel elements No. 5-9; power surges, the conditions of design basis accidents with loss of coolant (LOCA) and insertion of positive reactivity into the core (RIA);
- Postirradiation examinations.

The testing scheme for 10 experimental fuel rods with REMIX fuel is given in Fig. 16.8.

The research program provides for removal of one fuel rod for investigation of the operability characteristics in the course of testing at certain burnup values. It can be seen from the presented scheme that sufficient package of experimental data within the linear load range of 0 to 36 kW/m and the average fuel burnup of 0–50 MW·day/kg h.a. will be obtained upon completion of this program for validation of computer codes and assessment of the software calculation errors. Behavior of REMIX fuel in the normal operation mode, in case of any operational occurrences and design basis accidents will be also studied.

The results of studies (visual examination, profilometry, fission gas release measurement, gamma scanning, radiographic analysis) of the four removed fuel rods demonstrate absence of any abnormalities significant for operability.

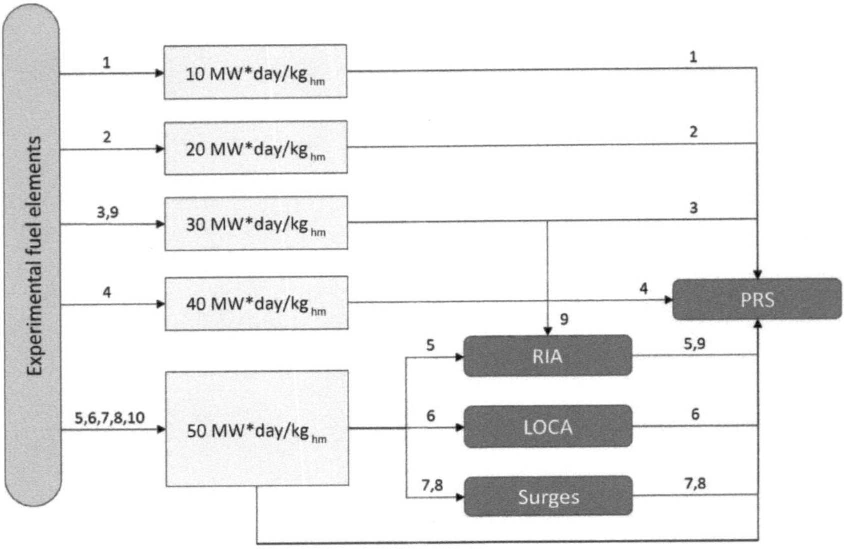

FIGURE 16.8 Testing scheme for experimental fuel rods with REMIX fuel (PIE, postirradiation examinations).

Visual examination of the fuel rod did not reveal any peculiarities of its state as compared to the regular VVER fuel rods with similar burnup. No abnormalities were detected in the fuel rod appearance including the state of weld joints. Reduction in the cladding diameter due to irradiation and thermal creep is 10–20 μm corresponding to the fuel burnup. No significant elongation of the fuel rod is registered. The fuel rod has retained its leak-tightness, the release of fission gas under the cladding is 1.38%. Distribution of fission products along the column length corresponds to the burnup profile. Gaps between the pellets registered according to the results of gamma scanning and radiographic analysis are due to displacement of the pellets and the fixture bushing in the course of fuel rod handling operations in the hot cells.

Thus the results of the ongoing reactor testing and examinations of the first removed fuel rod are indicative of operability maintenance for experimental fuel rods with REMIX fuel and the possibility for their operation up to higher burnup. Reactor testing of other fuel rods will end in 2019.

16.4 Reprocessing of irradiated REMIX fuel

Khlopin Radium Institute has accumulated extensive experience in hydrometallurgical reprocessing of SNF from different reactors (both research and power ones) in order to practice this technology for industrial implementation at radiochemical plants. More than 140 kg of SNF with the burnup of up to

70 GW-day/t have been already processed for almost 50 years of the plant operation.

The modern extraction technology for NPP SNF reprocessing emerged in the USA more than 60 years ago for reprocessing of irradiated natural uranium for military purposes [164]. It was called PUREX process (PUREX = Plutonium-Uranium Extraction). Improvement of the technology processes together with relaxation in the requirements for removal of fission products enables to simplify the SNF reprocessing flow chart structure reducing the refining units with their waste and transferring them to the production facilities for manufacturing of nuclear fuel from regenerated materials and at the same time - to establish the conditions for elimination of process ILW discharges as provided in the technology of the Pilot Demonstration Facility at MCP (Zhelezhogorsk) developed by Khlopin Radium Institute [63].

Upon completion of the reactor campaign in MIR reactor experimental reprocessing of REMIX SNF is planned in order to demonstrate the possibility for its recycling.

Chapter 17

Adaptation of uranium-plutonium fuel fabrication technologies

The main peculiarities of the manufacturing technologies for mixed nitride and oxide fuel as compared with the commercial uranium fuel production technologies implemented at the plants of JSC "TVEL," apart from mixing of fuel components itself, include mandatory aerosol protection of the equipment considerably complicating operation, maintenance and repair as well as enhanced radiation fields implying high level of production automation.

The flow diagram developed by JSC "VNIINM" (Fig. 17.1) is implemented at the pelletized fuel production facilities (metal-oxide fuel at MCC and mixed nitride uranium-plutonium fuel at Siberian Chemical Combine [SCC]). The technology is based on single-stage dry mechanical mixing of powdered components in the electromagnetic vortex mixer with subsequent traditional dry pressing and high-temperature sintering of pellets.

The technology provides for high plutonium content uniformity within the fuel pellet volume and sufficient solubility for reprocessing of spent oxide fuel assemblies, in particular with the use of water-based methods.

Transition to production of nitride uranium-plutonium fuel requires, first of all, availability of the carbothermal synthesis plant for mixed nitride powder. The necessity to maintain and control the inert atmosphere in the protective boxes and chambers so that to prevent oxidation of finely dispersed pyrophoric nitride powder and non-sintered pellets should be also considered as a peculiarity of nitride fuel production.

Manufacturing of oxide fuel on the nitride fuel production equipment is possible subject to minimum adjustment. But manufacturing of nitride fuel on the plants initially designed for oxide fuel is impossible.

The pellet fuel rod manufacturing technologies are identical for oxide and nitride fuel and based on the main joint sealing (welding) by the magnetic field method with remote control, but in each particular case they are determined depending on the equipment and fuel element design peculiarities. In particular, a spacing wire wound on the finished item or on the tube assembly in case

Closed Nuclear Fuel Cycle with Fast Reactors: Handbook of Russian Nuclear Power.
DOI: https://doi.org/10.1016/B978-0-323-99308-1.00031-3

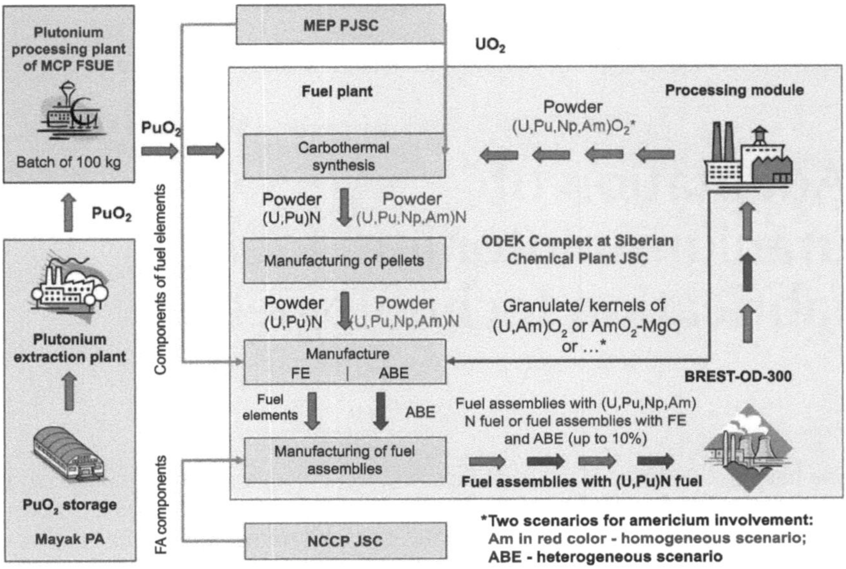

FIGURE 17.1 Mixed nitride fuel fabrication/refabrication scheme.

of manual manufacturing is used for the fuel elements of SFR (sodium fast reactors).

In spite of high level of equipment unification and identity of the technologies applied at the production facilities for fuel rods with any fuel including MOX and MN fuel the line may be used for manufacturing of the fuel rods of a particular reactor plant. Design of the equipment for the lines intended for plutonium-containing fuel has box protection as in case of pellet production, thus they are different from uranium lines in their arrangement and require special repair technologies. Fuel element assembly lines at the commercial production facilities are automated.

The technology for manufacturing of vibro-compacted fuel rods with oxide fuel is implemented only in JSC «RIAR» and based on pyrochemical production of the initial granulated material from molten salts. Automated plants (ALIT-1, ALIT-2) enabling to form vibro-compacted fuel ~~pellet~~ stacks and provide sealing of fuel rods in a hot cell through the use of the universal magnetic field method with remote control are arranged. This technology cannot be adapted for nitride fuel manufacturing, but it may be successfully used for involvement of minor actinides into the fuel cycle.

The manufacturing technologies for FR fuel assemblies with oxide and nitride fuel may be identical. In this case the process is fully automated as distinct from uranium production.

The remote technology developed on the basis of the TVS-2M assembly technology (for reactor plants of VVER type) at JNS «NCCP» is planned for commercial production of fuel assemblies with mixed nitride uranium-plutonium fuel at the fabrication/refabrication module (PDEC Complex of JSC «Siberian Chemical Plant»).

Analytical support of fuel quality control is different and defined mainly by the available capacities of the facilities. All types of control at the operating facilities actually depend on the basic fuel rod design and may be unified not only for the facilities, but also for fuel types.

Chapter 18

Usage of the industry-specific fuel infrastructure

18.1 FSUE "Mayak PA" ("Paket" on RT-1, RT-1)

"Paket" facility has been in operation since 1980s, the line is used to manufacture MOX fuel in support of manufacturing at FSUE «MCP». The facility includes the sections for manufacturing of pellets and fuel rods within the old line and a separate new complex for pellet manufacturing ("Paket-2") equipped with two advanced automatic presses. Production is manual with the maximum capacity of 3–3.5 t/year of fuel (up to 100 sets of fuel rods for fuel assembly) with subsequent FA assembly in JSC «SSC RIAR» or FSUE «MCP». The process based on vortex pulverization (VP-process) is implemented. The facility is designed for use of low-background plutonium (of weapon or energy origin). The products include fuel rods of SFR type with pelletized MOX fuel. There is no local FA assembly section.

18.2 FSUE "MCP" (MOX, pilot demonstration facility)

Production of MOX fuel at FSUE «MCP» was launched with plutonium. Implemented technology – VP-process. The production facility is fully automated and designed for both weapon-grade and energy-grade plutonium, for example produced from VVER SNF. The complex includes a plutonium re-purification plant, manufacturing sections for pellets, fuel rods and fuel assemblies. The products include BN-800 fuel assemblies with pelletized MOX fuel. Capacity of the plant is presented in Table 18.1.

TABLE 18.1 Capacity of MOX fuel production.

No.	MOX fuel production complex	Capacity
1	Plutonium repurification facility, t/year	2.15
2	Pellet manufacturing complex, t/year	12.44
3	Fuel rod manufacturing complex, thous. pcs/year	52.0
4	Fuel assembly manufacturing complex, pcs/year	400

Closed Nuclear Fuel Cycle with Fast Reactors: Handbook of Russian Nuclear Power.
DOI: https://doi.org/10.1016/B978-0-323-99308-1.00039-8

Research cells of the MCP pilot demonstration facility are suitable for R&D works with MOX SNF and require refurbishment for works with MN SNF.

18.3 JSC "Siberian Chemical Combine" (KEU-1, KEU-2, FRM)

KEU-1. KEU-1 complex of experimental facilities is an operating manual experimental production facility for fuel, fuel rods and fuel assemblies with pelletized MN fuel at the chemical and metallurgical plant of JSC «Siberian Chemical Plant» with the capacity of 200 kg of fuel and up to 5 sets per a year for fuel rods and fuel assemblies. Capacity of the manufacturing sections for fuel rods and fuel assemblies may be increased twice or thrice. The implemented technology – VP-process/carbothermal synthesis. Low-background weapon-grade plutonium is used at the complex. The complex is equipped with the intra-chamber atmosphere (nitrogen) maintenance and control system.

The products of the complex include experimental fuel assemblies of BN-600 type and BN-800 fuel rods with MN fuel. Manufacturing of fuel assemblies and fuel rods with MOX fuel without any significant equipment refurbishment is possible. Fuel elements with REMIX fuel for VVER-1000 and MIR reactors were manufactured at the plant.

KEU-2. The complex of experimental facilities at the chemical and metallurgical plant of MCP «Siberian Chemical Plant» expands the capacities of KEU-1 with regard to fuel pellet manufacturing. The section is arranged for testing of commercial equipment prototypes, practicing of industrial regimes and repair technologies with the possibility to manufacture experimental MN fuel. The section capacity is 400 kg of fuel per a year. The section is equipped with the improved intra-chamber atmosphere (nitrogen) maintenance and control system. The implemented technology – VP-process/carbothermal synthesis.

FRM. The newly arranged production facility for pelletized MN fuel, fuel rods, and fuel assemblies (the FRM within PDEC Complex) with the capacity of 14 tons of fuel per a year. Launch in 2020, the possibility for switching to re-fabrication within the CNFC – 2028 (subsequent to pilot operation of the start-up core and the first refueling FAs). Implemented technology – VP-process. The plant is designed for use of high-background energy-grade plutonium, as well as americium and neptunium from SNF of BREST-OD-300 in the re-fabrication mode, and equipped with the intrachamber atmosphere (nitrogen) maintenance and control system. The production is highly automated.

The products include fuel assemblies for BREST-OD-300. The option with manufacturing of pellets for BN-800 fuel rods with MN fuel with subsequent manufacturing of fuel rods and fuel assemblies at FSUE «MCP» is considered.

The option with refurbishment and increase of the existing capacities was proposed with regard to fuel supply for the IPC industrial facilities:

- Development of the FRM refurbishment project for timely supply of MN (or MOX) fuel to the power unit of the first of a kind IPC synchronized with the time limits for commissioning of Power Unit 1.

- Performance of R&D works for development of robotized process lines for uranium and plutonium fuel manufacturing and processing for the industrial power supply complex in 2019–2024. Subsequent to the development results the decision on the time limits for construction of the modular robotized nitride fuel (re)fabrication plant synchronized with the time limits for commissioning of the IPC Power Units 2/3 shall be made.

Pilot operation of the FRM at PDEC Complex since 2020 will enable to obtain the required data for the equipment improvement and optimization of the fabrication technology within the framework of R&D works at PDEC Complex. R&D works on the refabrication technology are in progress. The works on homogeneous involvement of minor actinides (particularly Np and Am) into MN fuel are the priority area of focus. Heterogeneous scenarios implying the possibility for involvement of actinides produced in the cores of thermal reactors into the cycle apart from own minor actinides are also considered.

18.4 JSC "SSC RIAR"

Experimental production of vibro-compacted MOX fuel from initial powders (pyrochemical granulated material) to fuel assemblies is implemented in JSC «SSC RIAR».

The pyroelectrochemical technology is applied at the production facility for production of granulated MOX fuel in molten chloride salts. The production facility can operate in partially automated mode, remotely and in hot cells. ALIT-1 plant enables to manufacture fuel elements of BN-600 type. The upgraded ALIT-2 plant will enable to manufacture BN-800 fuel elements in the automatic mode. The fuel element and fuel assembly manufacturing plants are the prototypes of commercial ones. Plutonium of any quality and origin is used at the complex.

The products include fuel assemblies of SFR type with MOX fuel. The FA assembly section is used for both vibro-compacted and externally supplied pellet fuel rods with MOX and MN fuel.

The equipment and arrangement of works at the fuel complex of JSC «SSC RIAR».enable to perform the works for manufacturing of MA-containing fuel rods.

The consolidated information on the existing experimental plants is presented in Table 18.2.

18.5 JSC "VNIINM"

Manual experimental production of uranium-plutonium fuel and fuel rods with the capacity of 10 kg of pelletized fuel per a year is arranged in JSC "VNIINM". There is no local FA assembly plant.

The fuel production process is based on VP-process; uranium-plutonium nitride is produced by the method of carbothermal metal oxide reduction.

TABLE 18.2 Operating experimental plants for manufacturing of mixed uranium-plutonium fuel.

Mixed fuel production plant	JSC "SSC RIAR"	JSC "VNIINM"	"Paket" PA FSUE "Mayak"	KEU of JSC "Siberian Chemical Combine"
Fuel type: pelletized/ vibro-compacted	Vibro-compacted	Pelletized	Pelletized	Pelletized
MOX fuel	Yes	Yes	Yes	Yes
MN fuel	No	Yes	No	Yes
Plant capacity, kg/year	Up to 1500	Up to 10	Up to 3500	Up to 600
	Manufacturing of fuel rods			
BOR-60	Yes	Yes	Yes	–
BN-600	Yes	Yes	Yes	Yes
	Manufacturing of fuel assemblies			
BOR-60	Yes	No	No	No
BN-600	Yes	No	No	Yes
	Plutonium dioxide use			
Low-background	Yes	Yes	Yes	No
High-background	Yes	No	No	No
Weapon-grade	Yes	No	No	Yes
	Possibility to produce initial powders			
Plutonium dioxide	Yes	Yes	Yes	Yes
$(U, Pu)O_2$	Yes	Yes	No	No
	Experience in manufacturing of fuel rods			
BOR-60 (MOX fuel)	Yes	Yes	Yes	No
BN-600 (MOX fuel)	Yes	Yes	Yes	No
BOR-60 (MN fuel)	No	Yes	No	Yes
BN-600 (MN fuel)	No	Yes	No	Yes

The process chains are equipped with the intra-chamber atmosphere (nitrogen) maintenance and control system and designed for low-background plutonium use. The products include fuel rods for irradiation devices of BOR-60, MIR, IVV-2M research reactors; fuel rods for experimental fuel assemblies of BN-600 and VVER-1000 reactors with pelletized MOX, REMIX and MN fuel.

Chapter 19

Structural materials for fuel rod claddings

19.1 Studies for substantiation of fuel burnup increase

Fuel burnup increase is one of the most efficient ways for improvement of the technical and economic parameters of NPPs with fast reactors.

For example, the task to achieve fuel burnup of $\geq$ 180,000 MW·day/t U corresponding to the damage dose of more than 180–200 dpa is set in the long run for fast sodium-cooled reactors. But at present the fuel burnup in fast power reactors does not exceed 100,000 MW·day/t U (burnup of 130,000–150,000 MW·day/t U was achieved in individual fuel assemblies). This is primarily due to the fact that the structural materials used for fuel rod claddings at different times turned out to be unstable and subjected to significant deformations in the neutron field. In this context attempts to improve austenitic steels were made: to reduce their irradiation swelling, irradiation creep and to increase their residual ductility.

Usage of steels modified due to introduction of alloying additives (titanium, silicium, etc.) and application of cold deformation were the first steps in solution of this problem. But the studies showed that irradiation swelling and creep of such austenitic steels remained quite high.

Selection of materials for FA components is determined by the set of their characteristics under irradiation – swelling, irradiation and thermal creep, short-term mechanical (σ_B, σ_T, δ) and long-term (creep rupture strength, stress rupture ductility) properties, corrosion resistance under the impact of both coolant and fuel as well as their manufacturability. Thus solution of the set of problems related to development of advanced structural materials for fuel assemblies providing for achievement of the optimum design operating characteristics of fast reactors is a complex scientific and technical issue.

In this regard programs were established in many countries for development of advanced structural materials for fast reactors including investigations of different steel grades and alloys representing the following three classes of materials: austenitic stainless steels, high-strength ferritic-martensitic steels, dispersion-strengthened alloys on the nickel-iron basis.

Closed Nuclear Fuel Cycle with Fast Reactors: Handbook of Russian Nuclear Power.
DOI: https://doi.org/10.1016/B978-0-323-99308-1.00025-8

Common tendencies characteristic for the programs of other countries may be specified through the example of the US National Program [184] for development of advanced structural materials for fuel assemblies of fast reactors launched in 1974. On the basis of studies including extensive reactor testing of 35 steel grades and alloys up to high doses austenitic (316 CW, D9), ferritic (HT 9, D-57), and dispersion-strengthened (D-21, D-66, D-68) steels were selected in the USA for further investigation.

At present the works on heat resistance enhancement for traditional ferritic-martensitic steels are in progress. Works on improvement of ferritic-martensitic steels and oxide dispersion strengthened steels (ODS steels) have been carried out in parallel. ODS steels were firstly considered in the USA as the material for the fast reactor fuel element cladding in 1960s. Anisotropy of properties due to extremely complex and expensive powder metallurgy was the main problem faced by the developers. At present new steels of 9–12% Cr type based on nanotechnologies and strengthened with titanium carbide or vanadium nitride with the working temperature up to 700°C are being developed. All these steel types with enhanced heat resistance require considerable investigations prior to the implementation stage.

In 1983, the first Integrated Program for Development of Radiation-Resistant Materials for Fast Reactor Cores [81] was developed in the Soviet Union; it was similar to the national programs of some other countries in many aspects. It was designed for development of materials providing for the burnup of at least 10 at%. The main techniques for irradiation swelling suppression were developed and tried in practice [80,183]:

- Optimization of austenitic steel alloying;
- Application of austenitic steels in the cold-worked state;
- Use of 12% chromic steels of ferritic-martensitic class for fuel assemblies wrappers;
- Development of high-nickel alloys of nimonic PE-16 type strengthened with γ'- and γ''-phases of Ni_3 (Ti, Al) and Ni_3Nb.

Extensive experience in testing and operation of fuel assemblies in BOR-60, BN-350, and BN-600 reactors enabled to implement the proposed options for structural material swelling reduction in the operating cores of fast reactors. Development of ChS68 cw austenitic steel and its use for fuel rod claddings and use of EP450 ferritic-martensitic steel for wrappers was the optimal variant. At present these materials provide fuel assemblies performance in BN-800 reactor up to the damage dose of ~95 dpa and fuel burnup of ~11.5 at%.

Further works are mainly related to improvement of austenitic steels in order to reduce their swelling and ferritic-martensitic steels in order to increase their heat resistance. Improved steels of austenitic (EK164 cw) and ferritic-martensitic (EK181) types including also oxide dispersion strengthened steels [80] are considered as promising materials for the fuel rod claddings in order to ensure step-by-step fuel burnup increase in SFRs.

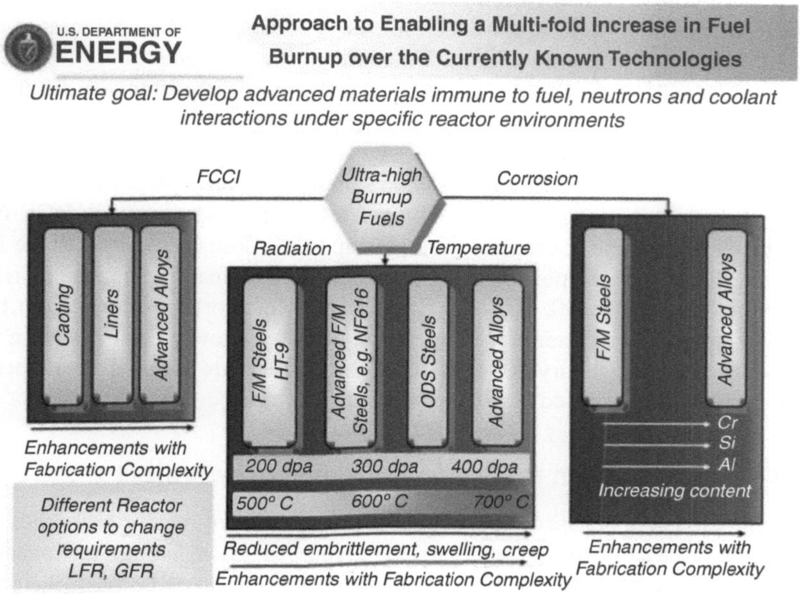

FIGURE 19.1 Areas of investigations carried out by the US Department of Energy aimed at multifold increase of fuel burnup [180].

Oxide dispersion strengthened ferritic-martensitic steels (ODS steels) with the use of powder metallurgy methods are developed in Russia, Japan, India, France, USA in order to achieve high damage doses (more than 180 dpa) and cladding temperature (~700°C). The required improvement of creep and rupture strength characteristics is achieved by the matrix strengthening with dispersive yttrium and titanium oxide particles.

In spite of positive results of PE16 high-nickel alloy irradiation as the fuel rod cladding in PFR reactor (UK) up to high doses this area was not developed in other countries, mainly because of failure to ensure acceptable ductility of high-nickel alloys.

Areas of investigations carried out by the US Department of Energy for multifold increase of fuel burnup are presented in Fig. 19.1. This is associated with development of promising (in view of reduced swelling, irradiation creep and embrittlement) irradiation-resistant steels for fuel rod claddings with the behavior independent from the type of fuel, coolant, and neutron irradiation.

The existing and developed steels comply (or shall comply) with the following limits: HT-9 steel – the maximum dose of 200 dpa, temperature of 500°C; improved ferritic-martensitic steel of NF616 type – the maximum dose of 300 dpa, temperature of 600°C; oxide dispersion strengthened steels and other promising alloys – the maximum dose of 400 dpa, temperature of 700°C. In order to enhance corrosion compatibility with metal fuel transition from traditional ferritic-martensitic steels to advanced alloys is accompanied with increase of Cr, Al, Si content with growth of their manufacturing problems.

Significant doses had been previously achieved in FFTF reactor with different items (hexagonal FA wrappers, fuel rod claddings, material research samples); for example, the maximum dose for the wrapper of ACO-3 assembly made of HT-9 steel was 155 dpa. Material research samples for subsequent irradiation in BOR-60 are made of this wrapper. The purpose of the experiment in BOR-60 is to irradiate previously irradiated samples up to the dose of 250 dpa (Stage I), 450 dpa (Stage II); samples of HT-9 and other ferritic-martensitic steels, irradiation temperatures: 370, 400, 450, 525, 600°C. The works are carried out by the US Department of Energy in cooperation with TerraPower Co. Irradiation in BOR-60 started in January 2014. In 2018, the works were stopped. The works on dispersion strengthened steels are in progress.

19.2 Studies within the framework of Proryv Project

Consolidation and analysis of the available data on the pre-reactor properties of EK164 and EP823 steels had been carried out in 2013–2014, and the sections for the second revision of temporary guidelines were developed on the basis thereof. Data on the short-term mechanical and physical properties, rupture strength, and thermal creep of EK181 and ChS139 steels planned to be used as fuel rod claddings at the main stage of BN-1200 reactor operation were obtained. The analysis demonstrated that data on the short-term mechanical properties at the temperature exceeding 700°C were insufficient or absent for all steels.

EK164-ID. Short-term mechanical properties were determined in 2015–2016 within the framework of the "Structural materials" project during elongation testing of cladding tubes of different dimensions made of EK164-ID steel at the temperatures of 750–1100°C. It was established that the strength and ductility values of all investigated tube sizes were similar. In case of the testing temperature increase from 750 to 1100°C the strength characteristics decrease: ultimate strength from 494–443 MPa to 35–37 MPa, the yield strength from 468–415 MPa to 30–33 MPa, and ductility increases from 15–30% to 122–132%. The obtained results on the short-term mechanical properties of cladding tubes made of EK164-ID steel at the temperature of 750–1100°C are included into the recommendations on «Initial data on EK164 Steel Properties» dated 30.11.2016).

In 2018, the rupture strength parameters of tubes Ø10.5 × 0.5 mm at 700°C and Ø9.3 × 0.5 mm at 800°C were compared. It was shown that the testing temperature increase by 100°C from the operating temperature of 700°C to the emergency mode operating temperature of 800°C resulted in significant changes of long-term mechanical properties: 3-fold reduction of the time to rupture, increase of the creep rate by an order; the ultimate rupture strength was reduced by more than 3 times, and the creep limit – 9 times.

EP823-Sh. The heat treatment parameters for manufacturing of tubes made of EP823-Sh steel were developed in order to ensure optimal structure of the tube metal. Comparative studies of short-term mechanical characteristics of the tubes of different dimensions made of EP823-Sh steel from the same cast performed in 2014 demonstrated that the ultimate strength and yield strength of tubes did not depend on their production flow diagram. Ductility of the tubes increases with the wall thickness reduction.

In 2016, the results of short-term mechanical testing of different metal product samples made of EP823-Sh steel were obtained for tubes of various diameter at the temperature of 20–900°C and rods at the temperature of 20–1150°C. Behavior of the mechanical characteristics of EP823-Sh steel depending on the temperature has similar pattern for all types of samples, and the strength characteristics reduce in the course of temperature increase. Ductility dependence on the testing temperature is more complex.

The strength parameters of EP823-Sh steel do not fall down to zero at the maximum testing temperature of 1150°C (σ_B = 14.0–16.7 MPa, $\sigma_{0.2}$ = 13.0–14.7 MPa) which is indicative of the possibility to use this steel under the emergency reactor operation but within a limited period of time.

The obtained data were included into "Validation Report for Substantiation of EP823-Sh Steel for BREST Reactor. Stage of 2016", dated 12.12.2016.

Experimental data on the rupture strength and thermal creep of a modern lot of tubes made of EP823 steel (particularly for the samples with modified surface) with the dimensions of 9.7 × 0.5 mm at the temperature of 750, 800, 850, 900, and 970°C were obtained. The up-dated analytical dependencies of the time to rupture and thermal creep deformation rate of EP823 steel depending on the stresses and temperatures are included into the third revision of the Temporary Guidelines.

ChS139, EK181. Rupture strength and thermal creep testing for the cladding tubes of a new size Ø9.3 × 0.5 mm made of ChS139 steel at the emergency (750, 800°C) and regular (700°C) temperatures of fuel rods operation was performed in 2018.

Analytical dependencies of the time to rupture and thermal creep rate on the stress and temperature as well as the ultimate rupture strength and creep limits were defined subsequent to the testing results. Analytical dependencies as well as the ultimate rupture strength and creep limits will be up-dated upon obtaining of any new data for the samples still being tested.

At present the available data on the rupture strength and thermal creep of EK164-ID, EK181-V and ChS139-Sh steels demonstrate higher heat resistance than for similar austenitic and ferritic-martensitic steels, but these data are insufficient to define the analytical dependencies of the time to rupture and thermal creep deformation rate (as a function of stress and temperature). Additional data on long-term mechanical properties of the modern lot of pressurized tubes with the BN-1200 dimensions made of EK164-ID, EK181-VI and ChS139-Sh steels at the temperature of 650–1200°C are required to define reliable dependencies.

Simulation studies of the physical – chemical interaction between the cladding steels and fuel fission products with regard to the "structural material – nitride fuel" system were performed within the framework of the project of structural materials in 2013–2015. Devices in the form of quartz vessels where samples and chemical agents were placed prior to sealing were used in the simulation corrosion testing. ChS68, EK164, EP450, EP450 ODS, EP900, EP823, ChS139, EK181 steel samples were studied. Iodine and cesium iodine were used as chemical agents. The nature of corrosion damage is different for austenitic and ferritic-martensitic steels. While the interaction of austenitic steels is of inter-crystalline (ChS68) or frontal (EK164) nature the interaction of ferritic-martensitic steels with iodine causes their pitting corrosion. Corrosion damage of ferritic-martensitic steels turned out to be lower than for austenitic steels.

Apart from cladding steels of the basic designs of BREST-OD-300 and BN-1200 fuel rods, steels considered as the candidates for usage in the cores with the average burnup of $\geq$12at. % are also studied.

EP900. EP900 is heat-resistant modification of EP823 steel distinguished for increased (~0.14% wt.) nitrogen content. It is established that it has enhanced structural stability and stability of mechanical properties in the course of thermal ageing within the temperature range of 450–670°C due to presence of dispersive chromium nitride particles in its structure. The rupture strength values of EP900 steel at the temperature of 650°C based on 10,000 h are approximately 20% higher in comparison with EP823 steel.

Bimetal. Development of bimetal cladding tubes with the coolant-contacting layer made of corrosion-resistant silicium-containing steel with increased chromium content and the base made of heat-resistant steel of EP900 type is a promising area in development of the cladding materials for fuel rods of BREST-OD-300 reactor.

Pilot batches of bimetal tubes with diameter 9.7 × 0.5 mm with different ratio between the layers made of 18CrSi1MoWVAl and EP900 steels were manufactured. It is shown that the optimal ratio between the layers is 1:3 with regard to a set of properties. It is demonstrated that the ultimate strength values of bemetal tubes at short-term and long-term mechanical tests are approximately 20% higher in comparison with EP823 steel.

Corrosion tests of bemetal tubes on the dynamic bench with lead coolant for 10,000 hours at the oxygen concentration in lead of $C_{[o]} = (1–4)\cdot 10^{-6}$ % wt. and the temperature of 650°C showed that dense protective oxide film was formed on the samples due to increased silicium content and aluminum alloying of 18CrSiMoWVAl steel.

Oxide dispersion strengthened steels. ODS steels of ferritic-martensitic class are developed for fuel rod claddings in order to increase MN fuel burnup up to $\geq$12 at%, temperatures up to 710°C. Enhanced creep and rupture strength characteristics are achieved through the matrix strengthening with dispersive oxide particles of different composition on the basis of yttrium oxide (yttrium-titanium, yttrium-titanium-aluminum, yttrium-titanium-silicium oxides, etc.).

Steels of various composition are used as matrix steels for production of ODS steels. Studies of ODS steel based on EP450 ferritic-martensitic steel were performed for BN-1200 reactor. It was shown that EP450 ODS steel had significant advantage in thermal creep characteristics and rupture strength as compared with the matrix steel. The samples rupture time for the samples made of this steel was equal to 15,000 h under testing at 700°C and the stress of 140 MPa

At present promising ODS steels for lead cooled reactors include EP823 ODS steel and ODS steel with increased chromium content (16% wt.), silicium content (~1% wt.) and aluminum content (~4.5% wt.).

The samples were subjected to corrosion tests on the dynamic bench with lead coolant at the temperature of 620°C for 4000 hours with the optimum oxygen concentration in lead of (1–4) × 10^{-6} % wt. It was shown that EP823 ODS steel was highly competitive with EP823 matrix steel in its corrosion resistance, and ODS steel with increased chromium and aluminum content showed almost no signs of interaction with the coolant under these testing conditions due to formation of Al_2O_3 protective oxide film on the surface.

In 2018, irradiation of EP450 ODS and EP823 ODS steel samples (including pressurized tubes) was carried out within BN-600 material assembly.

Technology works for acquisition of cladding tube manufacturing of ODS steel with increased chromium and aluminum content are in progress.

Vanadium alloys. They have the unique combination of heat resistance, corrosion and irradiation stability, mechanical strength making it possible to implement advanced reactors with different coolant types and extreme parameters. The principal results in development and technology acquisition of vanadium alloys:

- The physical and mechanical properties and the structural and phase state for V-4Ti-4Cr alloy are determined;
- The pilot facility for production of highly homogeneous ingots of (V-4Ti-4Cr) alloy with the weight of ~110 kg is arranged;
- Technical specifications for semi-finished products (plates, bars, tubes) are developed;
- Bimetal tubes "steel – V alloy" are manufactured;
- Extremely thin-walled tubes (6.9 × 0.4 and 9.3 × 0.6 mm) are manufactured of V-4Ti-4Cr alloy.

19.3 Bench testing of dummy fuel rods (dummy fragments) including spacing elements (small-scale liquid-metal benches) – corrosion in lead

The majority of corrosion tests for samples made of EP823-Sh steel were carried out in the flow of lead coolant with the dissolved oxygen concentration

TABLE 19.1 Estimated corrosion zone under long-term exposure of EP823-Sh steel to lead coolant.

Affecting parameter	Oxygen content in the coolant, %			
	10^{-5}		10^{-8}–10^{-7}	
Temperature, °C	~650	~550	~650	450
Time, h	5000	4000	3000	2000
Corrosion zone thickness, μm	<45	<30	<2	<3

of $C_{[o]}$ = (1–4) ×10^{-6} % wt. at the temperatures of 400–670°C on various experimental facilities.

Dependence of the gradual plateau reaching by the corrosion layer value on the testing time is noted through analysis of the obtained experimental data for all temperatures under consideration, and this value does not exceed 10 μm upon expiry of the operation period (35,000 h). As numerous experiments showed, the corrosive interaction zone thickness either reaches the maximum value as specified for the temperature of 500–670°C or has the tendency towards gradual slight increasing (as demonstrated at the temperatures of 400–420°C) with plateau reaching as early as within the first 1000 hours in the lead coolant flow. The estimated oxide film thickness for 35,000 hours at the temperature of ~650°C will be 2–3 μm, and at the temperature of ~540°C and ~420°C – 15–20 μm.

Thus EP823-Sh steel is corrosion-resistant within a wide temperature range of 420°C–670°C with the oxygen content of $C_{[o]}$ = (1–4) × 10^{-6} % wt. in the lead coolant flow. Growth or loss of oxide film in case of non-optimal oxygen content (1.5 × 10^{-8}–5 × 10^{-7} % wt.; (3–5) × 10^{-5} % wt.) can be predicted with a lower confidence level as the amount of experimental data is much less. As duration of the lead coolant impact in case of non-optimal oxygen content in lead is determined by the operating conditions of BREST-OD-300 reactor and shall not exceed 1000 h, estimation of the interaction zone growth (decrease) will enable to evaluate the material behavior under emergency conditions (Table 19.1).

It is possible to conclude from the presented data that optimum oxygen content in lead of (1–4) × 10^{-6} % wt. guarantees the permissible corrosion depth in tubular samples of EP823-Sh steel below 50 μm for a long period of time (~16,600 h).

In case of increased oxygen content in the coolant of ~0^{-5} % wt. the oxidation process is the most intensive, and the interaction depth can reach 60 μm as soon as in 3000 h.

Interaction clusters of up to 90 μm occur within 1000 h in case of reduced oxygen content in lead of ~10^{-8}–10−7 % wt. Thus it is necessary to maintain the optimum oxygen regime in order to provide optimal corrosion resistance of EP823-Sh steel.

Corrosion resistance testing of pressurized cladding tubes made of EP823-Sh steel, ODS steels and other materials in the lead coolant flow with the optimal oxygen content at the temperature of 650°C was carried out in 2019.

Chapter 20

SNF processing technologies

20.1 Requirements for the SNF processing technology in the CNFC

The SNF processing technology shall comply with the following requirements:

No pure Pu extraction in the course of SNF processing;
Processing of SNF with high (10% and more) content of fissionable materials;
Prevention of any trans-uranium element ingress into RW in the amount exceeding 0.1%;
Return of Am and Np into the fuel cycle and their transmutation/ burn-out with the possibility for curium extraction and storage;
Processing of SNF with high burnup (10% h.a. and more);

Processing of SNF with short period of storage (1–2 years) after completion of irradiation in the core.

None of the commercially available technologies complies with these requirements. Combined (pyro + hydro) processing of MNIT SNF which is to be checked at the PDEC processing module of Siberian Chemical Plant JSC and is based on nonaqueous primary operations enabling purification of the fissionable materials from fission products by approximately 100 times and hydro-metallurgical refining operations providing for deep purification and deep extraction of actinides from RW complies with these requirements to the fullest extent possible.

20.2 Existing capacities for processing of SNF from thermal and fast reactors

Assessment of the capacities for processing of SNF from existing thermal and fast reactors and those under construction in the RF is presented in Table 20.1 Analysis of the table shows that the industrial processing of about 1000 t/year of SNF from thermal reactors at radiochemical plants (RT-1, Mayak PA FSUE) and the pilot demonstration VVER-1000 SNF processing plant (pilot demonstration facility of MCP FSUE) will be possible after 2020.

Closed Nuclear Fuel Cycle with Fast Reactors: Handbook of Russian Nuclear Power.
DOI: https://doi.org/10.1016/B978-0-323-99308-1.00036-2

TABLE 20.1 Assessment of the existing and constructed capacities for processing of SNF from fast reactors.

Parameter	RT-1 (Mayak PA FSUE)	Pilot demonstration VVER-1000 SNF processing plant (MCP FSUE)	PDEC processing module (Siberian Chemical Plant JSC)
Status	Functioning (at least till 2035)	Under construction (planned to be started in 2025)	Under development (design commencement in 2021)
SNF type	Uranium SNF from TNRs, FNRs and transport and research RPs (U, REMIX, MOX and MNIT - optional) RBMK-1000, VVER-440, VVER-1000, AMB-100, AMB-200, BN-600, BN-800	Uranium SNF from TNRs (U, REMIX, MOX option mixed with TNR SNF) VVER-1000 BN-600 MOX, BN-800 - optional)	MNIT (MOX optional) from BREST-OD-300 (option - BN-800, transport and research RPs)
Maximum SNF processing capacity, t/year	Up to 400 t/year of TNR SNF with development to 600 t/year	250 t/year of TNR SNF with development to 450 t/year	10 t/year of MNIT*
Technology	Hydro-metallurgy (PUREX)	Hydro-metallurgy (simplified PUREX)	Combined (pyro+hydro, PH-process)
Possibility to separate Am and Cm	Refurbishment needed	Not provided	Provided
Possibility to extract Cs-Sr	Refurbishment needed	Not provided	Not provided**

**May be increased to 60 t/years in the design*
***May be provided in the design*
Note. TNR - thermal reactors; FNR - fast reactors.

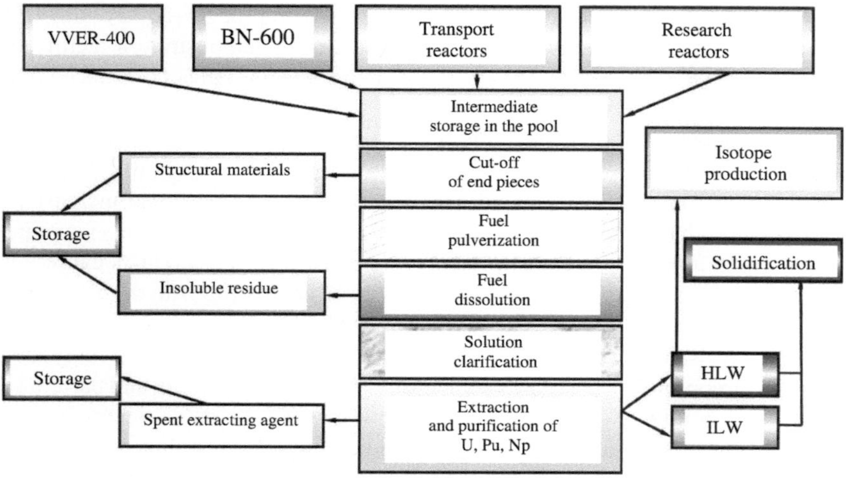

FIGURE 20.1 SNF processing flow diagram at RT-1.

Uranium SNF from BN-600 as well as SNF from transport and research plants is and will be processed at RT-1 till 2035. Refurbishment of RT-1 and the pilot demonstration facility or construction of the SNF processing module within the PDEC Complex at the site of Siberian Chemical Plant JSC is necessary for commercial processing of mixed uranium-plutonium oxide and nitride (MOX and MNIT) SNF from fast reactors and even only from BN-800 and BREST-OD-300.

20.3 Hydro-metallurgical technology for processing of SNF from thermal and fast reactors

At present single recycling of plutonium in BN-800 fast reactor is carried out in the Russian Federation. Industrial capacities for fuel production and SNF processing enabling multiple recycling of Pu are available [42].

Commercial SNF processing using hydro-metallurgical technology based on PUREX process has commenced in Russia since 1977. RT-1 plant and Mayak PA FSUE in Ozersk which was refurbished for the first time in 2016 are operating with the use of the hydro-metallurgical technology.

5600 tons of SNF have been processed at RT-1 plant for 38 years of operation. The final products of RT-1 are uranyl nitrate with enrichment up to 3% by ^{235}U, U_3O_8 with enrichment up to 76% by ^{235}U and recycled energy-grade PuO_2 [42]. The SNF processing flow diagram at RT-1 is presented in Fig. 20.1. The second extension of the pilot demonstration facility for processing of VVER SNF with production of initial oxides for MOX and REMIX (regenerated U and Pu mixture) fuel is planned to be commissioned at MCP in Zheleznogorsk by 2025 [47]. The hydro-metallurgical technology (simplified PUREX process) is also used at the pilot demonstration facility. The hydro-metallurgical technology based on the PUREX process has already proved its efficiency for processing of SNF from thermal reactors with the average burnup of 45–55 GW·day/t at RT-1. Economic parameters are improved at the pilot demonstration facility in comparison with RT-1 technology, and the environmental impact is reduced due to absence of liquid RW [157].

Disadvantages of hydro-metallurgical technologies include difficulties in processing of SNF with the burnup of more than 55 GW·day/t [140]. Approximately double burnup increase will result in an almost proportional increase in the fission product content in SNF, which is a serious problem from the viewpoint of activity and heat emission growth. Besides, many fission products cause the problem of deposit formation in the course of SNF processing with the use of traditional hydro-metallurgical methods based on the PUREX process [213]. An increase in the fission product content will inevitably cause deposition of molybdenum and zirconium in case of low acidity of solutions, or strontium and barium in case of high acidity of solution [157] as well as loss of fissionable materials with the deposit. The second problem of hydro-metallurgical technologies resides in the necessity to use voloxidation for predistillation of tritium, or crystallization purification in order to avoid usage of organic extraction agents with due regard for significant reduction in the number of operations during hydro-metallurgical processing of SNF [145,161]. And finally, the experience in commercial hydro-metallurgical processing of SNF with high content of fissionable materials is available only in France [134] and Russia [219]. Opportunities and limitations of the hydro-metallurgical technology are described in detail in [181,205,213]. It is obvious that the problems related to radiation and nuclear safety assurance encourages to consider combined (dry and water-based) SNF processing methods. It should be noted that hydro-metallurgical methods are suitable for operation with large amounts of fissionable materials but require application of diluted solutions or reaction vessels of certain geometry thus affecting the SNF processing process economy due to the increase both in the vessel volumes and in the amounts of process solutions. In this case MOX SNF of BN-600 reactor was processed in 2012 and 2014 at RT-1 plant. In total 8 fuel assemblies with the burnup of 73 to 89 GW/(day-tU) and the storage time of 17 to 21 years till the moment of processing were processed. Processing included the stages of dissolution, clarification, extraction and separation of uranium and plutonium with purification from any fission fragments, refining of uranium and plutonium at the relevant refining cycles. No increased plutonium content in the waste was registered in the course of BN-600 MOX SNF processing at the RT-1 plant [219].

20.4 Pyrochemical SNF processing technology

At present the development of the regeneration technology for spent fuel from fast reactors is highly promoted. The process design includes three stages according to the set of features [70]:

Nonchemical SNF processing with refining fusion of metal fuel;
Gas fluoride chemical processing of SNF;
Chemical processing of SNF using ionic or metal molten media.

The works for plasma processing of spent fuel have been resumed recently.

The most systematic studies among nonaqueous SNF processing methods mentioned above were carried out using the gas fluoride technology, which was primarily chosen due to the fact that the research was aimed at the development of enriched uranium production. However, it was stopped by the late 80s of the past century for various reasons of technological and material research nature. Nevertheless, the development of this technology still attracts attention, and the studies on the processes in the "solid fuel – gas" system appear in the published sources.

In the middle of the past century the Ural school of high-temperature electrochemistry established on the basis of the Institute of High-Temperature Electrochemistry (IHTE) of the Ural branch of the USSR Academy of Sciences and the Physical and Technical Department of the Ural Polytechnic Institute proposed the idea to use molten salts based on alkaline metal halogenides for electrochemical processing of FNR SNF. At present this method is known as the pyrochemical SNF processing technique.

One of the main advantages related to application of pyrochemical processes in the FNR SNF processing technology for the NFC closure resides in the possibility to reduce the SNF storage time to 1 year and less, as molten salts are radiation-resistant media and can be used for processing of SNF with high energy emission. Under high temperatures, the processes in molten salts proceed under the conditions close to equilibrium ones. It has a considerable effect on the selectivity of the developed processes in relation to the dissolved components.

Dry SNF processing technologies are also developed for processing of dense fuel from fast reactors [152,202]. It should be noted that some pyro-electrochemical processes under development [152] involve the oxide reduction to metal at the first stage, thus it proves that electrorefining technologies may be used in the future [117].

Pyrochemical technologies developed worldwide provide for processing of metal SNF based on U-Pu-Zr alloys or processing of MOX and MNIT SNF from thermal and fast reactors of different types [118].

The demonstration experiment on pyrochemical processing of irradiated MNIT SNF in the air chamber of NIIAR JSC was carried out in Russia at the end of 2013. However, the passivation of MNIT SNF pellets due to formation of UNCl surface film with low solubility in the molten salt electrolyte [179] and a possible UN/U_2N_3 overcharge resulting in loss of uranium were not taken into account in the above-mentioned experiment. At present the pyrochemical processing technology is being developed at the IHTE of the Ural Branch of RAS. The process flow including the following main operations is proposed for processing of MOX and MNIT SNF: oxidized MNIT or MOX SNF (the mixture of actinides oxides and oxides of rare-earth and noble metals) after voloxidation is supplied to metallization where actinides are electrochemically reduced to metal in molten salts, and the reduced cathodic product is supplied for refining flux melting. An alloy of actinides and noble metals is formed in the course of melting, and rare-earth metal oxides pass into the fluoride salt flux. This alloy is supplied for electrochemical refining. The produced cathodic deposit is supplied to hydro-metallurgical processing after electrolyte removal, and the anodic deposit containing noble metals and actinides is subjected to the "soft chlorination". For this purpose, it is dissolved in molten chloride using an oxidizing agent $CdCl_2$ or $PbCl_2$ then the target components are deposited in the form of oxides.

In order to obtain the product suitable for pure fuel manufacturing the actinides mixture is additionally purified after the electrochemical refinery operation – americium and curium are separated by controlled potential electrolysis.

Pyrochemical processing flow diagrams for FNR MOX and MNIT SNF. The SNF processing flow diagram for the PDEC module in the combined option is presented in Fig. 20.2 (pyrochemical processing). The proposed process flow enables the production of fissionable materials in the form of metal powder and decreases the SNF activity down to the level allowing usage of this product in the combined-scheme of hydro-metallurgical processing.

In case of the sole pyrochemical SNF processing technology implementation (in the IPC) the cathodic deposit shall be subjected to the controlled potential electrolysis with soluble anode after the electrolytic refining to provide a complete separation of fissionable materials and minor actinides. The resulting cathodic product is subjected to purification from electrolyte and supplied for refabrication. The flow diagram of sole pyrochemical SNF processing is presented in Fig. 20.3. The proposed process flow decreases considerably the process media material flows at the initial stage and simplifies the equipment for subsequent operations.

Theoretical and experimental substantiation of the main process operations. The actinide oxide reduction process was carried out using lithium generated on the inert cathode in the course of the molten LiCl-Li_2O electrolysis. Dissolved lithium reacts with the oxidized SNF particles within the molten material volume. The produced lithium oxide is dissolved in molten LiCl, it diffuses within the molten material volume and oxidizes on the graphite (sacrificial) anode with release of the CO-CO_2 mixture or on the inert (dimension stable) anode with release of gaseous oxygen.

Metals (Pt, Au, Ag), metal alloys (aluminum bronze, Cu-Fe-Ni), and oxide ceramics were studied as possible inert anodes. Subsequent to the results of the studies it was established that ceramic anodes based on NiO had the maximum corrosion resistance and thus were further used for the metallization process.

Dependencies of the reduction process for UO_2 powder and pellets (diameter – 5.0 mm, height – 8.0 mm) with the density of 97% of the theoretical value were investigated in the laboratory electrolysis unit, and the possibility of complete uranium dioxide reduction to metal was demonstrated. It was established that the UO_2 reduction process had the diffusion mode and was limited by the mass transfer in the reduced uranium pellet (particle) pores.

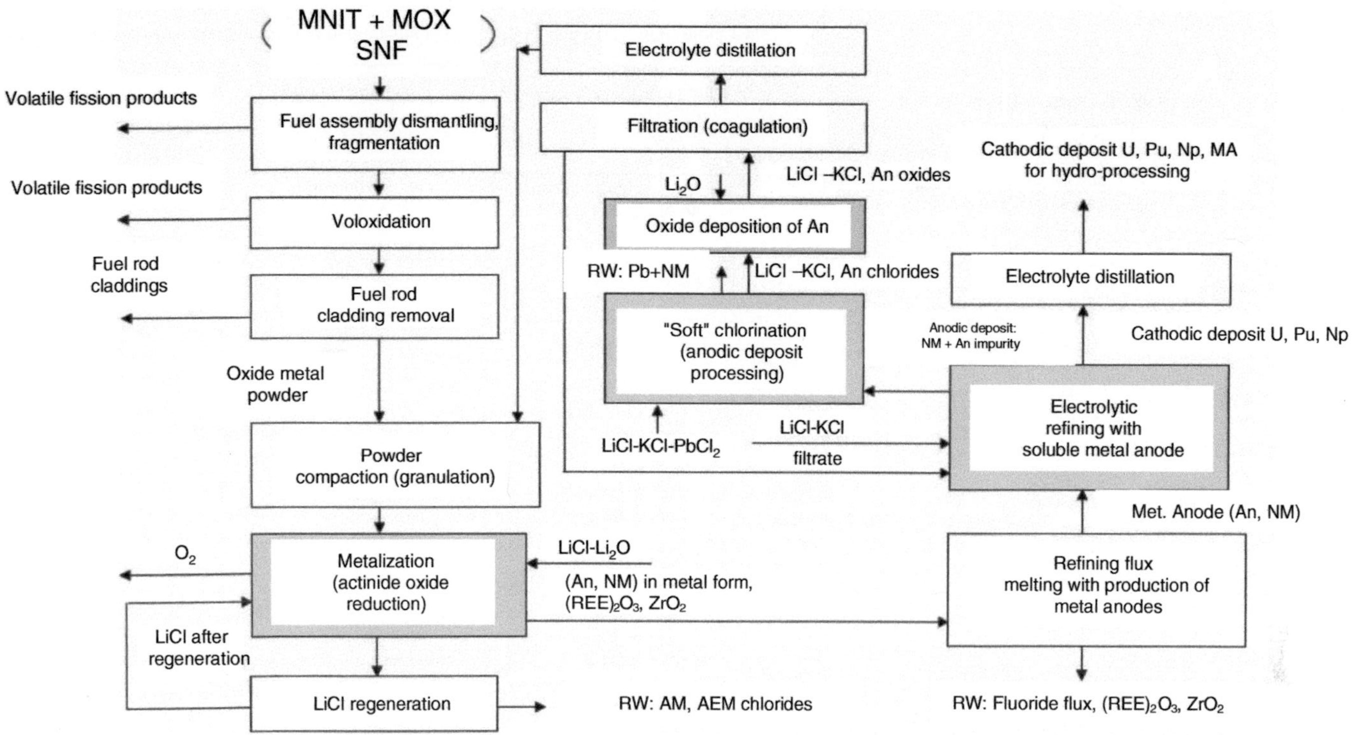

FIGURE 20.2 Flow diagram of the combined (pyro+hydro) FNR MNIT and MOX SNF processing.

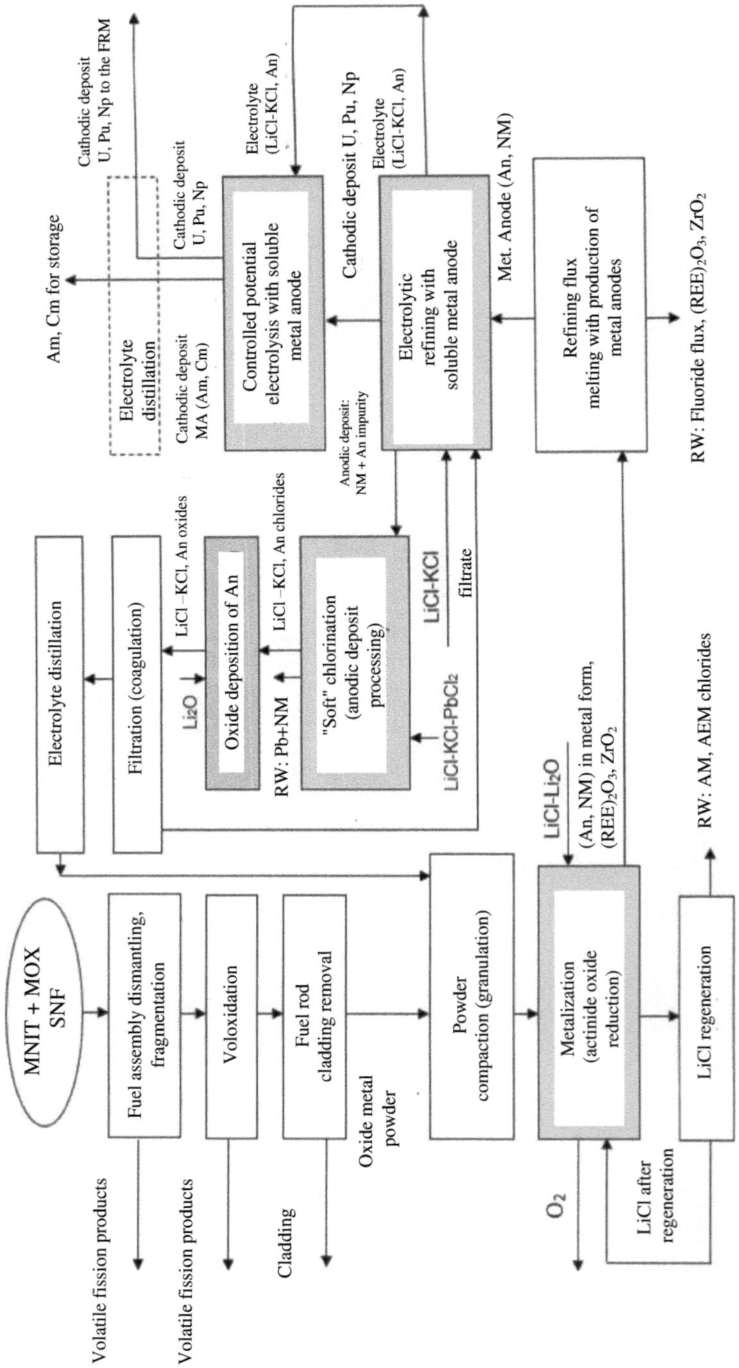

FIGURE 20.3 Flow diagram of sole pyrochemical FNR MNIT and MOX SNF processing.

FIGURE 20.4 Photos of UO_2 pellets after metallization: partially (A) and completely (B) reduced; (C) – micro-photo of the reduced pellet front.

Reduced UO_2 pellets maintain their initial shape and geometric dimensions (Fig. 20.4). Experimental verification of the model uranium nuclear fuel remelting (metal uranium with addition of rare-earth element (REE) oxide Nd_2O_3 and metallization products – uranium dioxide with addition of REE (Ce, La, Nd) oxides in the amount of 5.5% wt.) was performed.

Experimental studies of the process for electrochemical uranium deposition from molten 3LiCl-2KCl with added UCl_3 were carried out, and cathodic deposit of metal uranium was obtained on Mo-cathode. In arrangement of the process much attention was given to uniform current distribution across the solid cathode surface.

The process operation is based on chemical interaction between the oxidizing agent $PbCl_2$ ($CdCl_2$) and nitride fuel components. "Soft" chlorination (chlorination with bivalent metal chlorides) of uranium mononitride (UN) with cadmium chloride in the molten LiCl-KCl eutectic was investigated depending on the temperature and chlorinating agent content in the molten salt. It was demonstrated that the degree of UN → UC13 conversion increased almost linearly with temperature growth. The obtained results were confirmed by large-scale laboratory experiments on the benches of Siberian Chemical Plant JSC (Seversk) with the UN load up to 170 g. The application of lead chloride ($PbCl_2$) as the chlorinating agent is very promising for the "soft" chlorination process. Noble fission products do not react with lead chloride in the course of chlorination and remain in the metallic phase, and the generated metal lead representing the collector for noble fission products is removed from the process flow and sent for disposal (RW). Analysis of the thermodynamic data is indicative of the possibility to reduce the temperature by 50–100°C in case of the $PbCl_2$ usage as the chlorinating agent with maintenance of all other parameters and characteristics of the "soft" chlorination process.

Intermediate and final products of pyrochemical SNF processing include pellets or powdered metallized fuel, the cathodic actinide deposit after electrolysis contains 5–10 wt. % of electrolyte with admixture of fissionable materials and fission products. Purification of these products from electrolyte is necessary to

FIGURE 20.5 Photos of cathodic deposit before (on the left) and after (on the right) purification from chloride electrolyte by the distillation method.

remove fission products (chlorides of alkaline or alkaline-earth metals) from the process flows or to reduce their concentration as well as to decrease losses of the target components contained in the molten material in the form of chlorides. Distillation at the increased temperature and underpressure is one of the methods considered for electrolyte removal.

Experimental studies of the distillation of alkaline or alkaline-earth metal chlorides from their molten mixtures under various conditions were carried out. It was demonstrated that the distillation process enabled to remove 98.8–99.9% wt. of LiCl, KCl and CsCl. In this case uranium content in the distillate was negligibly low. The distillation process under these conditions decreases the $NdCl_3$ concentration in the remaining product after distillation by an order, and the concentrations of $BaCl_2$ and $SrCl_2$ – by two orders (Fig. 20.5).

Experiments on distillation of molten LiCl-Li_2O mixtures with the oxide content of 3.0% wt. are indicative of the fact that lithium oxide is scarcely distilled in contrast to chloride. The distillation regimes providing for the decrease in lithium concentration (within chloride and oxide) in the metallized uranium pellets and powders down to 0.01wt. % are determined. On the basis of the performed studies it may be concluded that the distillation method allows purifying metallized products only from alkaline metal chlorides. Other methods or techniques resulting in removal of involatile components should be used to remove Li_2O, chlorides or oxides of alkaline-earth and rare-earth elements.

REE oxides were deposited using neodymium trichloride. Lithium oxide and gaseous oxygen were used as the deposition agent.

Subsequent to the results of experiments it was also established that the particle size distribution of the produced neodymium deposit scarcely depended on the temperature and initial mole ratio $O^{2-} : Nd^{3+}$ within the investigated range of 0.5–2.0.

Experiments with cerium oxide were performed for experimental substantiation of the filtration process for molten chlorides containing REE oxides.

The filtration rate of the Al_2O_3 filters with the working porosity of 39% decreased from 45 to 40 $cm^3/h{\cdot}cm^2$ depending on the volume of molten chloride passing through the filter (from 10 to 70 ml respectively). The particle size distribution of the filtered Ce_2O_3 powder did not change, which confirms high efficiency of the filtration process.

The tested aluminum oxide filters may be recommended as filtering materials for molten media based on alkaline metal chlorides.

The characteristics of the process for the electrolyte (LiCl, LiCl-KCl) purification from CsCl and $SrCl_2$ impurities by the floating-zone refining method were determined. In all experiments the resulting salt ingot can be conventionally divided into two sections: the decrease in the impurity concentration as compared to its initial quantity takes place in the first (long) section, and significant enrichment by CsCl and $SrCl_2$ takes place in the second (short) section.

The degree of molten chloride purification from CsCl and $SrCl_2$ depended on the solvent salt composition, the initial concentration of impurities and the recrystallization front movement rate. During purification of LiCl containing CsCl and $SrCl_2$ mixture in the amount of 0.3 mol. % the content of impurities was reduced tenfold, and the purified salt volume was 80% of the total salt volume used in the experiment.

The results of experimental studies with uranium obtained in 2017–2018 confirmed the technology feasibility of the basic process operations for pyrochemical SNF processing – uranium dioxide reduction in the course of the LiCl – Li_2O electrolysis ("metallization"), refining flux melting of uranium, electrolytic deposition of metal uranium from molten chlorides using soluble metal anode and "soft" UN chlorination for the anodic deposit processing. the experimental verification is performed, and feasibility of the auxiliary process operations – oxide deposition, filtration of molten chlorides containing solid oxides of actinides and REE, as well as chloride electrolyte purification from chlorides of alkaline or alkaline-earth metals by the floating-zone refining method – is confirmed.

The obtained results allow developing the flow diagram for pyrochemical processing of FNR MNIT+MOX SNF.

20.5 PH-process is a combined (pyro + hydro) processing technology for SNF from fast reactors

The well-known hydro-metallurgical technology has a set of limitations hampering its application for processing of SNF from fast reactors with high burnup, high content of fissionable materials and short storage periods (see the section "Hydro-metallurgical processing of FNR and TNR SNF"). On the other side, pyro-electrochemical technologies [152,202] are known as lean, radiation-resistant, nuclear-safe, but not yet providing for high purification coefficients.

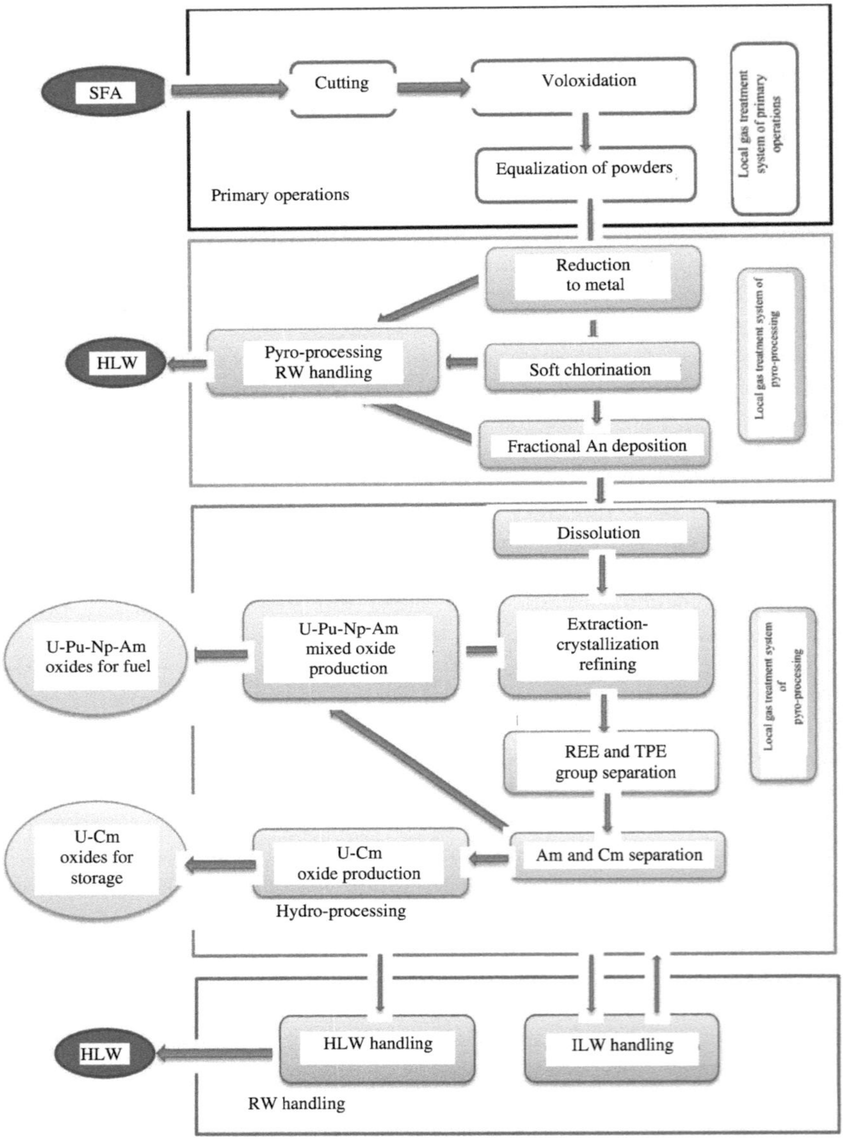

FIGURE 20.6 Flow diagram of the PH-process [102].

The developed combined (pyro + hydro) processing technology for SNF from fast reactors called PH-process [68,152] is based on application of non-aqueous primary purification of fissionable materials from fission products by approximately 100 times, and further hydro-metallurgical refining operations (Fig. 20.6).

The combined process flow allows:

Processing SNF with short period of storage, high burnup, with the content of fissionable materials up to 10-15% and the coefficient of fissionable material purification from fission products $K_p \sim 106$;
Maintaining the nonproliferation regime for nuclear materials;
Providing for loss of trans-plutonium elements $\leq 0.1\%$;
Manufacturing the products suitable for pelletized fuel fabrication;

Involving neptunium and americium into the fuel cycle after separation from curium.

Pyro-processing of MNIT SNF is considered in the section "Pyrochemical technology for processing of FNR and TR321 SNF." Operations of hydrometallurgical processing following pyrochemical one developed by the team of authors from VNIINM JSC, Radium Institute JSC with active involvement of VNIIHT JSC, Siberian Chemical Plant JSC, NIIAR JSC, Mayak PA FSUE as well as the RAS institutes including the Frumkin Institute of Physical Chemistry and Electrochemistry and the Institute of Geochemistry and Analytical Chemistry are described below.

The combined technology provides for extraction of americium within mixed (U, Pu, Np, Am) nitride fuel (homogeneous fuel). But processing of SNF with americium after-burning in separate fuel rods is also possible. The combined technology has been and is being developed in the two main areas: implementation of innovative processes and arrangement of test benches for equipment trial and verification of the adopted technological solutions.

Dismantling of spent fuel assemblies, fragmentation of fuel rods and subsequent voloxidation of SNF in the air flow at the increased temperature are the primary operations of MNIT SNF processing. Optimal voloxidation parameters (temperature, time, the oxidizing agent supply rate) and subsequent washing of fuel rod claddings were tested on the model nuclear fuel samples. It was determined that fuel was fully detached from the cladding, and losses of nuclear materials after washing were less than 0.01wt. %. Optimal voloxidation regimes were verified on the MNIT SNF samples from the BN-600 reactor plant. It was determined that more than 99.9% of tritium and 99.8% of ^{14}C were removed in the course of voloxidation due to complete destruction of fuel pellets.

Dissolution of the basic products in the combined process flow for processing of SNF from fast reactors. U-Pu-Np product after pyrochemical processing in the form of alloy grains with the size of 2–5 mm is supplied for hydrometallurgical processing. An actinide ingot (U – 84.5%; Pu – 15%; Np – 0.5%) was synthesized in order to verify the possibility for direct dissolution of actinide alloy in nitric acid. Subsequent to the experiments it was demonstrated that the solution containing 400 g/L of total actinides could be obtained within the technologically acceptable period using concentrated nitric acid. The content of nondissolved metal residue was less than 1.5 wt. %. It is proposed to dissolve this residue by the electrochemical method. Testing demonstrated that complete dissolution of the previously nondissolved residue was achieved with silver ions used as mediators. Total losses of nuclear materials at the dissolution stage amount to less than 0.05 wt. %.

The filtration (clarification) stage follows the SNF dissolution. It is proposed to use the cross-flow ultra-filtration method with ceramic membranes for this purpose. Integrated verification of the full-scale clarification plant model demonstrated the possibility to concentrate the pulp up to 100–103 g/L. Solid phase is completely removed from the clarified solution (the residual content is less than 0.1 g/L).

Extraction refining. In accordance with the published data the following uranium and plutonium purification coefficient is achieved in the course of the primary pyrochemical processing: from cesium – 10^3; from curium – up to 10^2; from rare-earth metals and yttrium – 40 [102]; from all other fission products – up to 10^2. Besides, the main bulk of fission products forming insoluble residue at the SNF dissolution stage including Zr, Mo [60,148], Rn-Pd [138,143] are separated from the fissionable materials in the course of the pyrochemical processing. It allows minimizing the number of purification operations and decreasing the amount of high-level liquid reactive waste (the primary extraction refined product) by more than 40%. The flow diagram of the primary processing cycle includes two processing stages:

Extraction of uranium, plutonium and neptunium and their purification from the main bulk of impurities, and extraction refining of the formed U-Pu-Np water flow with the specified Pu/U ratio;

Extraction of the uranium remaining after-purification and regeneration of the circulating extraction agent.

Verification of the extraction refining process flow options at the extended bench of Siberian Chemical Plant JSC demonstrated achievement of the declared purification parameters. Extraction of nuclear materials from the SNF solution exceeded 99.98%. New reagents for re-extraction of plutonium and neptunium as well as the equipment prototypes, particularly the liquid chromatography column for deep purification of uranium extract were tested in the course of the extraction process flow verification.

The coefficients of purification from γ-active fission products of at least 5–10^4, from americium – of at least 2–10^5 [102] were achieved in the described extraction refining process flow.

Crystallization refining. Crystallization refining of uranyl, plutonyl and neptunyl nitrate mixture was developed by the specialists of VNIINM JSC in cooperation with Siberian Chemical Plant JSC. The operation of plutonium oxidation to the hexavalent state with nitric acid performed in the course of the boiling solution hold-up is mandatory in preparation of the initial solutions for crystallization experiments. It is possible to oxidize 70–100% of plutonium to Pu(VI) and more than 99% of Np(VI) depending on the residual acidity.

It follows from the obtained experimental data that enrichment of the crystalline phase with uranium is observed (the uranium and plutonium cocrystallization coefficient is below 1). The achieved cocrystallization coefficients are technically acceptable, as the process flow provides for reflaxing of actinides by their processing at the extraction stage.

Experiments in the linear crystallization model demonstrated that purification of nonseparated U-Pu-Np mixture from the majority of fission products, particularly from technetium with the purification coefficient of 100–300 was possible depending on the crystallization mode. So, purification of nuclear materials by more than 106 times [102,145] is possible in the course of MNIT SNF processing with combined extraction and crystallization refining.

Production of actinide oxides. Direct denitration under microwave radiation is the most promising method for production of initial materials for nitride synthesis [72,73,182]. Both solid phase (for example, the material after crystallization refining) and directly mixed U-Pu-Np re-extract may be subjected to heat treatment. In the latter case the initial stage involves evaporation of the solution under the impact of microwave radiation.

It was demonstrated under laboratory conditions (in cooperation with the RAS Institute of Geochemistry and Analytical Chemistry) that UO_2 could be produced directly at the denitration stage in 5% hydrogen medium. UO_2 powders have the specified properties:

- More than 99% U in the oxidized state (IV);
- Particle size is within the range of 25–400 μm;
- Bulk density is 2.0–2.2 g/cm^3 and 2.3–2.5 g/cm^3 for tapped powders;
- Specific surface area is 2.0–2.7 m^2/g.

The results of experiments on the box laboratory unit with the total weight of nuclear materials up to 100 g confirmed the possibility to produce homogeneous powders of mixed actinide dioxides by the powder mixing in the course of denitration in argon flow with hydrogen content of 5%. Upon transition to testing of the microwave denitration technology on the pilot facility model (capacity – 7.5 kg of uranium) it was discovered that additional research and development works related to optimal use of the entire microwave emitter power were required for production of mixed dioxides. At present it may be stated with confidence that the process for evaporation of actinide nitrate solutions and their subsequent microwave denitration to mixed oxides U_3O_8-PuO_2 within a single stage is tested and can be implemented on the industrial scale. Losses of nuclear materials in the course of denitration are less than 0.01% wt.

Production of uranium and plutonium powders by the direct microwave denitration method will allow reducing the generation of liquid RW with high actinide content by 4.5 times.

Development and verification of the gas treatment technology for SNF voloxidation, oxidized SNF dissolution and actinide oxide production operations are completed. The following equipment is developed and tested:

FSGO-50, FARTOS-50, MTF, FyaS units for trapping of liquid and solid aerosols;
BRUNS-50, SMOG-50 for trapping of NO, NO_2;
The absorber column module for CO (^{14}C) trapping.

It was established that FSGO and FARTOS aerosol filters caught 70–80% of NO_2 in case of water injection into the purified aerosol flow; in this case FSGO filter accounted for the basic trapping. Then the gas stream is supplied to BRUNS unit (reagentless trapping in the fixed bed of aluminum gamma-oxide, efficiency of purification from nitrogen oxides – up to 90%) and two SMOG units installed in series (silicagel – urea gas treatment) filled with granulated aluminum gamma-oxide impregnated with water solution of carbamide and ammonium bicarbonate where deep (99.9% and more) purification from nitrogen oxides takes place. Verification performed on the cold test facility showed that efficiency of the entire system: BRUNS unit, FSGO+FARTOS, SMOG varied for purification from nitrogen oxides within the range of 99.80–99.97%.

20.6 Americium and curium extraction and separation

Laboratory studies and dynamic testing of various extraction systems where americium and curium separation was performed were carried out in 2012–2018. The results of the studies and analysis of published data substantiated selection of the N,N,N,'N'-tetraoctyldiglycolamide (TODGA)-F-3 system as the basic option and the carbamoyl-methylene-phosphine oxide (CMPO) – tributyl phosphate (TBP) – metanitrobenzotrifluoride (F-3) system and ATP-HDK-F-3 as redundant options for the technology development.

Dynamic testing of the fractionation process flow with the TODGA-based system was carried out in 2014 and 2015 at the laboratory mixer-settler facility of Mayak PA FSUE. The model solution imitated the refined product from the primary extraction block with introduction of ^{241}Am label. The extraction unit consisted of four blocks of mixers-settlers in the total quantity of 51 pcs

Subsequent to the testing results extraction of americium (^{241}Am label) from the solution imitating the refined product from PUREX process with the REE content of up to 4.2 g/L in the organic phase was at least 99.97%, and REE (except for lanthanum) – more than 99.99%. At the TPE (transplutonium element) re-extraction stage the degree of americium purification from REE reached 100% [153].

The effect of external ionizing radiation and internal alpha-irradiation on the degradation degree for extraction fractionation systems was studied in

2016–2017. ^{244}Cm, ^{238}Pu, ^{60}Co and accelerated electrons were used as the ionizing radiation source.

Check of fluoride ions release was performed in the double-phase system TODGA/F-3 – 3 mol/L HNO_3. The aggregate release of fluoride ions in the course of alpha-radiolysis was approximately 5.5 mg/L/kGy (2 at/100 eV), and in case of gamma-radiolysis – 2 mg/L/kGy (0.7 at/100 eV). Concentration of fluoride ions in irradiated water solutions containing dissolved F-3 did not exceed 4 mg/L [146]. It was noted in the course of the works that chelated TODGA was less susceptible to radiolytic destruction. This effect was manifested regardless of the dilutant used [212].

The test with the use of small samples (batch test) was performed in 2016 on the base of NIIAR JSC using number quantities of TPE. Verification of the TODGA-based extraction system in heavy polar fluorinated dilutant F-3 was performed for TPE extraction from the refined product after the first cycle of PUREX process. The model solution of the refined product from PUREX process containing 2 g/L of americium was used in the experiment [211]. The test demonstrated operability of the system and the possibility to achieve the target parameters.

The study of element distribution including macro-quantities of TPE (americium) in the extraction system TODGA – F-3 preliminarily subjected to radiation damage was carried out in 2017–2018. Extraction of the key elements (americium, REE, zirconium, palladium) with irradiated extracting agent at all basic stages of the fractionation process (TPE+REE extraction, TPE re-extraction, REE re-extraction) was checked. The dummy solution imitating the refined product from PUREX process and containing real quantities of americium (0.6 g/L) was used for verification. It was demonstrated that the radiation doses up to 50 kGy did not have any significant effect on distribution of metals.

In 2017, "hot" dynamic tests of the technology for americium and curium extraction from concentrated high-level refined product of BN-600 and VVER-440 SNF processing with the use of the TODGA-based extraction system were performed on the research facility of Mayak PA FSUE after a 30-year interval. Duration of the facility operation with real HLW in the continuous (24/7) mode was more than 70 hours. Extraction of TPE from HLW of more than 99.2% was achieved (99.9% taking into account after-purification with solid extracting agent). It was confirmed that the process ensured TPE and REE extraction at the extraction stage and separation of TPE and REE at the re-extraction stage. Impossibility to use mixers-settlers for this process was confirmed by experiments.

The studies on use of light dilutant (C8-C10 spirits) for the TODGA extracting agent were commenced in 2017. In 2018, recommendations on the explosion and fire safety conditions for the process were obtained, and bench-scale tests of the system were carried out.

In 2018, the extraction system based on multiradical DGA was tried. High resistance of the system to the 3-rd phase formation in the hydrocarbon dilutant (isopar-M, C-13, etc.) was demonstrated. Dynamic tests of the extraction system were carried out in 2019.

The process for acid removal from the salting-out agent of the extraction fractionation operation was developed. It was demonstrated that sufficient acid removal from the flow took place in case of electrodialysis for the salting-out agent reuse in the process. The proposed process enables to reduce the volume of ILW flow with high salt content (2.5 mol/L $NaNO_3$) by more than 700 times subject to complete replacement of the salting-out agent once per a month.

Pilot trials of the sorption technology for americium and curium separation with the use of cationite were carried out in 2015 at Mayak PA FSUE. The pressure in the system did not exceed 3 atm in the course of testing. Approximately 9 g of pure curium and about 65 g of pure $^{241,243}Am$ with the curium content of less than 0.8% wt. and $^{154,155}Eu$ – less than 0.1% by activity were extracted subsequent to the refining process [168,204].

Experimental verification of the dummy chromatographic unit (UHVD-100-2,1) designed for the pressure of up to 100 atm. with the dummy solutions (gram quantities of REE: holmium, europium, neodymium) was carried out in 2016–2017 with the use of fine-grain (50 μm) SAC 50 sorbent (8%).

The possibility to process REE mixtures with the total content of 45–70 g of metal was demonstrated. Subsequent to the chromatographic unit operation efficiency results test calculations were performed with the use of the mathematical process model. Discrepancies in the calculations and the experimental data were detected. The mathematical model was adjusted.

The set of works on safety analysis for chromatographic Am-Cm separation process was carried out in 2017–2018. It was demonstrated that the process ensured acceptable safety in the course of regular operations. It was also demonstrated that deviations from the sequence of process operations posed the main hazard [158].

20.7 SNF processing with the use of plasma separation

The above-mentioned data are indicative of the need for the CNFC arrangement and development of new SNF and RW processing technologies. Pyroelectrochemical technologies as well as plasma processing techniques including the approaches based on Ohkawa mass-filters, plasma-optical and resonant methods are developed [124,139,195,206]. Evaluation of energy efficiency for plasma processing of SNF from a reactor with the electric power of 1 GW gives the values below 0.5% of the own capacity which is considerably lower than the auxiliary power consumption of the reactor [195].

Let us consider certain results obtained by Archimedes Group [123,125,206] which has achieved significant progress in experimental demonstration of the

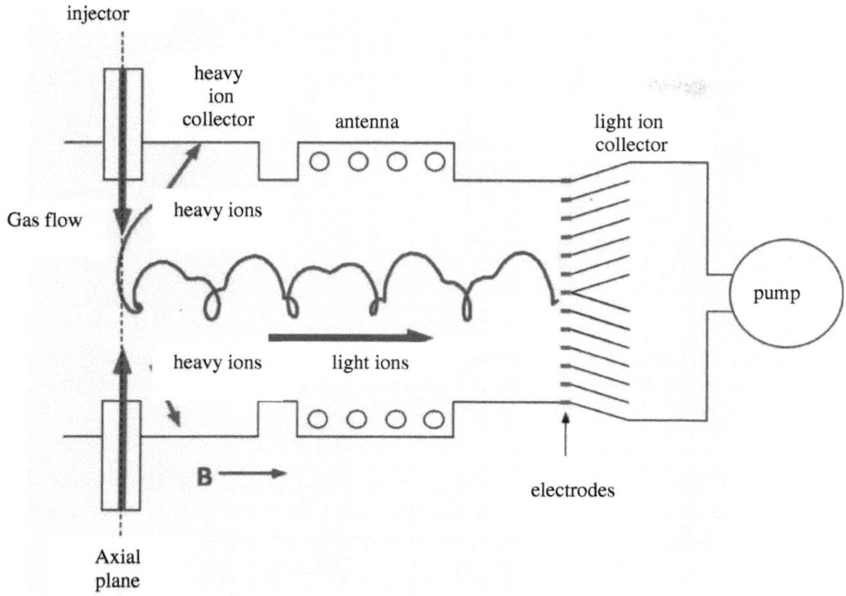

FIGURE 20.7 Sectional view of the filter [217].

possibility for plasma RW separation. Ions with different mass are separated in the following way: a helicon discharge is generated in presence of almost uniform magnetic field produced by current coils in a cylindrical vacuum chamber (the magnetic field is parallel to the axis). Ring electrodes submerged into plasma are installed at the chamber ends enabling to reproduce electric potentials of the electrodes [127] along the magnetic field lines and consequently to create a radial electric field. Plasma rotation starts in the resulting superposition of fields; in this case the radial size of the area where an ion can move depends on its mass [128,206]. Thus the conditions when light ions are drifting toward the ends, and heavy ones are thrusted to the side surface of the chamber by the centrifugal force can be arranged by selection of the field parameters and the separation chamber diameter (Fig. 20.7) [123].

In the case of Archimedes, the critical mass separating heavy ions from light ones was selected to be equal to 89 a.m.u. [124]. The following key experimental results obtained by the group may be distinguished [123,206]:

The possibility to generate multicomponent plasma with the concentration of approximately 10^{13} cm^{-3}, the electron temperature of ~2 eV, the ionization degree close to 100% and ionization multiplicity of 1 was demonstrated for the elements with the nucleus charge of more than 20;

The effect of high separation level for the elements with different mass was obtained;

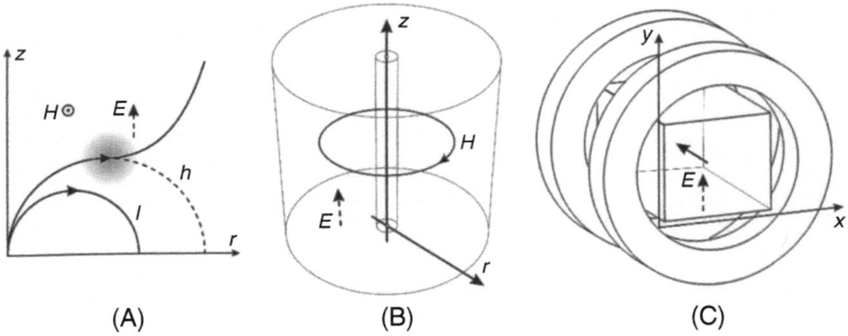

FIGURE 20.8 Plasma SNF separation mechanism (A) and separation chamber geometry options with azimuthal (B) and axial (C) magnetic field H for its implementation: the intensive electric field area E is darkened, l – the path of light ions, h – the path of heavy ions.

High efficiency of the method was demonstrated: 440 kg/day; in this case the ion price was about 500 eV corresponding to the values of (1 keV) taken for the calculation of the plasma separation method energy efficiency.

In spite of its incontestable advantages the approach implemented by Archimedes Group has some significant drawbacks. One of them is associated with the fact that the electrodes setting the electric potential in magnetized plasma are acting as light ion collectors at the same time. This complicates the process for the separated substance removal from the receivers and also poses a hazard of deviation from the RW component separation conditions due to distortion of the potential set by the electrodes because of the deposition products. The requirement for the point of the substance injection into the plasma volume is the second significant drawback (see Fig. 20.7). The problem is that ionization of separated products in Archimedes unit shall be carried out in the area close to the cylindrical chamber axis, otherwise the fraction of light ions entering the collector of heavy ones will increase sharply as their paths start to cross the chamber walls [206].

A different approach was proposed by V.P. Smirnov. His method involves the substance conversion into low-temperature plasma with subsequent spatial separation of ions according to their mass.

As distinct from electromagnetic separators, in this case ions move under the conditions of compensated volume charge [51] thus removing the limitations for the ion current value and increasing the capacity considerably. Separation efficiency is due to generation of the electric field in magnetized plasma [51,197] with the potential well catching heavy ions and leaving light ones almost untouched (Fig. 20.8). The accelerating potential is applied at the chamber inlet in order to overcome the initial scattering with regard to the ion speed value and direction. Taking into account the fact that the atomic mass of uranium fission products is ~35% less than the atomic mass of minor actinides the depth and position of this well can be adjusted in such a way so that the paths of heavy and light ions are separated spatially providing for SNF separation. The proposed concept enables to achieve higher capacity as compared with plasma-optical process flows [139,196,197], a more successful substance collection system than in Ohkawa mass-filters and also less sensitivity to fluctuations of the parameters than in resonant methods [218]. Computational modeling of the substance ion paths imitating SNF and RW was performed to substantiate the proposed plasma separation method, in particular ions with atomic mass of 240 and 150 were considered. The calculations were carried out for the collisionless mode in the single-particle approximation. Azimuthal and axial magnetic fields and several model configurations of the electric field reflecting different separation chamber geometry options were considered (see Fig. 20.8).

The calculations demonstrated the possibility for efficient spatial separation of SNF or RW substances within the plasma volume in combined electric and magnetic fields; in this case the initial energy of the separated substance ions can vary within wide ranges from 0.2 to 3 eV, and the initial scattering of ion speed according to the angle – from 0 to 45° (60° in some cases).

It was established that the experimental study of plasma separation of SNF modeling substances interesting from the viewpoint of technology development could be carried out at the experimental plant with the typical separation area size of approximately 1 m^3, the magnetic field induction of 0.1–0.2 T and electric potentials of approximately 1 kV.

The presented results enable to develop a laboratory facility with both azimuthal (see Fig. 20.7) and axial (see Fig. 20.8) magnetic fields. In this case the electric field potential profile necessary to separate the plasma flow within the chamber volume is created with the use of electrodes the magnetic field lines are based on. The final selection was made in favor of the plant with axial magnetic field from considerations of the bench ergonomics, the possibility for modular capacity increase as well as economic factors. Its appearance is shown in Fig. 20.9. The main parameters of the plant: the vacuum chamber diameter – 90 cm, length – 200 cm, pressure of up to 10^{-6} torr, the posxssibility to use different gases, the magnetic field of up to 2.1 kG with the degree of nonuniformity within the active volume of less than 10%, the high-frequency plasma heating power – up to 60 kW, the liquid cooling system with heat removal capacity of up to 160 kW, the electrostatic potential setting system with adjustable geometry of potential surfaces and the potential of up to 1250 V.

It should be noted that apart from the issues directly related to the plasma separation process arrangement of the high-capacity source for conversion of condensed SNF or RW substance into the plasma state is also an important aspect. The authors and their colleagues demonstrated the potential of vacuum arc discharge with diffusive cathodic tie-in as such a source. The modes when a vacuum arc was ignited on the cathode made of model metal were demonstrated; in this case the ionization degree of the resulting plasma was close to one, and the ionization multiplicity was about 100% [122].

FIGURE 20.9 Appearance of the experimental facility for trial of the plasma separation method for SNF modeling substances.

Thus plasma methods potentially comply with the requirements for the commercial SNF and RW processing technology.

The possibility for plasma processing with the capacity of up to several hundred kilograms per a day was demonstrated through experiments [123,124].

The approach to plasma separation of minor actinides and uranium decay products [206[was proposed; computational modeling demonstrating the possibility to implement this approach at the distance of about 1 m with the electric potential of approximately 1 kV and the magnetic field of about 1 kG was performed.

The plant enabling to perform experimental verification of the plasma separation method was developed and arranged [206].

Chapter 21

Radioactive waste management

Transportation of irradiated nuclear fuel (SNF) and high-level radioactive waste (HLW) is an inevitable and essential process stage of electric power generation at NPPs. Dozens of transportation operations for spent fuel assemblies (SFA) from various reactor plants sent for long-term centralized storage, radiochemical processing or material research are annually performed in our country alone. Hundreds of thousands tons of SNF have been accumulated worldwide at the stage of processing hold-up or long-term storage by 2019. Potential radiation risks for the operating personnel, the public and the environment are subject to comprehensive assessment in preparation for the above-mentioned operations and continuous control in the course of their implementation. These types of activities shall be carried out in strict compliance with the effective safety standards and rules.

21.1 SNF and HLW transportation

The International Atomic Energy Agency (IAEA) has taken the responsibility for development of the recommendations of safety assurance during transportation of radioactive substances (hereinafter – the IAEA Rules) since 1959 on the basis of the resolution of the UN Economic and Social Council. From then the IAEA Rules are taken as the basis for national requirements by other international organizations and many states including our country.

The first official rules for transportation of radioactive substances (nonfissionable radioactive materials) in the USSR based on the IAEA Rules were issued in 1973 (PBTRV-73 [10]), and the rules for transportation of nuclear materials based on the IAEA Rules were issued in 1983 (OPBZ-83 [11]). The first revision of the Russian federal rules and regulations in the area of atomic energy use regulating safety assurance in transportation of both fissionable and nonfissionable radioactive materials (NP-053-04 [13]) was published in 2004.

The IAEA Rules define the following objectives: "...to establish the requirements to be fulfilled in order to ensure safety and protection of people, property and the environment against radiation exposure in the course of radioactive material transportation" [14].

Closed Nuclear Fuel Cycle with Fast Reactors: Handbook of Russian Nuclear Power.
DOI: https://doi.org/10.1016/B978-0-323-99308-1.00017-9

The basic safety requirements for achievement of these objectives are limited to four items:

1. Containment of radioactive substance (RS) within the package.
2. Monitoring of the external radiation level around the packages.
3. Dissipation of heat generated inside RS.
4. Criticality control in case of fissionable RS.

In order to fulfill the above-mentioned requirements, it is recommended to apply the step-be-step approach to determination of the content limits for packages and vehicles, as well as reasonably sufficient regulatory requirements for the package design depending on the radioactive content and application of administrative control measures in case of necessity.

The priority component – the design safety concept – is the basis of the revision of the IAEA Rules recommendations effective in 2019. The design safety concept for transportation of radioactive materials is the requirement for minimization of the radiological consequences due to guaranteed integrity maintenance for the special-purpose transportation packages in case of any potential transportation incidents. So, any container with SNF shall represent a nuclear installation designed in accordance with the inherent safety principles. For this purpose, additional administrative and technical arrangements may be used in the course of transportation and/or process storage of radioactive materials in case of exceptional nonsystemic necessity.

The modern Russian Federal Rules for Safe Transportation of Radioactive Materials (NP-053-16) establish not only safety requirements for transportation of radioactive materials but also the requirements for operations and conditions related to their transportation and being a part of this process, such as design, manufacturing, maintenance and repair of packaging; preparation, loading, dispatch, transportation including transit storage; unloading and acceptance of the cargo and packages at the destination point.

In practice controlled fulfillment of both package design safety requirements and mandatory use of vehicles on the exclusive use basis in order to comply with the requirements for physical protection of fissionable material, i.e. arrangement of additional administrative measures is implemented in Russia in the course of SNF transportation. This approach enabled to prevent any radiological incidents during transportation of high-level radioactive materials within the entire period of the nuclear industry existence.

In accordance with the international classification all packages for handling of spent nuclear fuel are classified as B(U)F type packages. Packages of this type shall be designed in such a way so that the radiation level at the distance of 1 m from the package surface should not exceed 0.01 Sv/h (1 rem/h) after testing under emergency transportation conditions with the radioactive content of the maximum activity permitted for this package. Containers for transportation of SNF unloaded from thermal reactors are developed, certified and used in Russia (Fig. 21.1).

FIGURE 21.1 Modern TPS-153 at the power unit of a NPP with VVER-1000 RP.

All design solutions used in the transportation package set (TPS) for VVER SNF are tried and certified by the competent state authority and may be directly taken to develop the containers for spent fuel assemblies unloaded from fast reactors. However, there are significant differences in the parameters – the permissible fuel rod cladding temperature and the SFA storage period prior to loading into the TPS. Their effect on the applied design solutions shall be additionally studied through experiments in development of new TPS for SNF of fast reactors.

The unit cost of intermodal transportation (to the distance of 4000 km) of SNF from VVER reactors between sites is assessed to be about 30 USD/kg_{hm} which is an insignificant amount (less than 0.01 cent/kW·h) of the revenue from electric power sold by the NPP. Taking into account the difference in weight and dimensions and the required time parameters for transportation of spent fuel assemblies from fast reactors increase of these price values can be expected.

21.2 SNF and HLW storage

The entire preceding practice (both national and international) of SFA storage in the first years after their removal from the core of various reactor plants including the plants of BN type is limited to SNF placement into spent fuel pools with the use of water for residual heat removal and radiation protection. The main drawbacks of this technology for use in advanced reactor plants

FIGURE 21.2 SNF storage facilities on the basis of capital construction solutions.

FIGURE 21.3 Examples of the main types of ventilated SNF repositories.

with inherent safety – the necessity of active safety systems for design basis accidents and unacceptable consequences in case of any beyond design basis accidents -manifested to the full extent during the accident at the NPP in Japan in 2011. The modern rules prescribe location of spent fuel pools in the buildings capable of withstanding design-basis postulated external impacts of natural or human-induced origin for guaranteed protection of the radioactive content against any damage. All these aspects in the aggregate result in high costs of both construction and current operation of "wet" SNF storage facilities. "Wet" storage facilities are used for the process-related (particularly long-term storage of spent fuel assemblies in the countries where the effective SNF management concept defines its further destination clearly – in Sweden, Finland, France and also in Russia (for certain FA types).

Dry storage technologies are also well-tried and widely used for long-term storage of SNF and vitrified HLW. Storage facilities on the basis of capital construction solutions (Fig. 21.2) of the solid reinforced concrete type (Kurchatov Institute, IPPE JSC, Mayak PA, etc.) and chamber type (SFSF-1 at MCP Isotope Chemical Plant). The first ventilated repository of well type was commissioned in the USA in 2017. Various multi-site concrete ventilated storage systems are used to arrange near-plant storage (Fig. 21.3).

Popularity of multisite ventilated systems for near-plant storage of SNF from PWR and BWR reactor plants, primarily in the USA, is due to their unit cheapness in the absence of any final national strategy and clear decisions with regard to SNF destination. They are also popular for long-term storage of "cold"

FIGURE 21.4 Examples of SNF repositories with double-purpose containers.

SNF with low burnup from CANDU RPs (Canada, Romania, the Republic of Korea, and Argentina) due to large amounts of irradiated fuel annually unloaded from such reactors.

The main drawback of the systems intended for storage only resides in the necessity of high future expenditures for SFA transfer into transportation casks in order to transport them for final disposal or processing. Advantages include the expenditures for their purchase distributed over time and lower costs of the storage site operation in comparison with permanent buildings. Safety of small-scale multi-site ventilated systems is not fully compliant with the conservative requirements for postulated external impacts. Their acceptable safety level is usually substantiated by the possible single damage within the limited volume from the SNF stored in the "dispersed" manner, or other additional arrangements for physical protection of radioactive material.

Usage of TPSs is the safest solution for long-term SNF storage. In 2018, the IAEA completed development of the basic requirements for double-purpose (transportation and storage) containers (Fig. 21.4). More than 40 years of experience in various testing and practical usage of double-purpose and/or transportation packages for SNF and vitrified HLW in many countries around the world (England, Bulgaria, Germany, Italy, Lithuania, Russia, USA, France, Czech Republic, Switzerland, Republic of South Africa, etc.) were used as the basis for development of these international recommendations.

As mentioned in the previous section, containers developed in accordance with the international requirements for transportation of high-level radioactive materials are the most widespread nuclear installations designed on the basis of inherent safety principles for all possible emergency situations of natural and human-induced nature. In this case safety of this technology is confirmed by multiple conservative in-situ tests as well as by maintenance of the package integrity after real transportation incidents. For example, in 2011 some spent fuel assemblies at the NPP in Japan were located in metal containers; they were not damaged and maintained their leak-tightness in the course of an earthquake and subsequent tsunami.

Both outdoor sites and light buildings are used for location of double-purpose TPSs. In addition to guaranteed safety the main advantages of this solution in arrangement of SNF or vitrified HLW storage include the expenditures for TPS purchase distributed over time (in case of necessity) and the minimum possible current operating costs. The latter is due to the need to perform only lifting works with thick-walled transportation packages at the storage site, in contrast to concrete (permanent or multisite) ventilated storage systems where manipulations with thin-walled bottles containing SNF or vitrified HLW are necessary.

Usage of double-purpose containers for storage of spent fuel assemblies from VVER-1000 RP was recognized as more efficient than arrangement of sequential storage of this SNF in "wet" and "dry" chamber repositories according to the resolutions of the two sections of Rosatom Scientific and Technical Society in 2010. The unit cost of VVER reactor SNF storage for 60 years (deferred decision on processing) is assessed to be about 200 USD/kg_{hm}, that is, less than 0.05 cent/kW·h of the revenue from the electric power sold by the NPP. These evaluations include the expenditures for construction, maintenance and decommissioning of the location site for SNF packages as well as purchase and disposal of the required number of double-purpose TPSs.

In conclusion of this section it should be noted that the special-purpose regulatory document intentionally devoted to arrangement of process-related (particularly long-term) storage of SNF or vitrified HLW in double-purpose (for transportation and storage) containers have not yet been developed in 2019.

21.3 Radioactive waste generated in the course of NPP operation

About 6 keV of energy are released upon combustion of one molecule of organic fuel, but almost the same amount of energy (and 1.5-2-fold more in some cases) is required to mine, process, transport this molecule, convert it into electric power and dispose of the combustion products. Thus power engineering based on organic fuel is efficient only in case of the open fuel cycle when the majority of combustion products are released into the environment, and the nature itself disposes of the waste.

Radioactive waste (RW) is generated at all stages of nuclear fuel cycle (NFC) – from uranium mining to SNF processing. In accordance with the effective legislation [1] RW are subject to conditioning (conversion into the proper form suitable for final disposal) and removal for disposal in special-purpose burial sites providing for RW isolation from the biosphere for the entire period of potential hazard. The exception is provided by so-called special RW disposed at their generation site by arrangement of special-purpose engineering barriers at the path of potential radionuclide spreading in accordance with Federal Law No. 190 "On RW Handling…" (Article 4) due to the set of environmental and economic reasons.

Removable RW are divided into different categories according to the following characteristics from the viewpoint of processing and disposal:

- Short-lived and long-lived ones depending on the half-life of isotopes contained in the RW;
- Low-, intermediate- and high-level ones depending on the decay type and specific activity of the contained isotopes;
- Solid, liquid and gaseous ones depending on the physical state.

The environmental impact related to uranium ore mining is proportional to the amount of generated waste where some uranium and its daughter products (radium and radon) remain. Activity of radium and radon in the uranium mine dumps is at the average equal to 10^7 Bq and $7 \cdot 10^3$ Bq respectively in production of 1 kg of uranium concentrate. Escape of radioactivity with wastewater flowing through the dumps (tailings) and gaseous radon spreading are the main paths of radioactivity ingress into the environment.

Operation of any NPP is accompanied with:

- Gaseous radioactive releases into the atmosphere;
- Accumulation of liquid radioactive waste (LRW) partially discharged into the hydrosphere;
- Accumulation of solid RW.

Leakage of radioactive gases through the fuel rod claddings (particularly in case of any fuel rod cladding integrity loss) into the coolant is the cause of gaseous releases from NPPs.

These gases are removed from the coolant, stored for 3 months (for decay of short-lived isotopes) and released into the atmosphere via the filters catching radioactive particles and radioactive iodine.

Activity of radioactive noble gases at a power unit with VVER-1000 (with the capacity utilization factor of 0.70) is $\sim 7 \times 10^{13}$ Bq/year, $^{14}C - 2 \times 10^{11}$ Bq/year and $^{3}H - 3 \times 10^{13}$ Bq/year.

NPP LRW is associated with activated corrosion products ($\sim 4 \times 10^9$ Bq/year) and tritium ($\sim 10^{13}$ Bq/year) and also with so-called nonprocess LRW generated in the course of equipment decontamination and clean-up of the rooms. Upon their generation NPP LRW have low activity and are classified as low-level radioactive waste (LLW). The common approach is to evaporate LLW LRW (by some hundreds of times) to intermediate specific activity (ILW) (up to 10^5–10^6 Bq/L depending on salt content and isotopic composition) and to include concentrated solutions into the cement matrix (RW cementation). RW included into the cement matrix may be handed over to the National operator for their final disposal.

In the course of NPP operation 1 GW·year of generated electric power is at the average accompanied by generation of ~800 t of LLW and ILW and 20 t of high-level irradiated nuclear fuel.

The majority of waste – up to 90% – are classified as LLW. Their share in the aggregate activity does not exceed 1%. HLW account for 99% of the aggregate waste activity, but their share in the amount is less than 1%.

Approximately 300 various radionuclides are generated from uranium fuel in a nuclear reactor, and more than 30 of them enter the atmosphere. The major part of radioactivity in gas and aerosol releases is generated by short-lived radionuclides, and their impact on the biosphere may be neglected. The share of NPP releases in the total balance of natural radiation background amounts to as little as some fractions of percent.

Investigations of the effect of anthropogenic factors on plants and animals did not provide any credible evidence of specific NPP impact against the background of other human-induced sources. As of 2019 there is no convincing evidence of any negative impact of NPPs operating under normal conditions on the health of the public and the personnel.

Thus, staying within the bounds of reasonable arguments, a NPP operating under normal conditions does not have any negative impact on the human health and the environment subject to compliance with the sanitary and hygienic standards with regard to releases.

The situation changes drastically in case of any accident at the NPP, especially with any core damage. The aggregate activity of the core materials increases millionfold in the course of the reactor operation. In case of any severe accident with release of the reactor core materials into the environment the public, plants and animals are exposed to high activity. Such accidents are very rare but have extremely serious negative consequences for development of nuclear power although the integral impact (loss of human lives and health) of these consequences are is incommensurate with the impact of other human-induced factors (e.g., death of people in traffic accidents).

Development of reactor technologies as well as the technologies for fabrication of nuclear fuel and SNF processing shall be aimed at prevention of such accidents.

Besides, development of RW handling technologies including the gas treatment methods and equipment shall facilitate reduction of gas releases from NPPs into the atmosphere in the context of strengthened requirements for permissible exposure levels in design of modern NPPs [12].

21.4 Radioactive waste from SNF processing

A total of 1.0–4.5 m^3 of HLW, 50–150 m^3 of ILW, and up to 2000 m^3 of LLW are generated in processing of 1 ton of SNF from a VVER reactor. Radionuclides present in the RW from SNF processing may be divided into several groups:

- Volatile fission products; the main representatives of this group include tritium, iodine and in some cases – radioactive carbon C-14 forming volatile compounds with oxygen;

- Fuel composition residues – uranium and plutonium (and also neptunium) isotopes non extracted in the course of processing;
- Minor actinides (americium and curium);
- Radioactive noble metals; the main representative of this group is Ru-106 isotope which firstly forms a volatile compound with oxygen like all ruthenium isotopes and secondly releases significant amounts of energy in the course of decay (via Rh-106), and high specific heat emission is provided in the matrices used for RW immobilization due to absorption of this energy;
- Alkaline and alkaline-earth fission products; the main representatives of this isotope group include relatively long-lived Sr-90 and Cs-137 with significant decay energy accounting for the main radiation burden on the materials used as the matrices for RW disposal;
- Rare-earth elements.

The modern practice for handling of high-level RW from SNF processing at RT-1 plant (Mayak PA) involves their vitrification without any preliminary separation of radionuclide mixture, temporary storage of vitrified blocks (in order to reduce heat emission from the vitrified HLW) and their subsequent disposal. Aluminum-borophosphate glass is used as the RW immobilization matrix at Mayak PA. Use of aluminum-borophosphate glass has historical reasons, as the vitrification line at Mayak PA was used for conditioning of LRW from processing of SNF from power reactors (both thermal and fast) and also for conditioning of LRW from processing of SNF from transport reactors with high aluminum content.

Borosilicate glass is planned to be used for RW vitrification at the pilot demonstration facility arranged at MCP for processing of VVER-1000 SNF and at the processing module for nitride SNF from BREST-OD-300 fast reactor with heavy (molten lead) coolant arranged at Siberian Chemical Plant within the framework of Proryv Project.

It is obvious that the percentage of fission products included into the glass matrix has limitations related to chemical and, more importantly, radiation resistance of glass. These limitations are specified in the regulatory documents [17,18].

In order to assess potential hazard of generated RW it is reasonable to apply the radiation equivalence principle involving evaluation of the time period needed for the environmental hazard of generated RW to become equal to the environmental hazard of raw materials extracted for manufacturing of nuclear fuel producing these RW in the course of its reactor irradiation and processing [98]. It is obvious that the shorter the radiation equivalence achievement period is the less potential hazard generated RW pose. Assessment demonstrated that the fraction of minor actinides had the maximum effect on the radiation equivalence achievement period from among all above-mentioned fractions of radioactive isotopes present in RW from SNF processing. The requirement for extraction of at least 99.9% of minor actinides in the course of SNF processing is introduced in

Proryv Project Framework. Extracted minor actinides are planned to be included into the fuel composition subsequent to controlled interim storage and after-burnt in fast reactors or specially developed after-burning reactors. Elimination of minor actinides from the RW composition sharply reduces potential hazard of generated RW, but does not solve the problem associated with the need for deep underground disposal of RW from SNF processing.

Extraction of other fractions present in SNF does not result in any considerable changes of the disposal conditions. However, significant reduction in the final amounts of disposed RW is possible. This is due to the fact that cesium and strontium isotopes with the half-life of about 30 years provide for the main radiation burden on glass (after extraction of minor actinides). Thus two-stage RW processing can be arranged. Glass-like matrix with the content of cesium and strontium isotopes exceeding the permissible value by 4 times is made at the first stage. The matrix is sent for controlled storage, and in 60 years the matrix is processed without any volume changes after four-fold reduction of the main dose-generating isotope activity and sent for final disposal.

21.5 HLW vitrification equipment

EP-500 melters. EP-500 furnaces with direct resistance heating of molten glass are traditionally used at Mayak PA for vitrification of RW from SNF processing. The capacity of such furnaces is up to 500 kg of glass per hour. The fifth EP-500/5 furnace is in operation in 2019. The drawback of the furnaces used resides in impossibility to replace them in case of any assembly failure due to the design and dimensions of these furnaces.

Induction melter with cold firepot. An induction melter with cold firepot is planned to be used for HLW vitrification at the SNF processing module developed within the framework of Proryv Project Framework. The capacity of the developed unit is small (up to 7 kg of glass per hour) but sufficient taking into account small capacity of the processing module. The cold firepot advantages in comparison with similar equipment of different design reside in the fact that the produced molten glass mixed with radioactive products does not have any direct contact with the parts of the unit, as there is a layer of unmelted glass (skull) between the molten glass and the metal firepot basket. This fact facilitates long-term operation of the firepot.

The prototype vitrification unit has been developed (up to the basic design stage) by 2019, the full-scale model of the pilot equipment for vitrification of HLW from hydro-metallurgical and pyrochemical SNF processing stages has been manufactured, melting of glass with various compositions has been tested, the unit testing in different modes has been carried out, and continuous operation of the unit for 8 days has been checked.

Peculiarities of the vitrification unit design:

- Remote control (automated instrumentation and control system);
- Remote replacement of assemblies with exhausted life span;

- Unified design of equipment – single-type units for hydro- and pyro-processing HLW; difference in the dosing and gas treatment systems, different temperature regimes of melting;
- Use of borosilicate glass for the HLW from hydro-metallurgical processing and borophosphate glass for pyrochemical processing HLW;
- Capacity – up to 7 kg per hour.

21.6 RW from the production facilities with increased plutonium content (as exemplified by PDEC nitride nuclear fuel fabrication module)

Analysis of the isotope composition and activity of liquid radioactive waste (LRW) generated at the mixed nitride uranium-plutonium fuel fabrication module constructed within the framework of Proryv Project Framework at the site of Siberian Chemical Plant JSC demonstrated that subject to concentration of trans-uranium elements (TUE) in the LRW at the level of 1 mg/L the final forms of generated RW (SRW produced by cementation of these LRW) were classified as intermediate-level long-lived TUE-containing RW. This class of RW is subject to deep underground burial [6]. The method of alkaline deposition of TUE from the LRW solution with subsequent purification of the solution from suspended matter through cross-flow filtration was proposed for LRW purification from TUE in the fabrication module. Experiments performed in VNIINM JSC, RI JSC and Mayak PA demonstrated that purification of the solution from TUE up to the concentration of less than 1 μg/L for plutonium and less than 0.2 μg for americium could be achieved by the initial solution acidity adjustment. The produced residue after decantation is heat-treated and supplied to the storage facility for controlled storage under nuclear-safe conditions and subsequent processing with return of plutonium into the process. LRW purified from TUE are cemented and sent for disposal (sub-surface).

21.7 Disposal of radioactive waste

In accordance with the effective legislation RW shall be converted into the form suitable for disposal [16] and buried at the special-purpose burial sites (except for the above-mentioned special RW subject to local disposal). The RW burial sites shall provide their reliable isolation from the environment within the time period when RW are hazardous for the environment, that is, the unstable materials present in the RW are classified as RW in accordance with their physical state and activity [99]. The classification introduced in the document [6] divides the RW suitable for disposal into six classes. RW classes are specified in Table 21.1.

The multibarrier system combining natural barriers (these barriers are usually represented by the geological environment and sub-surface soil) and engineering barriers (the primary RW package; the container material; additional engineering barriers provided by the storage facility design) is used to isolate the radionuclides contained in RW from the environment.

TABLE 21.1 Classification of disposed RW.

RW class for disposal	RW types sent for disposal	Type of disposal facility
Class 1	Solid HLW subject to disposal in the geological RW repositories with preliminary storage for reduction of their heat emission (residual heat not exceeding 1 kW/m^3)	Geological repository
Class 2	Solid HLW and solid long-lived ILW subject* to disposal in the geological RW repositories without any preliminary storage for heat emission reduction, including radionuclide SSS of hazard classes I and II**	Geological repository
Class 3	Solid ILW and solid long-lived LLW subject* to disposal in the subsurface RW repositories located at the depth of up to 100 meters from the ground surface, including ionizing radiation SSS of hazard class III in accordance with NP-067-11.	Sub-surface RW repository
Class 4	Solid LLW subject* to disposal in the sub-surface RW repositories located at the ground surface level, including radiation SSS of hazard classes IV and V**	Sub-surface RW repository with relaxed requirements
Class 5	Liquid ILW and LLW subject* to disposal in the geological RW repositories (NIIAR JSC, MCP FSUE, Siberian Chemical Plant JSC)	Deep well injection land
Class 6	RW generated in the course of uranium ore mining and processing and also in mining and processing of mineral and organic raw materials subject* to disposal in the sub-surface RW repositories	Subsurface RW repository with relaxed requirements at the RW generation site

Note. SSS, spent sealed sources.
**In accordance with the acceptability criteria established by the federal regulations and rules governing RW management.*
***Hazard classes are established in accordance with the federal regulations and rules in the area of atomic energy use.*

Final disposal in the geological formation is considered the most acceptable technique for isolation of high-level waste. But the majority of countries have faced many difficulties (primarily of political and social nature) in practical implementation of this concept.

The site for the federal repository in Nevada (Yukka-Mountain group) was selected in the USA within the framework of the national program for removal of irradiated fuel and high-level waste. But the repository commissioning deadlines were shifted several times due to antagonism of the local authorities. The US Department of Energy planned to commence storage of irradiated NPP fuel in the repository in 2010 and HLW vitrification – in 2015. According to assessment of the US Department of Energy the works for potential RW disposal site selection in the USA alone will cost 6 bln. USD.

In France public outcry in 1989–1990 resulted in suspension of the site selection for disposal of high-level waste. The final decision was postponed for at least 15 years.

In Germany the works for arrangement of the HLW geological repository in the salt dome in Gorleben were suspended under the pressure of public opinion in 1991. Initially it was planned to open the repository in 1998, and then this date was shifted to 2008 and further.

In Japan the site for HLW repository on the island of Hokkaido was selected in 1984. But objections of the local residents prevented the government from issuance of the repository construction permit.

In Russia, the site in the Nizhnekansk granitoid massif (Krasnoyarsk Territory) was selected for geological burial of RW from SNF processing. Long-term investigations of the massif geological structure, tectonic peculiarities, and underground water travel patterns preceded the final site selection. The set of works for predictive mathematical modeling in relation to migration of radionuclides within the disposal zone was carried out. At present the integrated safety analysis program for development, operation, and decommissioning of the geological RW repository in the Nizhnekansk massif is implemented (the term "repository decommissioning" means termination of any RW acceptance in the full repository and transition to the repository sealing stage with subsequent monitoring of its state) [6]. In accordance with this program construction of an underground research laboratory was commenced in 2018 and planned to be completed in 2025. In case of the positive results with regard to safety of the arranged geological repository it is planned to commence construction of the first stage in 2030, and beginning of operation (acceptance of RW for disposal) is expected not earlier than 2035.

Of course, the requirements for the disposal conditions of RW with lower potential hazard for the environment (having low specific activity or consisting of short-lived isotopes) are less strict (subsurface burial sites with relaxed requirements for the engineering barriers in the radionuclide migration paths are permissible). The regional network of RW burial sites accepting RW of classes

3 and 4 for disposal exists in Russia. Besides, such RW repositories are arranged at the majority of large nuclear industry enterprises.

Underground injection of liquid radioactive waste was carried out in three places in Russia. This is done at the production sites of chemical plants in Zheleznogorsk, Seversk and at the disposal site of NIIAR near Dimitrovgrad.

Liquid waste is injected under pressure to the depth of several hundreds of meters. The preliminary geological survey makes it possible to predict that LRW disposed in this way will not pose any hazard for the environment and people within the foreseeable future – up to 1000 years.

Despite absence of any serious incidents associated with underground RW injection in the USSR and later in the Russian Federation this RW disposal technique is not supported by international organizations, in particular the IAEA.

Part IV

Advanced reactor technologies and the nuclear power engineering infrastructure

Chapter 22

New generation reactor technologies within the framework of Generation IV International Forum

Generation IV International Forum (GIF) was arranged at the initiative of the USA with the declared aim to organize international cooperation between the countries developing nuclear power in the area of R&D works carried out to substantiate the designs of generation IV nuclear power systems and exchange of the obtained results. De facto it was the mechanism for control on the part of the USA and appropriation of innovative developments from other countries. To a certain extent GIF was the response to INPRO project arranged at the initiative of the Russian Federation and initially aimed at practical cooperation in acquisition of the new technology platform based on reactors with inherent safety and the CNFC. More than 10 countries joined the project. Under the prohibition for NPP SNF processing this meant falling out of the global promising development trends for the USA. Later INPRO project was restructured into the analytical variant fruitless on the practical level under the pressure of the USA and with submission of Russia.

Six innovative reactor technologies were defined within the framework of GIF:

1. Sodium-cooled fast reactors;
2. Lead-cooled fast reactors;
3. Gas-cooled fast reactors;
4. Water-cooled reactors with supercritical parameters;
5. Molten salt reactors;
6. High-temperature gas-cooled reactors.

Sodium-cooled fast reactor (SFR). Developments of SFR reactors are carried out with both traditional three-circuit heat removal scheme with the steam-and-water third circuit and Brighton gas cycle. Integral, looped, and modular concepts of the reactor plant layout are considered. Moderate and familiar

Closed Nuclear Fuel Cycle with Fast Reactors: Handbook of Russian Nuclear Power.
DOI: https://doi.org/10.1016/B978-0-323-99308-1.00015-5

parameters of the coolant temperature (up to 550°Cs) are adopted for the designs. Great attention is paid to safety enhancement through maximum utilization of the inherent safety properties of fast reactors and safety systems based on the passive operating principles.

The main characteristics of individual SFRs considered within the framework of GIF are presented in Table 22.1. Presentation of the Chinese concept of CFR1200 is planned for the nearest future.

Examples of the integral and looped SNF concepts implemented in the foreign projects within the framework of GIF are presented in Figs. 22.1–22.3. Reactor plants for the modular concept are developed in the integral layout.

Basic areas of R&D work for SFRs:

- Safe operation, development of the design safety criteria for generation IV SFRs;
- Improved fuel;
- Design of equipment and the power conversion unit;
- International demonstration of the global actinide cycle;
- System integration and assessment of SFRs for compliance with the requirements for generation IV installations.

In September 2017, the concept of the Russian BN-1200 reactor was officially adopted and approved by all members of the Governing Board as compliant with the GIF requirements. It will be described in detail below.

Lead-cooled fast reactor. Three designs are considered as reference ones within the framework of GIF:

- Lead-cooled fast reactor ELFR (Euratom);
- BREST-OD-300 reactor (Russia);
- Modular lead-cooled reactor SSTAR (USA).

Development of ALFRED experimental reactor with the thermal power of 300 MW planned to be located in Romania is carried out by FALCON Corporation with support of Euratom. MYRRHA fast-neutron research ADS installation with lead and bismuth coolant is constructed in Belgium.

Layout options for the lead-cooled reactors considered within the framework of GIF are presented in Fig. 22.4. The main design characteristics of ELFR, BREST-OD-300 and SSTAR are given in Table 22.2.

Gas-cooled fast reactor (GFR). The Project agreement on the conceptual design and safety is effective within the framework of the GIF systemic arrangements with regard to GFRs. The high-capacity GFR concept (2400 MW_t) is considered as the reference one (Fig. 22.5).

Crisis is observed in this area of activities due to lack of financing, and thus the works on the fuel and core materials are suspended.

At present the primary focus is on development and implementation of ALLEGRO experimental reactor project.

TABLE 22.1 Key parameters of the generation IV SF reactor concepts announced within the framework of GIF.

Design parameters	JSFR (Japan)	KALIMER (Korea)	SMFR (USA)	ESFR (Europe)	BN-1200 (Russia)
Electric power, MW_e	1500	600	50	1512	1220
Thermal power, MW_t	3570	1500	125	3600	2800
Primary circuit configuration	Looped	Vessel	Vessel	Vessel	Vessel
Power unit efficiency factor, %	42	40	~38	42	43.5
Coolant temperature at the core outlet, °C	550	545	~510	545	550
Coolant temperature at the core inlet, °C	395	390	~355	395	410
Main steam temperature, °C	503	503	480	490	510
Main steam pressure, MPa	16.7	16.5	20	18.5	17.0
Campaign duration, years	1.5–2.2	1.1	30	1.35	Up to 6
Number of refuelings	4	5	1	5	5–6
Core diameter, m	5.1	4.2	1.75	4.72	4.18
Core height, m	1.0	0.89	1.0	1.0	0.83/1.0
Type of fuel	MOX+MA	Metal (U-Pu-10%Zr)	Metal (U-Pu-10%Zr)	MOX	MNIT/ MOX
Structural material	ODS	HT9M	HT9	ODS	AS/FM/ODS
Enrichment by Pu, %	13.8	25.2	15.0	15.7	Up to 20
Burnup depth, GW·day/t	150	139	–87	150	Up to 100/125 (average)
Breeding ratio	1.0–1.2	0.74	1.0	1.0–1.2	1.35/1.2

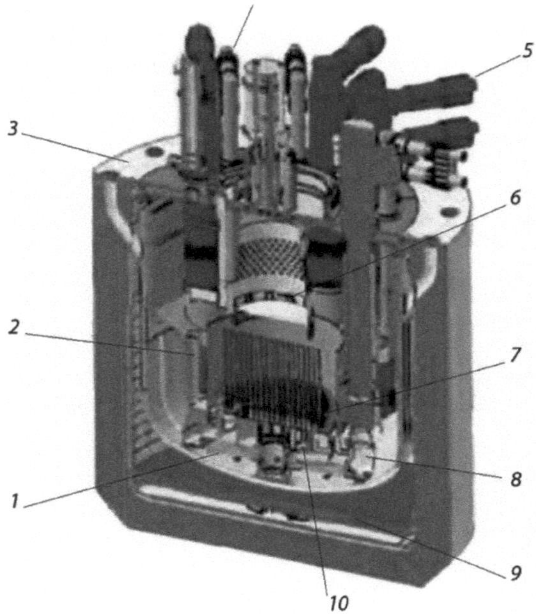

FIGURE 22.1 Integral SFR concept (ESFR): 1 – internal core catcher device; 2 – in-vessel refueling device; 3 – cellular formed cover; 4 – emergency heat exchanger; 5 – intermediate heat exchanger; 6 – central pivotable column; 7 – base plate; 8 – pump; 8 – external core catcher device; 10 – stiffeners.

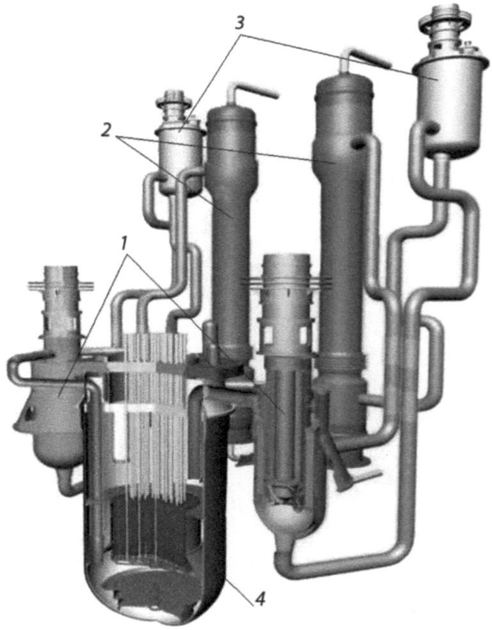

FIGURE 22.2 Looped SFR concept (JSFR): 1 – RCP-1/ IHE; 2 – SG; 3 – RCP-2; 4 – reactor tank.

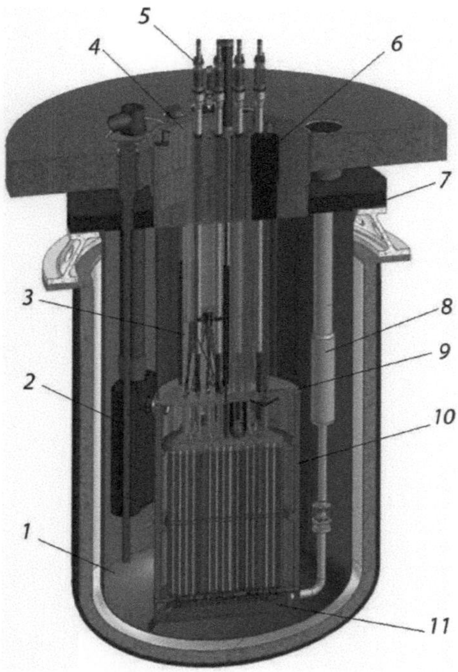

FIGURE 22.3 Modular SFR concept (AFR-100): 1 – reactor tank; 2 – IHE; 3 – central pivotable column; 4 – rotating plug; 5 – operating mechanisms of the control rods; 6 – removable plug; 7 – reactor tank cover; 8 – electromagnetic pump; 9 – core cover; 10 – core shell; 11 – pressure chamber.

Supercritical water-cooled reactor (SCWR). R&D works in two areas are considered within the framework of GIF in relation to SCWRs:

1. Thermal hydraulics and safety;
2. Materials and chemistry.

Three main conceptual SCWR options are considered within the framework of GIF:

- The concept of the European pressurized three-way-entry SCWR (HPLWR);
- The Canadian SCWR concept with pressurized pipes (SCW CANDU, Fig. 22.6A);
- The Chinese SCWR concept with pressurized vessel (CSR1000, Fig. 22.6B).

Molten salt reactor (MSR). Developments in the two main areas are carried out within the framework of GIF in relation to MSRs:

1. Reactors with molten salt fuel (fuel as the circulating liquid molten material);
2. Reactors with solid fuel and coolant on the basis of molten salts.

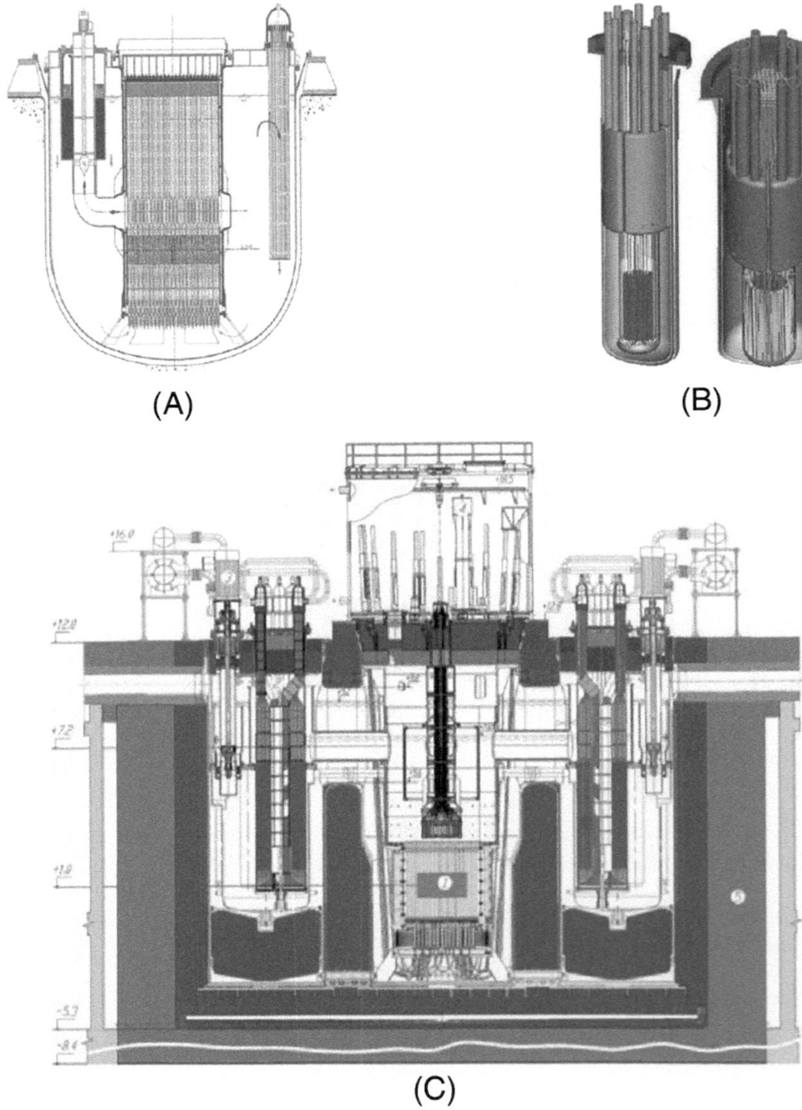

FIGURE 22.4 Layout of the primary circuit in the designs of lead-cooled fast reactors: (A) – ELFR; (B) – SSTAR; (C) – BREST-OD-300.

The following MSR concepts are considered within these two areas:

- The MSR concept with fast neutron spectrum (MSFR) developed by France and Euratom (Fig. 22.7A);
- The Russian concept of a molten salt MA burn-out reactor (MOSART) with and without uranium-thorium make-up (Fig. 22.7B);

TABLE 22.2 Main characteristics of the generation IV lead-cooled fast reactor concepts considered within the framework of GIF.

Design parameters	ELFR	BREST-OD-300	SSTAR
Electric power, MW_e	600	300	20
Thermal power, MW_t	1500	700	45
Primary circuit configuration	Vessel	Vessel	Vessel
Power unit efficiency factor, %	42	42	44
Coolant temperature, °C at the core outlet at the core inlet	480 400	540 420	567 420
Secondary circuit working medium	Steam-water	Steam-water	CO_2
Temperature at the turbine inlet, °C	450	505	553
Secondary circuit working medium temperature at the SG/TI inlet, °C	335	340	402
Pressure at the turbine inlet, MPa	18.0	18.0	2.0

FIGURE 22.5 Layout concept for high-capacity GFR.

- The concept of a high-temperature reactor with solid fuel cooled by molten fluoride salts (FHR) considered in the USA and China;
- The concept of a thorium MSR (TMSR) with graphite moderator (China, Japan and the Republic of Korea).

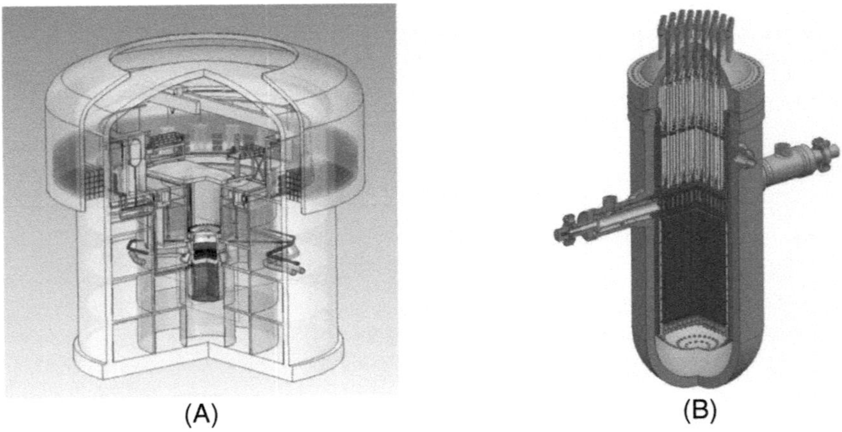

FIGURE 22.6 Conceptual SCWR options: (A) – SCWR reactor building (Canadian SCWR concept); (B) – SCWR vessel (Chinese concept CSR1000).

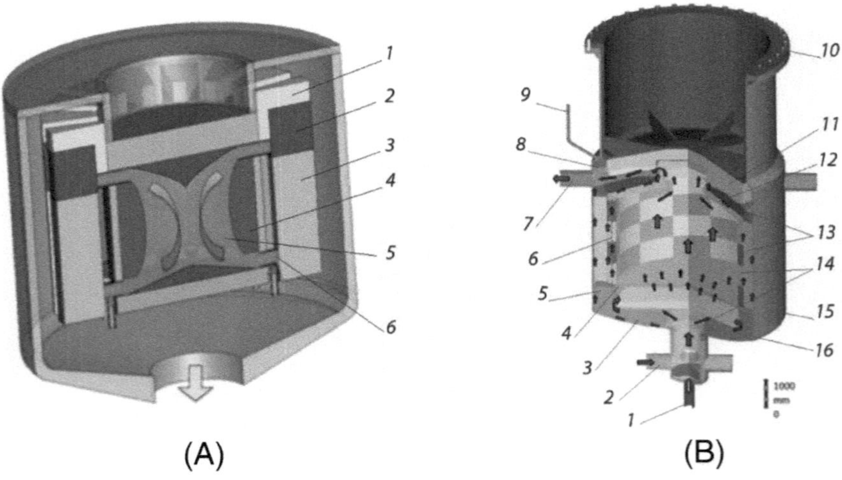

FIGURE 22.7 Concepts of MSR with molten salt fuel: (A) (MSFR): 1 – fuel salt processing unit; 2 – pumps; 3 – heat exchangers; 4 – blanket salt; 5 – core salt; 6 – release of bubbles; (B) (MOSART): 1 – fuel salt drain line; 2 – fuel salt inlet; 3 – protection; 4 – distribution grill; 5 – support ring; 6 – reflector cooling system; 7 – fuel salt outlet; 8 – fuel salt level; 9 – gas utilities; 10 – cover; 11, 13 – protection; 12, 14 – reflector; 15 – vessel; 16 – fuel salt inlet window.

High-temperature gas-cooled reactor (HTGR). The following areas are considered within the framework of GIF R&D works in relation to HTGRs:

- Hydrogen production;
- Fuel and fuel cycle;
- Materials;
- Validation and benchmarks of the calculation methods.

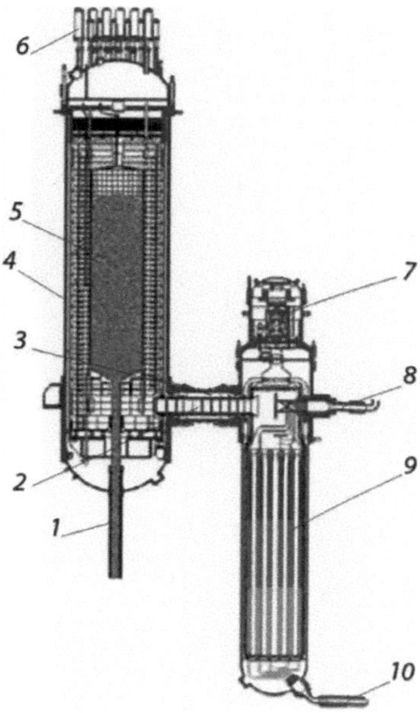

FIGURE 22.8 Flow diagram of the HTGR primary circuit: 1 – fuel discharge pipe; 2 – hot gas pipe; 3 – cold helium flow; 4 – reactor pressure vessel; 5 – core with spherical fuel rods; 6 – CPS control mechanisms; 7 – helium pump; 8 – steam outlet; 9 – steam generator; 10 – feedwater.

The basic flow diagram of the HTGR primary circuit is presented in Fig. 22.8.

At present there are two operating HTGR plants in the world: HTTR plant in Ōarai Research Center (Japan) and HTR-10 experimental plant in Tsinghua University (China).

Chapter 23

Development of technologies based on fast reactors

23.1 Fast reactor development stages in Russia

Russia has the maximum competencies in fast reactor (FR) technologies continuously developing with accumulation of positive experience in design, construction and operation of sodium-cooled fast reactors (Fig. 23.1).

At present two fast power reactors with sodium coolant (BN-600 and BN-800) as well as BOR-60 experimental reactor are operated in Russia. These reactors enabled to gain the experience in design, construction, and operation of this reactor type offering the possibility to proceed with commercialization of the fast reactor technologies and to develop two-component nuclear power.

In general, the program for development and commercial application of the fast reactor technologies consists of the following successive stages:

- Accumulation of the operation experience and acquisition of decommissioning technologies for BN-600 RP;
- Acquisition of mixed uranium-plutonium fuel and basic CNFC technologies with the use of BN-800 RP;
- Development of the multipurpose fast research reactor (MBIR) as a new experimental facility for substantiation of materials and fuel cycles for fast reactors (in replacement of BOR-60 reactor);
- Development of the commercial power unit with BN-1200 RP competitive in comparison with thermal reactors;
- Development and practising of the technology for lead-cooled fast reactors implementing the inherent safety properties to the full extent on the basis of BREST-OD-300 pilot demonstration RP;
- Development and implementation of the complete closed nuclear fuel cycle within the pilot demonstration experimental energy complex;
- Development of the industrial power supply complex design, including 1–3 FNR-1200 (both sodium- and lead-cooled) and the CNFC facilities as the industrial reference sample of the new nuclear power technology platform capable of competing with traditional and renewable energy sources.

Closed Nuclear Fuel Cycle with Fast Reactors: Handbook of Russian Nuclear Power.
DOI: https://doi.org/10.1016/B978-0-323-99308-1.00028-3

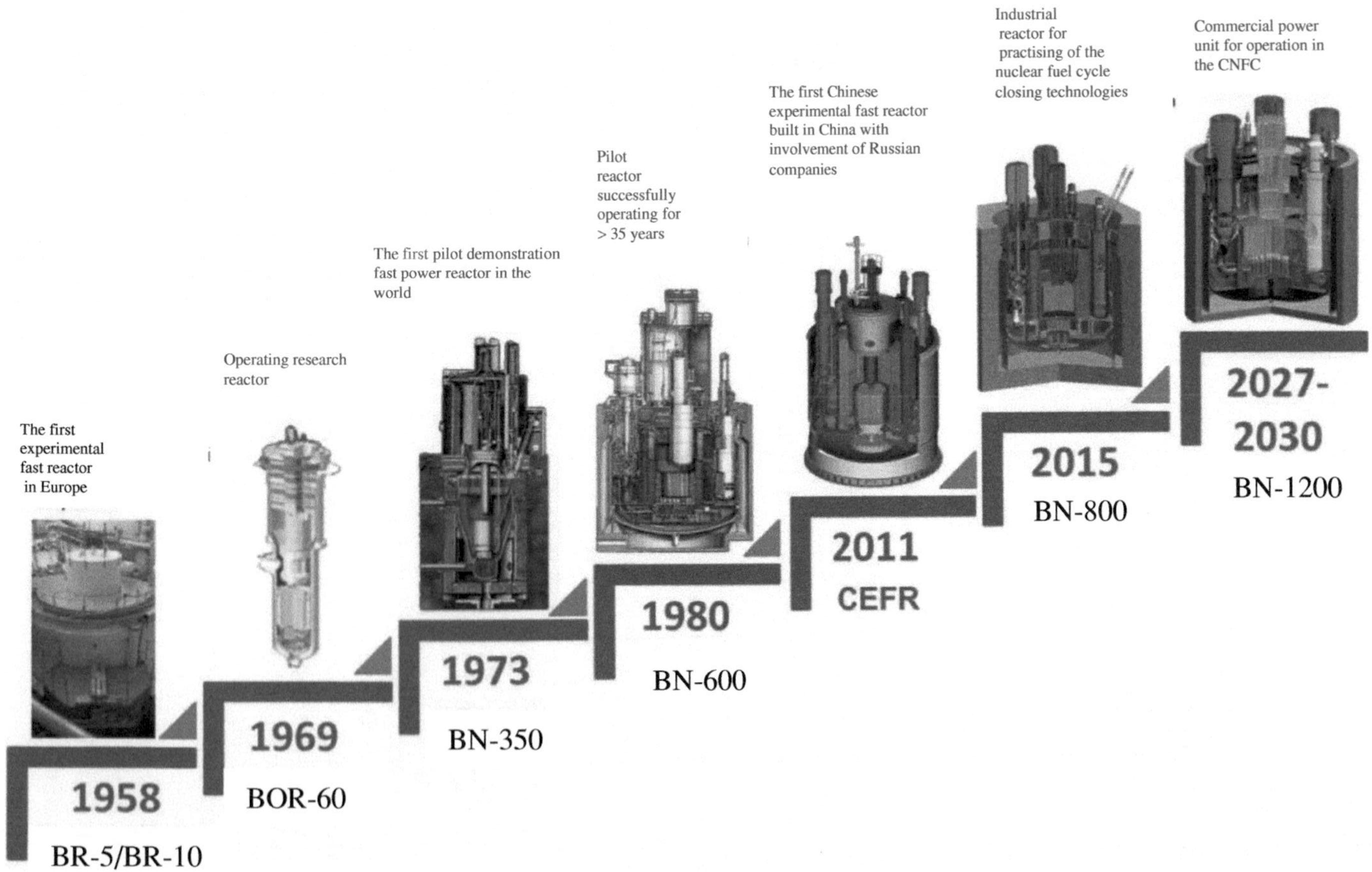

FIGURE 23.1 Stages of the fast reactor technology development.

The power unit with a BN-1200 RP is considered as the first commercial power unit in the world intended for operation within the CNFC and comparable with the best counterparts based on thermal reactors with regard to the LCOE values subject to achievement of the safety level compliant with the requirements for generation IV power units.

The pilot demonstration experimental energy complex should demonstrate stable operation of all CNFC technologies aimed at the short fuel cycle typical for fast reactors for the first time in the world. It includes BREST-OD-300 RP, the fabrication/refabrication module for mixed nitride uranium-plutonium fuel (MNIT) and the fuel processing module.

The third major project includes development of the commercial power unit with lead-cooled BR-1200 reactor plant within the industrial power supply complex where additional reduction of capital costs and construction period by 20% in comparison with the parameters achieved in VVER-TOI is planned.

The analysis results with regard to the integral LCOE parameter for BN-1200M and BR-1200 power units (at 2018 values) taking into account the influence of risks are presented in Fig. 23.2.

The results prove that the LCOE value is comparable or even exceeds the parameters for thermal reactors with steam generator units, subject to the optimal characteristics of power units that can be achieved with due regard for successful implementation of the lead- and sodium-cooled FNR technology development programs.

For comparison, LCOE of VVER-TOI and VVER-S (disregarding the influence of risks) amounts to 1882 RUB/MW–h and 1750 RUB/MW–h respectively, and LCOE of CCGT-410 at the site in Kursk region is 1980 to 2330 RUB/MW–h with the capacity utilization factor of 65 and 50% respectively (at 2018 values).

23.2 BN-800 reactor and establishment of the closed NFC

The maximum success in operation of fast reactors worldwide for approximately 40 years was achieved at BN-600 reactor of Beloyarsk NPP (Fig. 23.3A). It definitively confirmed acquisition of the sodium technology with high parameters of operational reliability.

The mission of new BN-800 reactor (Fig. 23.23B) commissioned in 2015 is to practice Russian innovative CNFC technologies (Federal Special Program "Development of the Russian Nuclear Power Generation Complex for the Period of 2007–2010 with Further Extension till 2015" approved by Resolution of the Government of the Russian Federation No. 605 dated October, 6, 2006). BN-800 provides the unique opportunity for Russia to practice the key physical and technology CNFC elements and to save time as one of the main resources for technical implementation of the CNFC within the framework of Proryv Project [45].

It is time that represents one of the main restricting factors (together with financial and technology aspects) for commercialization of the CNFC

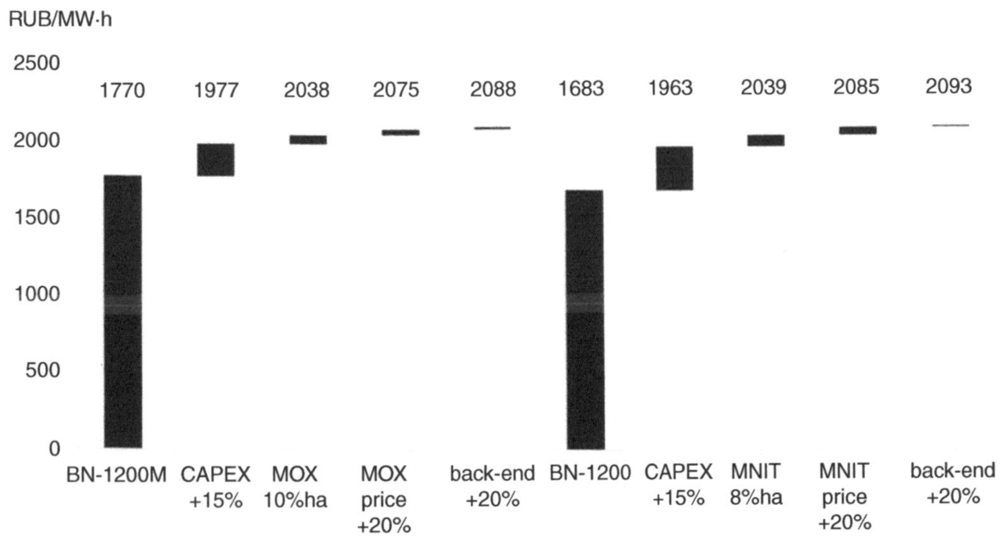

FIGURE 23.2 Integral LCOE parameter for BN-1200M and BR-1200 power units with due regard for influence of any risks related to changes of the FNR technical and economic parameters.

(A) (B)

FIGURE 23.3 Power units No. 3 with BN-600 RP (A) and No. 4 with BN-800 RP (B) of Beloyarsk NPP.

technologies taking into account duration of the reactor fuel testing and practicing of its multiple recycling.

The main objectives of BN-800 with regard to acquisition of the CNFC technologies are the following:

- The first transition to mixed uranium-plutonium fuel (MOX and MNIT) in Russia on the basis of plutonium from SNF of VVER reactors and gaining of the operating experience for transition to the two-component structure of nuclear power eliminating SNF accumulation;
- Demonstration of the multiple nuclear fuel recycling from FNRs with BR>1 as the key element of fuel supply for large-scale nuclear power within any required time period;
- Substantiation of the concept and practicing of "equilibrium fuel" with MNIT fuel with BRC of ~ 1 as the key element for nuclear safety enhancement in the new generation reactors;
- Substantiation and demonstration of minor actinide transmutation in FNRs as the key elements of the CNFC environmental safety.

Thus the relevant R&D program with the use of BN-800 reactor will enable to solve the postponed problems with regard to SNF, to substantiate and implement the optimal technique for disposal of long-lived RW without any additional special-purpose after-burning plants and with maintenance of the natural radiation balance in the course of their disposal.

But BN-800 is not capable of the CNFC establishment for two-component nuclear power due to commercial noncompetitiveness in comparison with thermal reactors and alternative generation.

Chapter 24

Fast reactors within the framework of Proryv Project framework

24.1 Power unit with BREST-OD-300 pilot demonstration reactor plant

The power unit with BREST-OD-300 RP with dense nitride fuel and high-boiling lead coolant is developed within the framework of Proryv Project Framework for establishment, implementation and demonstration of the innovative inherent safety nuclear power technologies based on fast reactors and their closed nuclear fuel cycle. BREST-OD-300 RP is considered as the prototype of future commercial fast reactors of BREST type for large-scale nuclear power capable of accounting for the basic electric power generation increase and solving the problem of energy supply for sustainable development of the humankind for many years ahead.

The power unit is planned to be operated within PDEC Complex with the near-plant closed nuclear fuel cycle. Practicing of the full process cycle is the second main objective of its arrangement. The main technical characteristics of the power unit and the reactor plant are given below.

The power unit, the reactor plant, and the main equipment are innovative and have no counterparts in operation. Selection of the main technical characteristics and engineering solutions in BREST-OD-300 RP including the power level of 700 MW(t), the specific lead coolant circulation system, the normal operation and emergency cooldown systems, etc. was determined not only by the objective to demonstrate the inherent safety properties of this reactor technology but the requirements for succession of engineering solutions in the future high-capacity RP designs were also taken into account.

Closed Nuclear Fuel Cycle with Fast Reactors: Handbook of Russian Nuclear Power.
DOI: https://doi.org/10.1016/B978-0-323-99308-1.00010-6

FIGURE 24.1 BREST-OD-300 reactor plant.

Main characteristics of the power unit and BREST-OD-300 RP

Rated thermal/ electric power, MW	700/300
Capacity utilization factor	0.8
Duration of operation between refuelings, ef. Days	300
Coolant temperature at the core inlet/outlet, °C	420/535
Steam temperature/pressure, °C/MPa	505/17
Efficiency factor, %	43.5
Primary coolant volume, m^3	900
Seismic resistance, DBE/SSE, points	7/8
Service life, years	30
Industrial production personnel size	316

BREST-OD-300 reactor plant is the main innovative solution within the power unit. Its scheme is presented in Fig. 24.1. The peculiarity of the reactor unit resides in its design with integral layout of the lead circuit containing the core with reflectors and CPS devices, steam generators (SGs), pumps, lead purification systems and oxygen regime maintenance systems, as well as other auxiliary equipment located in the steel-lined central and four peripheral (according to the number of lead coolant circulation loops) cavities of the metal and concrete casing with the cooling system (Fig. 24.2). The temperature of the external casing surface at the border with the civil structures is maintained within the acceptable limits through the use of the cooling system with natural air circulation.

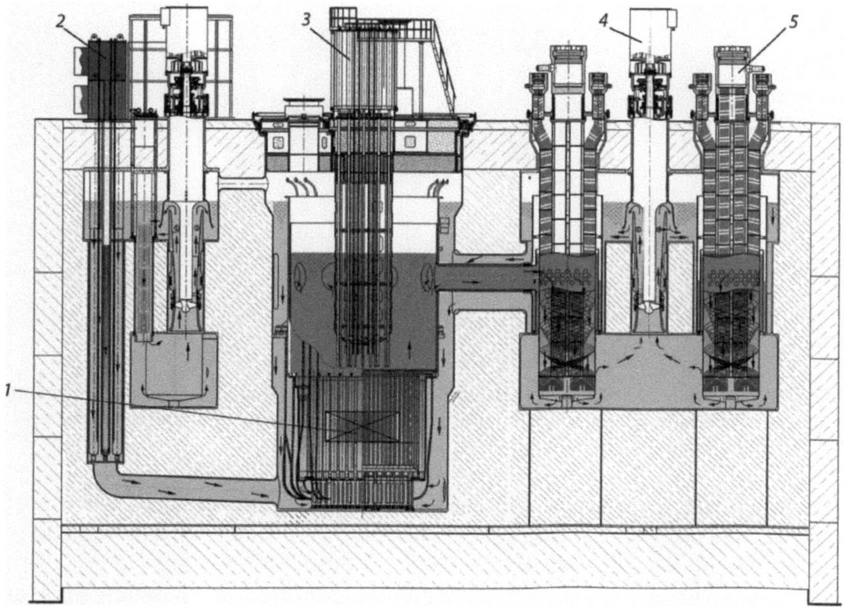

FIGURE 24. 2 Coolant circulation scheme in BREST-OD-300 RP: 1 – core; 2 – cooldown heat exchanger; 3 – casing of the CPS actuators; 4 – RCPS; 5 – steam generator.

Forced lead circulation through the core is ensured due to "cold" and "hot" coolant level difference provided by the pumps. Lead cooled in the SG is supplied to the upper (pressure) level through the use of pumps and then is supplied to the core inlet via the downcoming ring section of the circuit within the central cavity; in the core it is heated in the course of flowing through the core from the bottom upward. Then lead is supplied to the SG and flows to the suction chambers through the shell side giving up the heat to the secondary circuit water and steam, and then is lifted to the empty upper level by the pumps. Lead coolant contacts with the gas circuit at the pump outlet where steam separation can be arranged in case of any SG pipe leakage.

The secondary circuit feed water heated with main steam in the high-pressure mixing heater to the temperature of 340°C is supplied to the inlet of each SG with the pressure of 18.5 MPa. It prevents the lead coolant temperature drop below its melting temperature (327°C) in the course of reactor start-up modes and in emergency situations. The secondary circuit peculiarity resides in the fact that it does not perform any safety functions for emergency heat removal from the reactor in contrast to the existing NPP designs.

The core of BREST-OD-300 RP consisting of 169 hexagonal fuel assemblies of jacketless design is arranged in the form of two radial zones – the central one (CZ) and the peripheral one (PZ) (Fig. 24.3). Spacing and fastening of

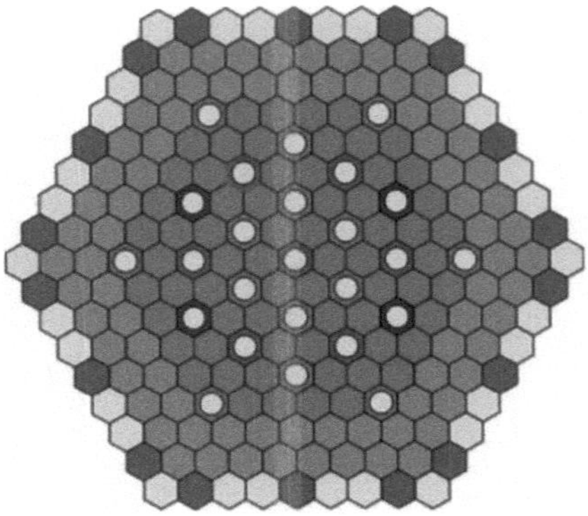

FIGURE 24.3 Core map: ● – CZ fuel assemblies; ●– PZ fuel assemblies; ○ – fuel assemblies with automatic control rods; ○ – fuel assemblies with shim rods; ○ – fuel assemblies with emergency rods; ● – passive feedback system; ○ – reflector.

rod-shaped fuel rods in the fuel assemblies are arranged through the use of grates. Fuel composition, the number and interval of fuel rods are the same in all fuel assemblies. Equalization of power, maximum temperatures of fuel rods, and coolant heating among the fuel assemblies is ensured due to radial profile of fueling and lead flow rate through the use of fuel rods with smaller diameter in the CZ fuel assemblies and larger diameter – in the PZ fuel assemblies. Use of the same fuel composition in all fuel assemblies provides for stability of equalized distributions in the course of the campaign subject to complete breeding of fissionable nuclides in the core (BRC of ~1).

Main characteristics of the BREST-OD-300 core

FA quantity in the reactor core, pcs	169
Maximum reactivity margin at power, βef	0.54
Temperature-power effect, %	–0.57
Average heat density, MW/m^3	175
Maximum linear load on a fuel rod, W/cm	410
Maximum fuel burnup, % h.a.	6*/9
Start-up MNIT fuel loading weight, t	20.8
Plutonium mass in the start-up loading, t	2.67
Fuel weight at refueling with the maximum burnup of 6%/9% h.a., t	7.2 / 4.8
Breeding ratio (BR and BRC)	1.05

* *For the initial stage of operation.*

Apart from fuel rods some CZ fuel assemblies contain control rods of the control and protection system (CPS CRs). The set of CPS CRs forms two independent reactor shutdown systems; one of them consisting of EP CRs represents the emergency protection system, and the other consisting of the compensation and automatic reactivity control rods (shim rods and automatic control rods) forms the second system. The CPS CR drives are located on the upper rotating plug, and the control rods themselves in the withdrawn position are situated under the core. In the course of refueling the drives are detached from the control rods, and they float up into the core and remain there under the impact of Archimedes force providing for deep sub-criticality of the reactor plant.

Usage of jacketless fuel assemblies ensures higher level of safety in comparison with jacketed ones. Heat removal from jacketless fuel assemblies in case of the coolant flow cut-off at the inlet is provided due to the coolant overflow from adjacent fuel assemblies. Calculations demonstrated that postulated coolant flow cut-off at the inlet of seven fuel assemblies of jacketless design in the central part of the core did not result in any fuel rod surface temperature increase exceeding the adopted acceptability criterion of 800°C.

The core is surrounded with rows of replaceable lead reflector blocks and steel protection; there is no blanket in the design. The reflector blocks are arranged in the form of a hexagonal dense steel casing filled with lead coolant with low circulation flow rate. Some lead reflector blocks are arranged in the form of devices similar to a gas bell where the lead column level "monitors" the coolant pressure (flow rate) and affects neutron leakage. Channels with lead columns representing the components of the passive feedback system (PFS) provide for relation of the reactor power (reactivity) to the coolant flow rate (pressure) in the core, enable to reduce the in-process reactivity margin on the control rods and represent an important safety factor through insertion of negative reactivity in case of any forced circulation reduction or termination.

Absence of the uranium blanket traditional for fast reactors and its replacement with lead reflector blocks prevent from production of weapon-grade plutonium (the technology measure for enhancement of the nonproliferation regime), provide for equalization of the spatial distribution of the neutron field and power in the core and eliminate any positive reactivity insertion in case of the lead level decrease in the circuit because of the coolant leakage or pump-out.

Substantiation of the main innovative equipment and system. Investigations of lead spreading along the metal-concrete border under the realistic conditions with regard to loads were carried out in order to substantiate the RP casing. Subsequent to the results of experiments the main lead spreading mechanism was discovered. The performed evaluation of the potential volume of cracks and void spaces in the drying system enables to predict that the molten lead ingress volume in case of any emergency leakage to the concrete casing filler (Fig. 24.4) does not exceed the permissible level.

Subsequent to the testing results supplements to the verification reports on the thermal analysis and strength calculation methodologies for the RP casing were developed. Besides, experimental works for determination of concrete properties: compression strength at 900°C, thermal conductivity, the thermal linear expansion coefficient, and temperature shrinkage were performed. These results were used for further substantiation of the reactor plant casing.

In order to substantiate the SG a series of fretting corrosion tests for the "pipe – spacer" joint and studies of the thermal and hydraulic characteristics of the 18-pipe model with lead coolant in the evaporation mode were performed for safety substantiation in case of any emergency feedwater supply cut-off.

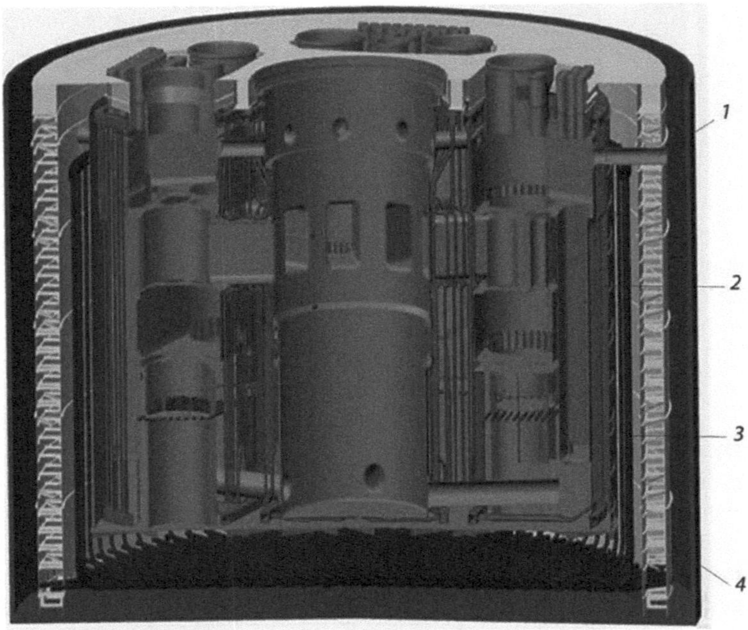

FIGURE 24.4 Scheme of the metal-concrete reactor casing: 1 – pipelines of the SG leakage confinement system; 2 – pipelines of the heat-up system; 3 – pipelines of the cooling system; 4 – structural shaft.

Corrosion and mechanical testing of the model weld joint samples as well as testing with model water and steam environment for determination of the effect on the corrosive and corrosion-mechanical properties of the SG base metal and weld joints were performed in order to substantiate the SG design (Fig. 24.5). Subsequent to the testing results the estimated value of the total surface corrosion allowance for the wall thickness of heat exchange tubes within the design lifetime of the SG was determined. The testing results confirm absence of any susceptibility to corrosion cracking of metal.

Experimental and theoretical study of fretting corrosion wear of the heat exchange tube friction pairs with their subsequent material research after testing was carried out. The experimental data were used to assess the parameters of wear rules and to adjust the developed physical and chemical model of wear contact interaction between a heat exchange tube and a spacer in the lead coolant flow.

The steam generator leakage confinement system (SLCS) (Fig. 24.6) is intended to protect the reactor unit against any exceedance of the permissible pressure limits in case of any operational occurrences caused by loss of the SG tube integrity. The SLCS consists of four branches connected to the gas cavity of the reactor unit. The gas cavity of the reactor unit includes the central gas

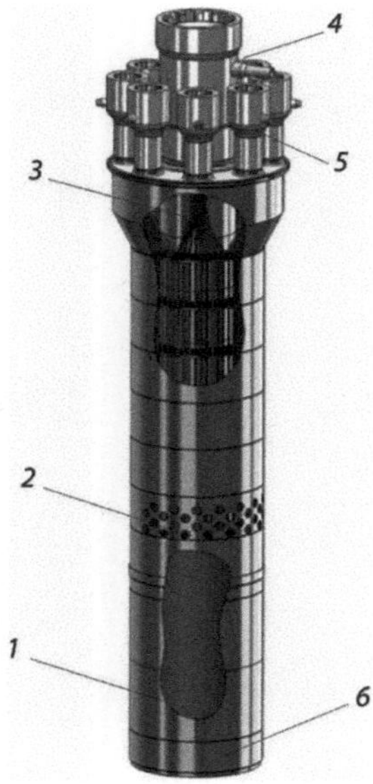

FIGURE 24.5 Steam generator scheme: 1 – heat exchange tubes; 2 – coolant inlet; 3 – mounting flange; 4 – feedwater chambers; 5 – steam chamber; 6 – shell.

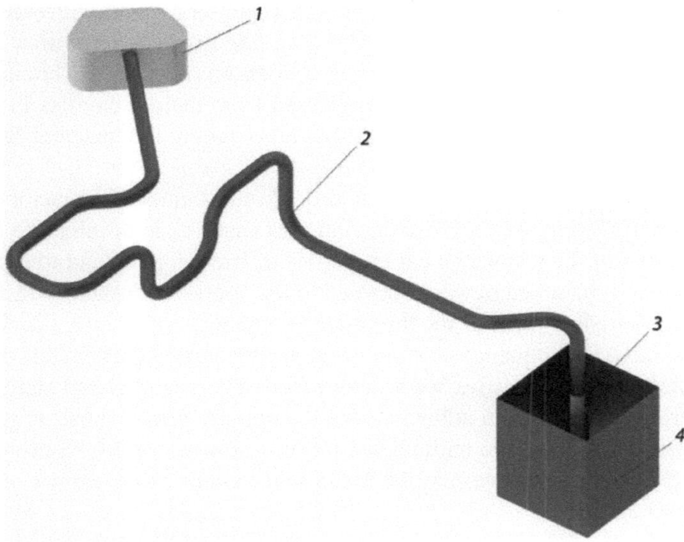

FIGURE 24.6 SLCS scheme: 1- peripheral cavity of the reactor unit; 2- SLCS pipeline; 3- circulation and bubbling device; 4- pressure suppression pool.

FIGURE 24.7 Scheme (A) and experimental model (B, C) of the reactor coolant pump: 1- casing sections; 2 – discharge; 3 – cooler; 4 – radial/axial bearing; 5 – motor; 6 – shaft; 7 – impeller.

cavity above the core and four peripheral gas cavities where the SG modules are installed. Each system branch includes a steam-and-gas mixture discharge pipeline (a SLCS pipeline) and a bubbler tank filled with water with a circulation and bubbling device where steam is condensed. Four SLCS branches are designed for condensing of the steam released from the gas cavities in case of simultaneous guillotine rupture of eight SG tubes taking into account failure of one SLCS branch.

The bearing tests were carried out in order to substantiate the reactor coolant pump set (RCPS, Fig. 24.7). Experimental data enabling to develop the method for evaluation of the tribotechnical properties of friction pairs and to assess the fatigue strength parameters were obtained in the course of durability testing.

Testing for adjustment of the dummy flow channels of different design was also continued. Subsequent to testing of the next experimental models pressure and cavitation characteristics were determined. The accumulated statistics on the experimental data from adjustment of the dummy flow channel of different design enables to select the optimal variant for testing of the RCPS prototype.

The following solutions may be noted *with regard to the power unit with BREST-OD-300 RP:*

- A double-circuit RP cooling and energy conversion system (the primary circuit is the RP lead coolant circuit, and the secondary one is the steam-and-water circuit for the medium supplied to the turbine);
- A full-speed condensing turbine of K-300-15,7/50 type;

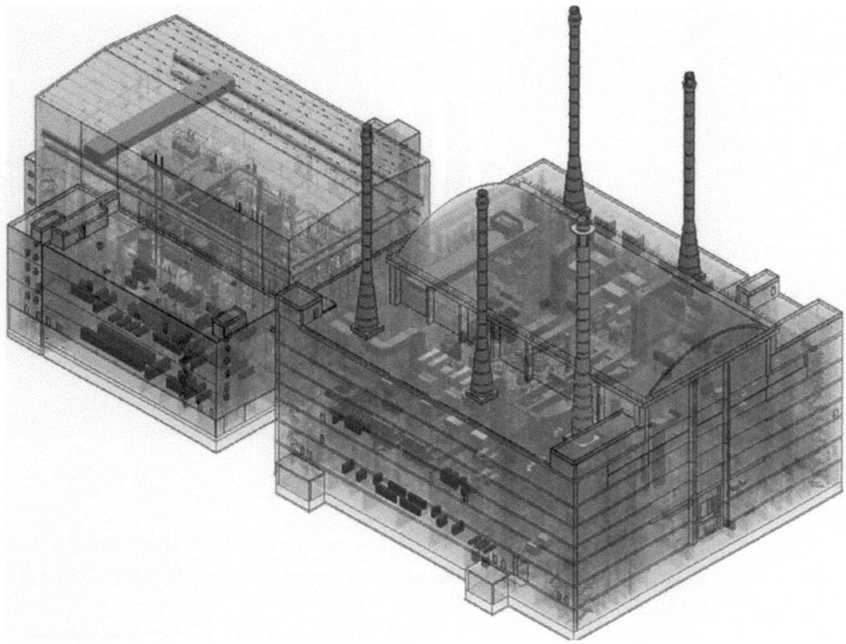

FIGURE 24.8 Three-dimensional scheme of the reactor building and the turbine hall.

- The secondary circuit scheme without any deaerator and a feedwater mixing heater providing the feedwater temperature of at least 340°C at the SG inlet;
- Recycling service water supply with a cooling tower.

The power unit design has no dedicated "nuclear island" and special building and provides for location of the main and auxiliary RP equipment in the same common building. The building design provides for "free" location of the equipment in the central reactor hall thus simplifying maintenance and ensuring the space required for practicing of the handling and repair operations at the first pilot demonstration power unit (Figs. 24.8 and 24.9).

Safety of the power unit with BREST-OD-300 RP. Combination of the properties of dense and thermally conductive nitride fuel (γ = 14.3 g/cm^3, X~ 18 W/m·deg.) with the properties of radiation-resistant, low-activated, high-boiling lead coolant (T_b >1700°C) passive in contact with water and air and not requiring high pressure in the circuit prevents any accidents with fires, steam and hydrogen explosions and enables the following in the design:

- To provide complete breeding of fissionable nuclides in the core, low ($\Delta\rho < \beta_{eff}$) and stable reactivity margin in the course of the micro-campaign;
- To prevent lead boiling with occurrence of the positive void reactivity effect and uncontrolled power and temperature increase in the core;

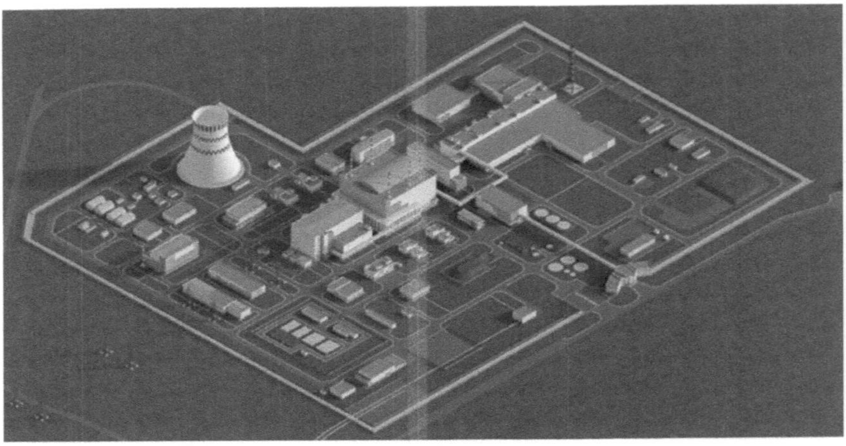

FIGURE 24.9 Layout of the power unit with BREST-OD-300 RP.

- To prevent coolant losses from the circuit in case of any RP casing damage due to integral arrangement of the circuit and self-curing of the cracks in concrete with solidified lead;
- To use high heat capacity of the lead circuit and moderate coolant circulation rate in order to slow down transient processes thus enabling to relax the requirements for fast response of the safety systems, to reduce temperature jumps and the probability of any fuel rod and SG tube damage;
- To use high albedo properties of lead in reflectors in order to equalize distribution of power and consequently maximum temperature of fuel rods and coolant heating in the fuel assemblies and also to reduce the critical mass of the loaded fuel.
- Low reactivity margin in the course of the RP operation ($\Delta\rho < \beta_{\text{eff}}$) radically enhances safety in case of any reactivity-related accidents by preventing the reactor transition into the critical state on prompt neutrons and catastrophic consequences of rapid and uncontrolled runaway even in case of full reactivity margin insertion.
- Low moderation of neutrons with heavy lead enables to expand the FE grid without any deterioration of the physical properties of the RP core and to increase the coolant flow passage and the level of power removed from the core through natural lead circulation. High heat capacity of the lead circuit and passive residual heat removal through natural air circulation via the air heat exchangers submerged into the lead coolant without any time limitations and with heat release into the atmosphere make accidents with fuel and lead overheating highly unlikely. In general, the aggregate probability of the BREST-OD-300 core damage from all initiating events on the basis of 24 h does not exceed 9×10^{-9} 1/year.

- The circulation scheme, high density and low velocity of lead in the downcoming section provide for high separation level of steam and gas bubbles and eliminate their ingress into the core in the volume capable of causing any uncontrolled increase of reactivity and power. In addition, in case of any disabling of one or several pumps nonuniformity of the coolant flow rate is reduced, and in case of rapid stop of all pumps within the period of more than 20 seconds inertia (run-down) of the flow rate in the core is ensured due to the different levels in the downcoming (cold) and raising (hot) lead sections. In case of disabled pumps heat removal from the core is arranged by natural lead circulation; in order to improve this circulation, the SG and the air heat exchanger tubes of the normal operation and emergency heat removal system are located above the core. Several auxiliary lines are provided in the circulation circuit (the RCP bypass for supply of filters and mass transfer equipment, the SG bypass for "underheated" lead from the reflector periphery, a circular header of the RCP pressure chambers, the core bypass for the "cold" coolant flow to the downcoming line of the SG tube bundle) in order to improve the circuit characteristics under normal operation conditions and in case of any transient processes.

In the limiting case of heat removal loss from the reactor core with disabling of four RCPs and feedwater supply cut-off in the course of operation at the rated and partial power levels it was demonstrated through experiments that no melting of fuel rod cladding and fuel took place, and the circulation circuit integrity was maintained. Release of fission products from the RP due to partial loss of fuel rod integrity amounts to not more than $6.14 \cdot 10^{10}$ Bq for the first day (does not exceed the reference level of daily releases under normal operation conditions). Probability of this scenario implementation is negligibly low ($\sim 2.8 \times 10^{9}$ 1/year).

The integral looped arrangement of the primary circuit is used in order to mitigate the consequences of any emergency situation with SG tube ruptures where the core with the adjacent structural components (in the central cavity) and the SG with the coolant circulation pumps (in the peripheral cavities) are spatially separated and hydraulically connected with pipelines forming four loops for heat removal from the core. Such arrangement together with the selected lead circulation scheme and steam discharge from the reactor vessel into the confinement system via hydraulic locks and then to the atmosphere via filters prevent any ingress of dangerous steam amounts into the core and overpressure in the reactor vessel.

Impossibility of any single SG tube rupture development into multiple one was demonstrated through experiments. The internal burst pressure in the SG tubes at 700°C is 185 MPa (53 MPa for the tubes with the residual thickness of 1 mm). The SG tube burst pressure value exceeds 40 MPa at the coolant temperature of 950°C even in case of weak spots (a patch with the depth of 2 mm).

Thus the risk of the most severe accidents with loss of heat removal and uncontrolled power increase resulting in unacceptable radioactivity releases requiring evacuation of the public may be drastically reduced or even eliminated naturally without any enhancement of expensive engineering barriers and safety systems, only due to peculiarities of the chain fission reaction neutron balance in the fast reactor, natural patterns, properties and qualities of its main components – lead coolant and fuel as well as the engineering solutions facilitating their implementation. This approach to self-protection in case of extremely severe accidents represents the essence of the inherent safety principle embodies in the characteristics and design of BREST RP.

As of 2019 development of the basic design of BREST-OD-300 RP including the engineering, computational and experimental works, substantiation of the adopted engineering and technology solutions and also adjustment of the engineering and substantiating documentation for the equipment and the entire RP is completed within the framework of the Federal Special Program "New Generation Nuclear Technologies" and Proryv Project Framework. The basic design of BREST-OD-300 RP and the design materials for the power unit with BREST-OD-300 RP underwent expert appraisal of Rosatom.

The positive statement of the RF State Expert Evaluation Department FAI was obtained for the capital construction facility "Construction of a pilot demonstration power unit with a lead-cooled fast reactor at the site of the restricted administrative unit of Seversk."

It is planned to solve the following tasks after commissioning, low power testing and pilot operation of BREST-OD-300 RP:

- Optimization and improvement of the main RP equipment (the RCPS, the refueling complex, etc.) and the core items (fuel assemblies, reflector blocks. protection, etc.);
- Optimization of the control algorithms for the processes and systems supporting the reactor plant operation;
- Enhancement of accuracy and verification of the design codes for the RP safety analysis;
- Confirmation and adjustment of the equipment parameters in support of the lead coolant technology, practicing of the lead coolant purification regimes, control and maintenance of the process parameters under various RP operating conditions;
- Practicing and optimization of the transportation and handling operations in the course of refueling;
- Optimization of the technical diagnostic means for the RP equipment and the diagnostic algorithms;
- Demonstration of operation in the equilibrium mode with the low reactivity margin and change ($\Delta\rho < \beta_{\text{eff}}$) within the period between refuelings;
- Transmutation and burning of long-lived actinides recycled with the fuel within the closed NFC in the core;
- Updating of the regulatory documentation for reactor plants with lead coolant.

TABLE 24.1 Main technical characteristics of power units.

Reactor	BN-600	BN-800	BN-1200
Rated thermal power, MW	1470	2100	2800
Gross electric power, MW	600	880	1220
Number of heat-generating loops	3	3	4
Primary coolant temperature, °C (at the intermediate heat exchanger inlet/outlet)	535/368	547/354	550/410
Secondary coolant temperature °C (at the SG inlet/ outlet)	505/318	505/309	527/355
Third circuit parameters: – Main steam temperature, °C – Main steam pressure, MPa – Feedwater temperature, °C	 505 14 240	 490 14 210	 510 17 275
Gross/net efficiency factor, %	42.5/40	41.9/38.8	43.5/40.7

24.2 Power unit with sodium-cooled BN-1200 reactor

The power unit with BN-1200 reactor as the first commercial marketable fast neutron power unit is developed within the framework of the Federal Special Program "New Generation Nuclear Power Technologies for the Period of 2010–2015 and for Further Extension to 2020," the investment resources of Rosatom and the R&D program of Rosenergoatom JSC. The main areas of improvement in comparison with the previously developed projects include improvement of the technical and economic parameters of the power unit with BN-1200 RP, safety enhancement due to development of self-protection properties, adaptation for the CNFC and two-component nuclear power.

Development of engineering solutions and improvement of the technical and economic characteristics of the reactor plant. Proven reference engineering solutions implemented and having confirmed their efficiency in BN-600 and BN-800 are applied in the design of BN-1200 RP. These solutions include the three-circuit cooling and energy conversion scheme and integral arrangement of the primary circuit. In general, the adopted parameters of the circuits are also traditional, except for the increased sodium temperature at the core inlet (410°C) and higher efficiency factor (Table 24.1.).

But the project contains many innovative solutions aimed to reduce the capital costs due to decrease of material consumption, elimination of some systems and operating costs by extension of the equipment life span (Fig. 24.10).

The most significant technical changes in BN-1200 RP aimed to enhance the technical and economic parameters in comparison with BN-600 and BN-800:

- The new core with reduced average power density (from 450 to 230 MW/m^3) and the increased diameter of fuel rods (from 6.9 to 9.3 mm) using both MOX and MNIT fuel (Fig. 24.11);

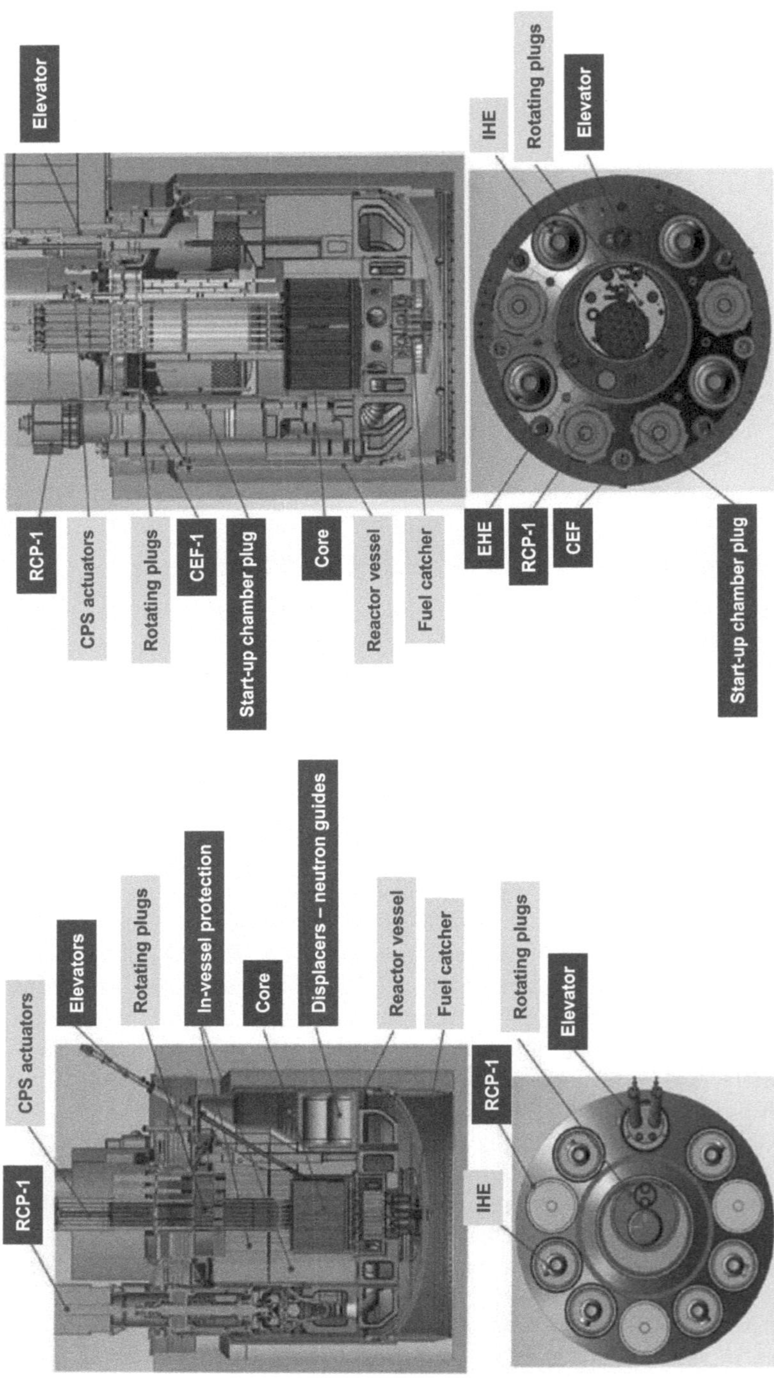

FIGURE 24.10 Comparison of the main equipment of BN-800 and BN-1200 (the components excluded in BN-1200 are highlighted in red color, innovative solutions – in dark blue, and reference solutions – in light blue).

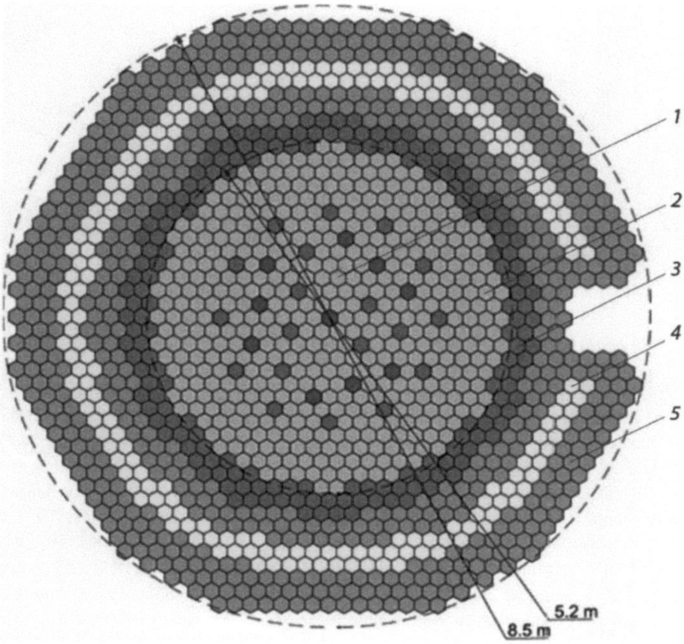

FIGURE 24.11 Core map of BN-1200 RP (the number of components is specified in brackets): 1 – CPS rods (31); 2 – core fuel assemblies (431); 3 – fuel assemblies of the side blanket or steel protection assemblies (174); 4 – in-vessel storage assemblies (198); 5 – boron protection assemblies (573).

- Boron protection within the core enabling to reduce the in-vessel protection by ~30 times;
- A pressurized vertical direct-flow steam generator (8 modules) instead of sectional and modular SGs for BN-600 (72 modules) and BN-800 (60 modules);
- The new refueling system with direct unloading of the spent fuel assemblies, elimination of the sodium drum (with auxiliary subsystems) for spent fuel assemblies and expansion of the in-vessel storage of spent fuel assemblies.

The new core requires considerably lower reactivity margin for fuel burnup enabling transition to the full-year operation cycle and increase of the capacity utilization factor up to ~0.9 and higher.

Enlargement of the core elements reduces their annual consumption ~3–6 times thus considerably decreasing the fuel factor in electric power generation. Besides, the single-zone arrangement with one type of fuel assemblies was used for the first time (in the previous designs of BN-600 and BN-800 there were three zones containing fuel assemblies with different enrichment).

The service life of the main equipment was extended to 60 years due to optimization of the in-vessel protection and reduction of the radiation impact on

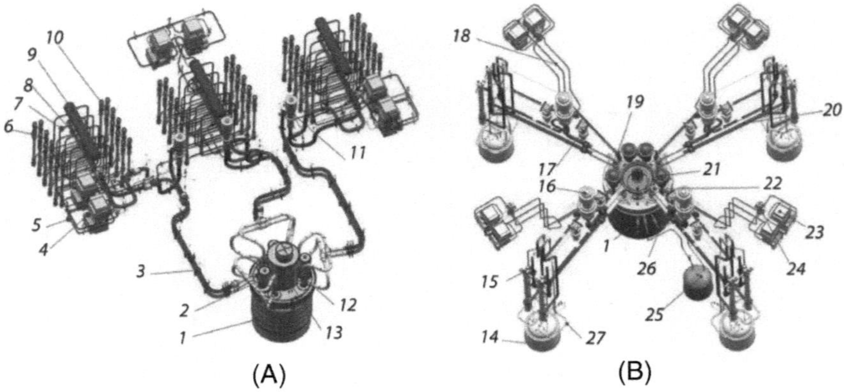

FIGURE 24.12 Scheme of the secondary circuit with the SG for BN-800 (A) and BN-1200 (B) RPs: 1 – reactor; 2 – primary circuit RCP; 3 – secondary circuit pipelines; 4 – air heat exchanger; 5 – ECS pipelines; 6 – steam superheater; 7 – SG piping; 8 – distribution header; 9 – sodium buffer tank; 10 – evaporation unit; 11 – secondary circuit RCP; 12 – intermediate heat exchanger; 13 – protective cap; 14 – V250 emergency discharge tank; 15 – SG modules; 16 – secondary circuit RCP; 17 – secondary circuit cold entrainment filter; 18 – EHRS pipelines; 19 – independent heat exchanger; 20 – secondary circuit pipelines; 21 – primary circuit RCP; 22 – intermediate heat exchanger; 23 – air heat exchanger; 24 – sodium expansion tank; 25 – compensator tank; 26 – overflow and gas compensation system pipelines; 27 – discharge pipelines.

the nonreplaceable reactor components as well as application of new structural material for the vessel components and pipelines.

The effect from application of the new solutions for the SG (Fig. 24.12):

- The specific metal content in the SG is reduced from 1.48 to 0.33 t/MW (by 78%) and the specific construction volumes of the SG boxes are reduced from 32 to 16 m^3/MW (e) (by 50%);
- The SG life span is extended from 23 to 30 years (by 30%) due to application of new structural material – 07H12NMFB steel.

New engineering solutions for the refueling system (Fig. 24.13) defined considerable reduction in the auxiliary systems, pipeline lengths and valves. As a result, the total weight of the refueling equipment was reduced by 780 tons, and the total refueling time per a year was reduced ~1.5 times.

Besides, the following improvement of the technical and economic characteristics was achieved due to new engineering solutions for the RP:

- The life span of the replaceable equipment was extended ~3 times (the replacement frequency was reduced from 10 to 3) particularly due to optimization of the RCP-1 flow channel in order to enhance cavitation resistance and RCP-2;
- The pipeline length was reduced 1.9 times, particularly due to application of bellows joints;

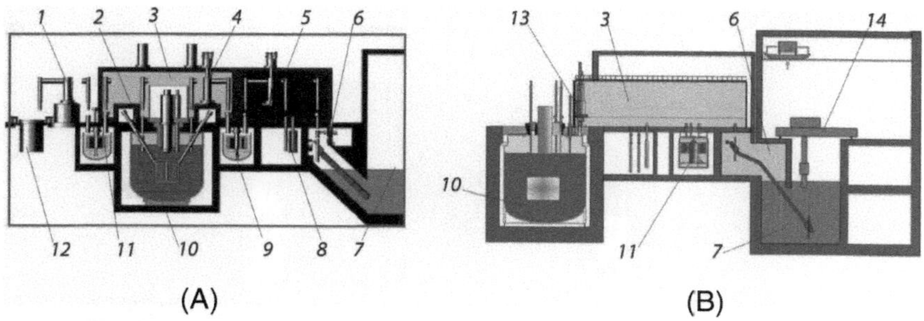

FIGURE 24.13 Comparison of the refueling systems of BN-800 (A) and BN-1200 (B) RPs: 1 – fresh fuel assembly transportation mechanism; 2 – elevator; 3 – reloading box; 4 – reloading machine of the reloading box; 5 – washing box; 6 – delivery shaft; 7 -spent fuel pool; 8 – reloading machine of the washing box; 9 – spent fuel assembly drum; 10 – reactor; 11 – fresh fuel assembly drum;12 – in-plant transportation package; 13 – refueling machine; 14 – reloading machine of the spent fuel pool.

- The number of sodium valves was decreased by ~80% (valves were excluded from the main pipelines of the secondary circuit for BN-1200 RP).

Development of engineering solutions and improvement of the technical and economic characteristics of the power unit. Engineering solutions for the reactor building (Fig. 24.14) are considerably changed in the design materials for the power unit with BN-1200 RP (2014) and BN-1200M RP (2017) in order to improve its technical and economic characteristics. The reactor building option of the cylindrical shape with the RP cavity in the center and diametrical arrangement of the RP equipment was adopted for the first time in the Russian practice.

Some other significant new solutions are listed below:

- A light spherical dome above the central hall in combination with application of advanced high-strength concrete and reinforcement bars united in the reinforcement blocks thus providing for reduction of the time period for commencement of the main reactor building equipment installation;
- The building of the fresh fuel storage facility and the spent fuel pool is separated as an individual module enabling to use it both for the centralized fuel production option and the near-plant option without any significant engineering and layout modifications;
- Reinforced external civil structures along the building periphery without any change of the internal volume due to arrangement of the expansion joint along the outer diameter of the reactor building walls (in case of strengthened requirements for consideration of human-induced accidents);
- The number of equipment in the emergency power supply system is considerably reduced in comparison with BN-800 due to application of passive safety systems and also due to connection of the emergency heat removal system directly to the reactor.
- The turbine hall was also affected by changes. The basic design of K-1200–16,0/50 turbine with the capacity of 1221 MW in the nominal mode was

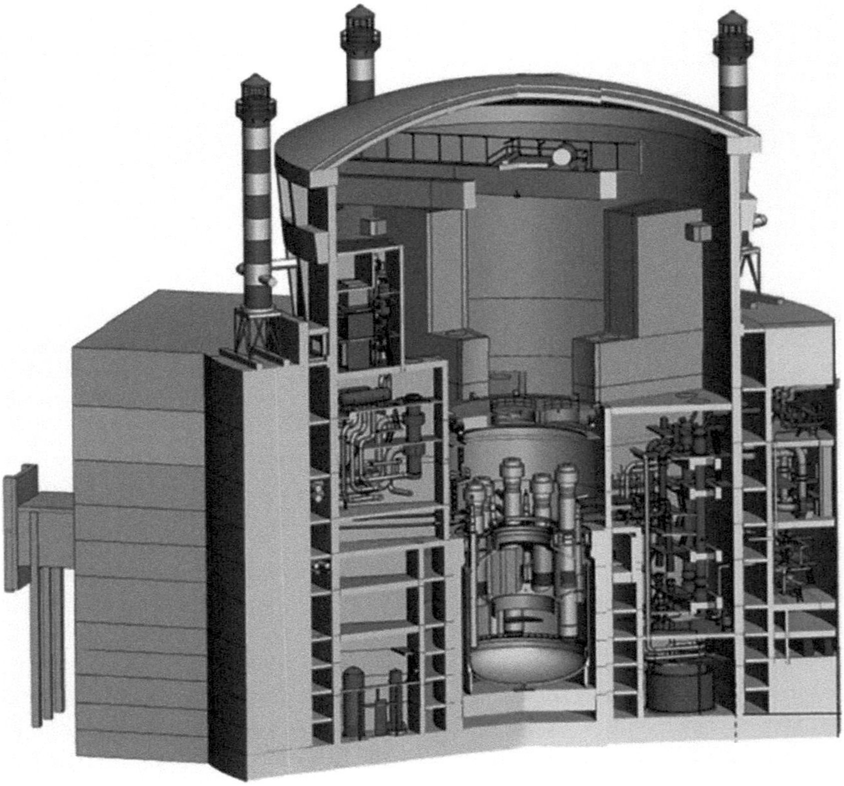

FIGURE 24.14 Layout of the reactor building.

TABLE 24.2 Comparative analysis of the turbine hall characteristics.

Parameter	BN-800	BN-1200	VVER-1200M
V_{vol}, m^3/MW	283	225	222
Heat efficiency of the turbine, kcal/kW·h	2040.3	1970.5	2364
Overall length of the turbine, m	42.07	42.16	52
Turbine weight, t	1850	2050	2100

developed enabling to improve the technical and economic parameters of the turbine hall (Table 24.2).

The set of the applied design and engineering solutions provided for considerable reduction of the construction volumes, primarily for the reactor building, in particular:

TABLE 24.3 Comparison of the structural volume.

Characteristics	BN-800	BN-1200	VVER-TOI	AR-1000
Reactor building footprint area, m^2	11,806	6658	6390	3000
Reactor building structural volume with auxiliary buildings, m^3	678,100	409,410	375,900	158,000
Reactor building specific structural volume, m^3/MW	766.215	327.528	299.52	131.67

- The specific volume of the reactor building was reduced by ~30%;
- The number of site welds was reduced ~4 times.

More detailed numerical data are provided in Table 24.3 and illustrated with Fig. 24.15. In general, it may be noted that the power unit of much higher capacity has much smaller building footprint.

Application of the developed optimized solutions for the first time ever provides the opportunity to ensure the required competitiveness of the power unit with BN-1200 RP as compared with the existing thermal reactors and CCGTs (Fig. 24.16).

There are uncertainties of technical and economic characteristics related to the current stage of the project as of 2019. They will be eliminated at the subsequent project development stages taking into account search for new engineering solutions and their substantiation by the developers.

Safety of the power unit with BN-1200 RP. In the course of BN-1200 reactor development great attention was paid to development of the inherent safety properties enabling to reduce the risk of any emergency situation occurrence and the self-protection properties based on the passive principles. The core of BN-1200 RP has considerably lower reactivity margin, especially in case of mixed nitride fuel application thus significantly reducing the probability of any reactivity-related accidents and limiting their consequences. Reactivity change in the course of fuel burnup demonstrating reduction of the required reactivity margin by 4–10 times in comparison with BN-800 is shown in Fig. 24.17.

This also simplifies the problem of compliance with the nuclear safety requirements in the context of reactivity balance assurance. Nevertheless, hydraulic suspended PAZ-G rods with passive activation in case of the coolant flow rate reduction below the specified level applied in BN-800 for the first time as well as new PAZ-T rods with passive activation upon reaching of the temperature of 800°C are included into the design.

The requirement for maximum reduction of the sodium void reactivity effect is implemented in the design in order to prevent reactivity-related accidents in case of sodium boiling (Fig. 24.18). For this purpose, the sodium cavity is arranged above the flattened (H/D ~0.2) core.

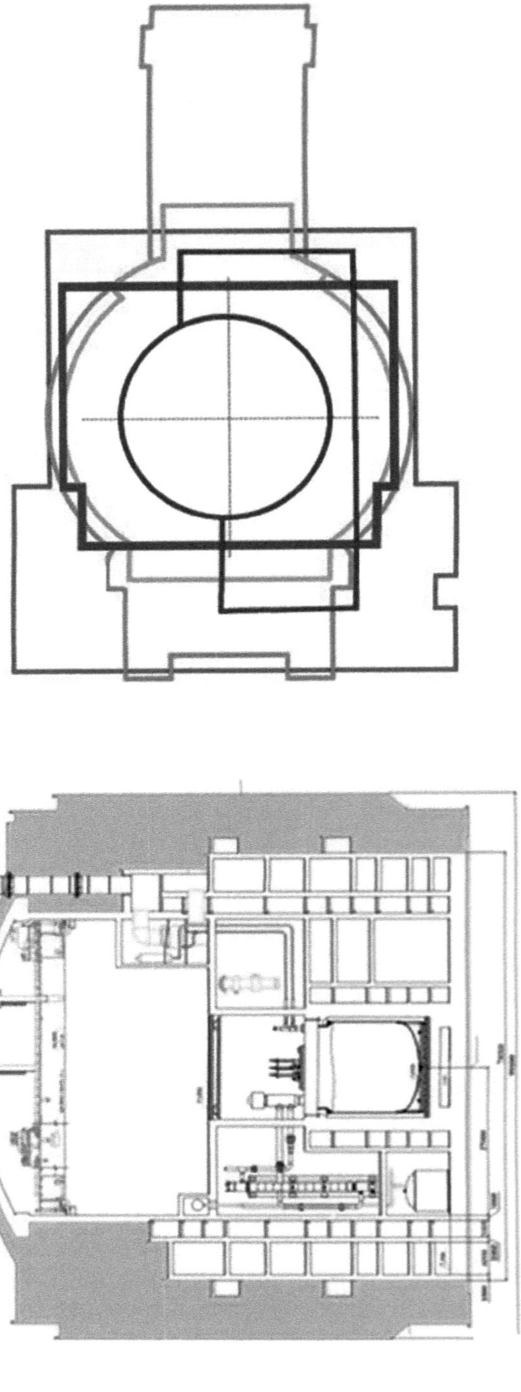

FIGURE 24.15 Comparison of the reactor building volumes for BN-1200 RP (green), BN-1200M RP (blue), BN-800 (red) as well as VVER-TOI (dark blue), and AR-100 (violet).

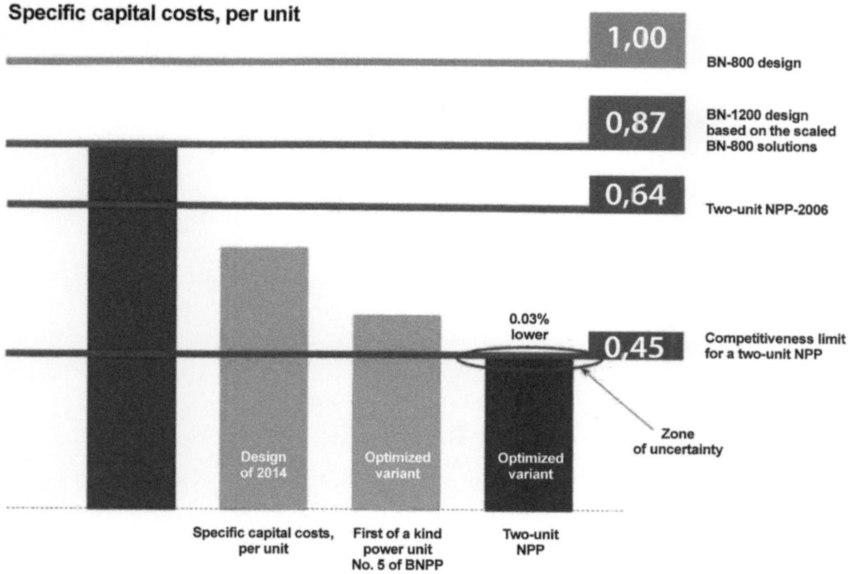

FIGURE 24.16 Competitiveness of the power unit design with BN-1200 RP (2014) and the optimized design with BN-1200M RP (2017) with regard to specific capital costs.

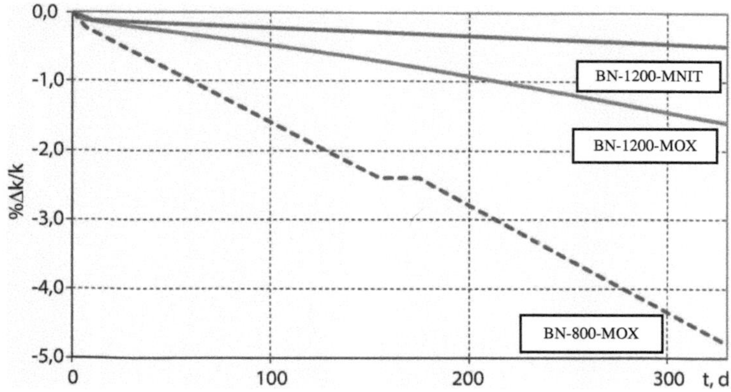

FIGURE 24.17 Reactivity change in the course of fuel burnup in BN-1200 RP (annual micro-campaign) and BN-800 (extrapolation).

Calculations show efficiency of the above-mentioned solutions for prevention of uncontrolled reactor power increase observed in case of any core dimension deviation from the optimal values and in the absence of the sodium cavity. From the physical viewpoint the effect is associated with increased neutron leakage from the core with sodium density decrease at the core periphery.

The specific energy emission in the fuel was reduced considerably from ~6 to 2 W/cm^3 due to reduction of power density in the core and in-vessel storage

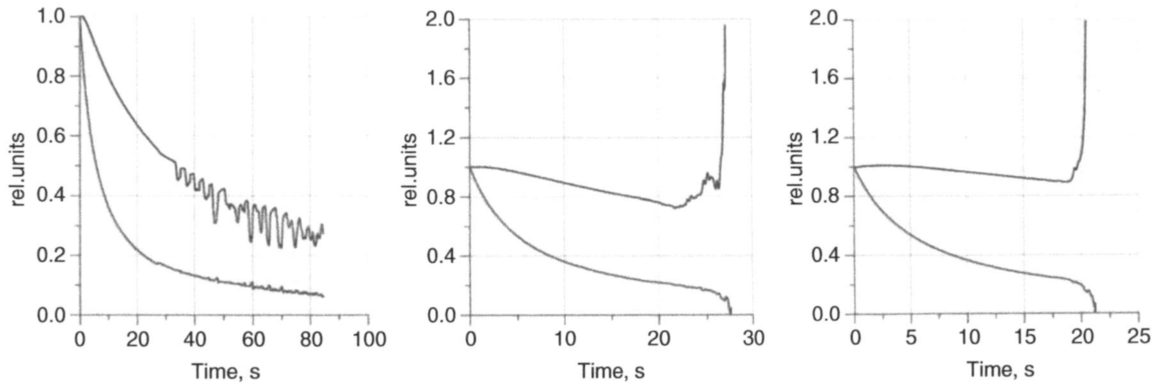

FIGURE 24.18 Changes of power (blue curve) and coolant flow rate (red curve) in the course of ULOF accident (two figures on the right present the curve run in case of flattening reduction and in the absence of the sodium cavity).

providing for long-term storage of spent fuel assemblies (2 years instead of *1/2* year in BN-600 and BN-800). This provides the possibility for direct safe transportation and washing of spent fuel assemblies and enables to exclude the spent fuel assembly drum from the RP design as a source of potential accidents with radioactive sodium.

A significant step toward sodium safety assurance was made in BN-1200 RP design due to complete integration of the main sodium systems and the primary circuit equipment inside the reactor vessel. All primary circuit pipelines with sodium outside the reactor vessel are excluded thus enabling to prevent any radioactive sodium leaks into the RP rooms and, consequently, eliminate any possibility for sodium-induced fires with generation of radioactive aerosols.

Thus sodium safety has technical and not radiological nature.

The arrangements for minimizing of the nonradioactive sodium leakage probability are also implemented in BN-1200 design in order to enhance technical safety of the plant. This is ensured due to considerable (approximately 2.8 times in comparison with BN-800) reduction of the secondary circuit pipeline length by transition to pressurized steam generators, application of bellows joints and complete enclosure of pipelines into casings.

The emergency heat removal system with arrangement of natural circulation along the circuits directly from the primary circuit is a major innovation. Independent heat exchangers of this system are located directly in the reactor vessel while in BN-800 design they are connected to the secondary circuit.

Considerable reduction of the probability of severe core damage for any internal events in the course of power operation of BN-1200 reactor up to $\sim 5 \times 10^{-7}$ is achieved due to the internal self-protection and inherent safety properties, availability of various passive means and improvement of these systems which is significantly lower than the relevant values for BN-600 $\sim 10^{-5}$ and BN-800 $\sim 1.2 \times 10^{-6}$.

Subsequent to analysis of severe beyond design basis accidents (ULOF and UTOP types) it was demonstrated that the exposure doses for the public outside the NPP site did not exceed the limit of 50 mSv within the first year of accident even with the worst accident development scenario eliminating any need for evacuation, let alone relocation of the public.

In September 2017, the Governing Board of the GIF agreement on system integration and assessment of sodium-cooled fast reactors officially accepted and approved the BN-1200 concept as compliant with the requirements imposed on generation IV fast reactors.

Operation of the BN-1200 within the closed nuclear fuel cycle. The fuel cycle of a BN-1200 RP provides for the reactor start-up with the use of plutonium extracted after processing of SNF from thermal reactors (VVER in Russia) with subsequent transition to regenerated fuel manufactured from its own SNF products. Start-up of a new power unit requires 10–13 t/GW (Table 24.4) depending on the fuel type (MOX or MNIT) and the implementation stage (at

TABLE 24.4 MOX and MNIT fuel balance for BN-1200 RP.

Parameter	Core with MOX fuel		Core with MNIT fuel	
	Initial stage	Main stage	Initial stage	Main stage
Initial reactor fueling, pcs FA	432	432	432	432
Initial fueling, t				
– Fuel	39.1	39.1	51.1	51.1
– Plutonium	8.4	8.4	8.95	8.95
depleted uranium:				
• In FAs	34.4	34.4		
• In the side blanket	48.4	48.4		
Refueling frequency	4/5/6	6/7/8	4/5/6	6/7/8
Campaign for the basic set of fuel assemblies, ef. days	1320	1980	1320	1980
Maximum fuel burnup, % h.a.	13.8	20.2	11.1	16.4
Average fuel burnup, MW-day/kg	91.3 (58.1)***	128	72.4	103
Annual consumption of				
– Fuel, t/year	8.8	6.1	11.6	7.92
– Pu, t/year	1.9	1.3	2.07	1.40
Consumption of Pu from the storage facilities for the reactor start-up, t*	14.1	12.3	15.2	13.1
Consumption of Pu from the storage facilities for FNR start-up, t/GW	11.8	10.3	12.7	10.9
Breeding ratio (BR)**	1.12/1.25	1.12/1.25	1.07/1.37	1.06/1.36
Excessive plutonium production, kg/year	110/220	110/220	44/294	42/292
Specific excessive Pu production, kg/year/GW	92/183	92/183	36.7/245	35.0/243
Consumption of MAs from thermal reactors, t	1.4	1.2	1.5	1.3

*With the fuel cycle duration of 3 years.
**The data for the configuration without and with the side blanket are presented via a slash for MOX, and for the configuration without and with breeding blankets – for MNIT.
***With due regard for the material breeding.

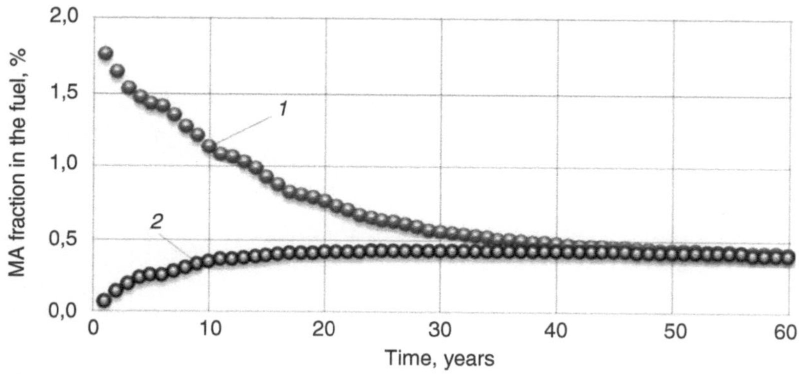

FIGURE 24.19 Changes in the MA concentration in the course of fuel burnup without MAs and with inclusion of MAs: *1* - start-up with VVER MAs; *2* - accumulation of own Mas.

the initial FNR implementation stage reduced burnup parameters are adopted). In the CNFC mode the power unit consumes only ^{238}U.

The breeding ratio may in its turn vary from 1.06 to 1.36 depending on the fuel type and the core configuration: without blankets, with the bottom and side blankets and provide for production of excessive plutonium for start-up of subsequent FNRs at the amount of up to ~240 kg/GW·year.

Minimization of RW amounts due to transmutation of MAs and long-lived fission products is the second key objective. The reactor can burn its own MAs (neptunium and americium) generated in the course of fuel irradiation to the full extent. Concentration of own MAs in the fuel is low and does not exceed 0.5–0.6% h.a.

MAs should be included into the start-up loading fuel together with plutonium from thermal reactors in the same amount as in VVER SNF (10–20% in relation to plutonium). In this case the MA concentration in BN-1200 RP fuel at the initial stage can amount to 1.8% with gradual decrease up to the equilibrium value (Fig. 24.19). Thus 1 power unit with BN-1200 RP can utilize ~1.5 t of MAs from VVER thermal reactors at the average. In case of necessity this value may be increased to 2 tons.

Inclusion of MAs in the specified amounts does not reduce safety and can be used for the reactivity margin stabilization at the equilibrium level (Fig. 24.20). MAs can be used in the similar way to compensate for the difference in isotope composition of plutonium thus ensuring "indiscrimination" of BN-1200 with regard to plutonium.

Optimization of the power unit design with BN-1200 RP is in progress as of 2019. The approximate time limits for completion of the main R&D works and substantiations – 2021. Start-up of the first power unit with BN-1200 RP may take place in 2031.

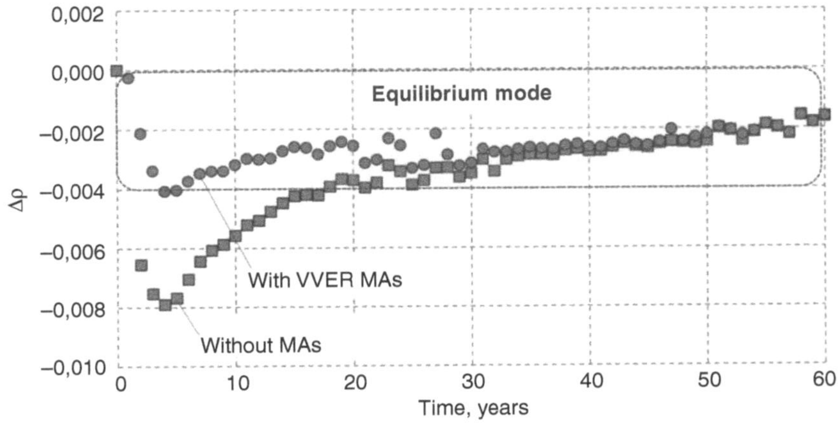

FIGURE 24.20 Reactivity changes in the course of fuel burnup without MAs and with inclusion of Mas.

24.3 Conceptual design of the IPC with BR-1200

The concept of multivariant search for the optimal design solutions at the initial facility development stage is actively applied in Proryv Project Framework. This is primarily implemented through the use of conceptual (front-end engineering) design – minimal development of the facility with fixation of only the key parameters enabling to predict its basic final technical and economic characteristics. 3D model is an integral part of the conceptual design.

The main design stage of any nuclear facility is characterized by issuance of the design and detailed design documentation. The conceptual design is the pre-design stage necessary to begin development of principally new and knowledge-intensive nuclear facilities. This stage does not involve high degree of detail which makes development of variants quite labor-intensive and expensive. The main limitations preventing from achievement of the competitiveness criteria are identified particularly at this stage. In some cases, review of the relevant standards and regulations on design of nuclear facilities is required.

The feasibility study for NPPs with BREST-1200 RPs was performed in 2000. Development of the conceptual design for the industry power supply complex (IPC) based on NPPs with BR-1200 and nuclear fuel production/processing plants commenced in 2015 in the Innovation and Technical Center of Proryv Project. Top-Level Technical Requirements for the conceptual design of the IPC based on BR-1200 RPs were developed. The Technical Assignment for development of the conceptual design of the IPC based on BR-1200 RPs for the works in 2016–2019 was developed and approved in 2016. The fundamental technology and space planning solutions were defined in 2015–2017, the cost of the IPC facilities was evaluated, and the possibility to achieve competitiveness was demonstrated. The IPC conceptual design provides for location of two power units of the NPP with BR-1200 RP at the site of South Ural NPP and optional location of the fuel production facilities at the sites of Siberian Chemical Plant JSC, MCP JSC, Mayak PA FSUE (the centralized NFC) and South Ural NPP (the near-plant NFC).

The IPC conceptual design is structured in accordance with the requirements for the design documentation composition; it covers all basic technology, architectural, construction and arrangement solutions, has the cost estimate section and may serve as the basis for further development at the design stage.

As far as power units with BR-1200 RPs are concerned the IPC conceptual design is based on both the experience in development of the lead-cooled fast reactor design BREST-OD-300 which passed the state expert appraisal in 2018 with the positive result and the experience in design of NPPs with VVER reactors.

BR-1200 reactor plant with the double-circuit heat removal scheme where water- steam with subcritical parameters is used as the working medium of the secondary circuit is developed as a commercial reactor plant for large-scale nuclear power with inherent safety. The aim of development is to arrange the reactor plant with BR-1200 lead-cooled fast reactor with the electric power of at least 1200 MW subject to maximum application of the engineering solutions provided for BREST-OD-300 pilot demonstration reactor. The expected technical characteristics of the RP are specified below.

Main technical characteristics of BR-1200 RP

Design thermal power, MW	2930
Design electric power, MW	1260
Steam capacity, t/h,	at least 6120
Maximum neutron flux density, $cm^{-2}\ s^{-1}$	5.7·1015
Primary coolant	Lead
Principle of heat removal from the core	Forced circulation
Primary coolant movement through the core	From the bottom upward
Coolant flow rate in the primary circuit, t/s	173.0
Average coolant temperature at the core inlet/ outlet, °C	420/535
Gas (argon) overpressure above the coolant level, MPa: average/ maximum	(0.003–0.005)/0.1
Reactor operation duration between refuelings in the partial refueling mode, ef. days	330
Secondary coolant	Water-steam
Secondary coolant temperature at the SG inlet/ outlet, °C	340/505
Secondary coolant pressure at the SG inlet/ outlet, MPa	18.4/17
Design service life of the reactor unit, years	at least 60

Purpose of BR-1200 RP:

- Thermal energy generation for subsequent conversion into electric energy within the power unit;
- Full breeding of plutonium in the core, arrangement of the nuclear fuel cycle closing with only depleted or natural uranium make-up.

The problems solved in the course of development and construction of BR-1200 RP for the commercial NPP power unit within the IPC complying with the following basic requirements of Proryv Project:

- Prevention of any NPP accidents requiring temporary evacuation and even relocation of the public;
- Assurance of the nuclear power competitiveness in comparison with alternative generation;
- Establishment of the closed NFC for complete utilization of the energy potential of natural uranium raw materials and burning of minor actinides;
- Consistent approach to radiation-equivalent (in relation to natural raw materials) disposal of radioactive waste;
- Technology enhancement of the nonproliferation regime (departure from weapon-grade plutonium production in the blanket and its extraction in the course of SNF processing, consistent departure from uranium enrichment for nuclear power, reduction of nuclear material transportation);
- Reduction of the capital construction costs for NPPs with fast reactors at least to the level of the best NPPs with thermal reactors due to process, design and engineering solutions intrinsic only for fast reactors;
- Environmental safety of the power supply complex facilities at all stages of the closed NFC for the public and natural ecosystems.

The works for adjustment and optimization of the design and engineering solutions aimed at improvement of the technical and economic parameters are going on continuously. The optimization solutions for the following were presented in 2018:

- The weight of the main BR-1200 RP equipment with reduction by 675 t and the lead coolant by 500 t – cost reduction by 4.5 bln. RUB;
- The secondary circuit heat balance scheme with substantiation of the gross power increase to 1275.6 MW;
- The reactor building volumes with reduction to 317 thous. m^3 and the IPC general layout with the possibility to decrease the site size to 58 ha;
- The floor spaces and the personnel size through the use of robotized production lines of the closed NFC.

Besides, the possibility to use Techa cascade of water bodies at Mayak PA FSUE for service water supply to the NPP based on BR-1200 RP at the site of South Ural NPP.

Chapter 25

Thermal reactors

25.1 Light-water reactors

The VVER technology development program is based on the accumulated experience in operation of reactor plants and aimed at solution of the main systemic problems of nuclear power, enhancement of competitiveness and reduction of the risks associated with development of alternative technologies, the possibility for efficient operation in two-component nuclear power generation system with the closed NFC. The considered basic designs are based on the evolutionary and innovative technology solutions enabling to ensure smooth transition to establishment of two-component nuclear power with thermal and fast reactors.

The following main candidates for promising light-water reactor technologies for two-component nuclear power are considered:

- Optimized VVER-TOI reactors;
- Improved VVER reactors with spectral regulation;
- Innovative VVER reactor technologies with the supercritical coolant parameters (this technology requires significant amounts of R&D works and is not considered in the scenario investigations).

The key parameters of the improved light-water reactor series and the expected time of their commissioning commencement are presented in Table 25.1 [9].

The medium-term horizon of the VVER technology development within the period of up to 10 years involves development of the reactor plant design with spectral regulation (VVER-S). Implementation of the spectral regulation concept in VVER reactor plant will enable to depart from usage of boron for excessive reactivity compensation, to reduce the capital and operating costs, to decrease the natural uranium consumption and to increase efficiency of the reactor plant considerably within the closed NFC with 100% uranium and plutonium fuel loading. The main changes in VVER-S design are related to refurbishment of the reactor plant and emergency safety systems. Evolutionary development means preservation and further development of the main engineering solutions related to the secondary circuit equipment, the turbine and the reactor building.

Closed Nuclear Fuel Cycle with Fast Reactors: Handbook of Russian Nuclear Power.
DOI: https://doi.org/10.1016/B978-0-323-99308-1.00032-5

TABLE 25.1 Key parameters of the improved light-water reactor series.

Parameter	VVER-1200	VVER-TOI	VVER-S	VVER-SKD
Electric/thermal power, MW	1200/3200	1255/3300	1255/3300	1200/2800
Specific CAPEX, %	118	100	80	75
Duration of construction, years	6	6	6	5
Implementation, year	2017	2023	2036	2046
evolutionary solutions				
	new technology			

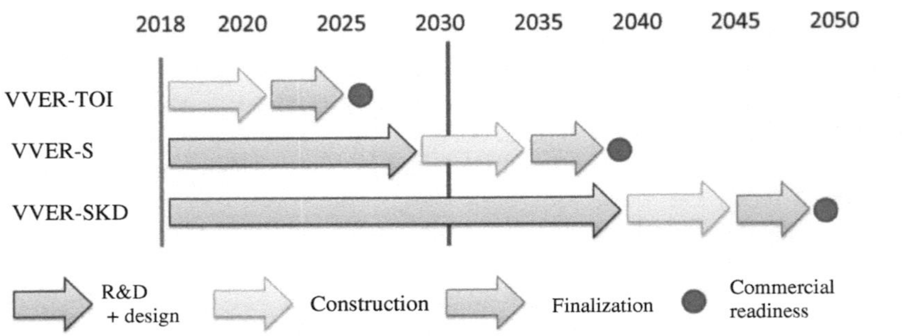

FIGURE 25.1 Assessment of the time limits for commercialization of VVER reactor technologies.

In particular, it is proposed to consider the potential for reduction of the number of steam generators and relocation of the spent fuel pool beyond the NPP.

The main aim of the VVER-S reactor technology development is to create a new competitive proposal for the Russian and foreign market capable of efficient operation at the transient stage of the multicomponent nuclear power system establishment with thermal and fast reactors.

It is expected that implementation of new technology solutions will enable to reduce the power unit cost by 20% in comparison with VVER-TOI. More detailed assessment may be performed in the course of the design works. Commissioning of the first power unit may be carried out in 2036. Estimated time limits for commercialization of VVER reactor technologies are presented in Fig. 25.1.

The parameters specified in Table 25.2 are assumed for analysis of the scenarios with functioning of light-water reactors in the open NFC.

The parameters specified in Table 25.3 are assumed for analysis of the scenarios with functioning of light-water reactors in the closed NFC [9].

The possibility to use regenerated nuclear materials in thermal reactors is determined by the scenario conditions. Use of regenerated uranium (multiple) enables to reduce natural uranium consumption in the VVER NFC. The possibility to arrange 100% loading of the core with MOX fuel is taken into account in VVER-S design. This decision shall be made with due regard for reduction of the plutonium stock available for the FNR fleet development.

TABLE 25.2 Characteristics of fuel campaigns for light-water reactors in the open NFC.

Parameter	VVER-1200	VVER TOI	VVER-S
Electric/thermal power, MW	1200/3200	1255/3300	1255/3300
Availability factor	0.93	0.93	0.90*
Cycle duration, ef. days	500	500	330
Initial loading, t_{hm}	77	77	125
Initial loading enrichment, %	3.3	3.1	2.9
Annual loading, t	22.6	22.7	21
Annual loading enrichment, %	4.69	4.77	3.5
Burnup, MW-day/kg_{hm}	48.4	49.4	50
AC**	0.42	0.43	0.35

**To be adjusted.*
***Accumulation coefficient of the fissionable material is the ratio between the mass of all fissionable isotopes upon fuel unloading to the mass of loaded ones.*
Note: t_{hm}, kg_{hm} - ton of heavy metal, kilogram of heavy metal.

TABLE 25.3 Fuel campaign characteristics for light-water reactors in the closed NFC.

Parameter	VVER TOI	VVER-S	VVER-SKD
Electric/ thermal power, MW	1255/3300	1255/3300	1200/2800
Availability factor	0.93	0.90	0.90
Cycle duration, ef. days	500	330	330
Initial loading, t_{hm}	77	125	40/60
Initial loading enrichment/ Pu* content, %	3.9/-	2.9/-	0.2/14.0
Annual loading with UOX/U-PuOX, t_{hm}	13.7/9.7	-/21	6.2/13.8
Annual loading enrichment/ Pu* content, %	4.86/9	-/8.2	0.2/16.5
UOX/U-PuOX burnup, MW-day/kg_{hm}	49/46	-/50	15/62
Accumulation coefficient for fissionable materials	0.44/0.72	0.66	0.04/0.92

**To be adjusted.*

Analysis of the effect of risks on the integral LCOE parameter for VVER-TOI and VVER-S power units (at 2018 values) is presented in Fig. 25.2. Comparison is performed with regard to the optimal characteristics of power units that can be achieved taking into account successful implementation of the evolutionary VVER technology development programs.

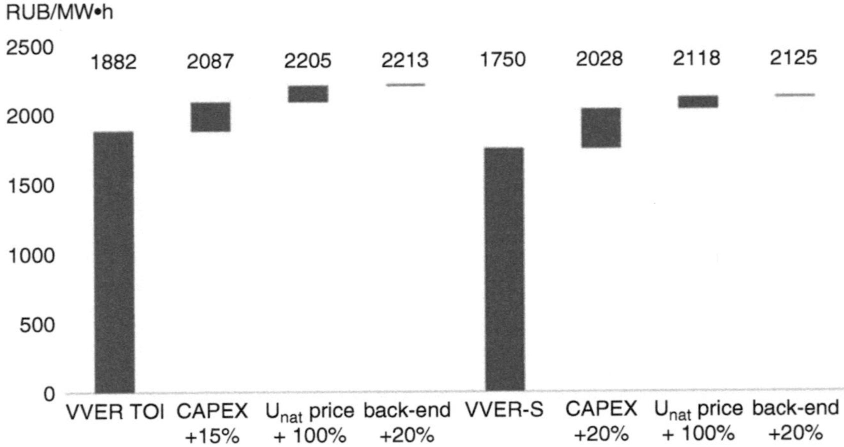

FIGURE 25.2 Effect of the risk of changes in the technical and economic parameters of VVER on the integral LCOE parameter for VVER-TOI (commercial) and VVER-S power units. (For reference: in 2018 values. LCOE of CCGT-410 at the site in Kursk region is 1980 RUB/MW ● h with the capacity utilization factor of 65% and 2330 RUB/MW ● h with the capacity utilization factor of 50%).

The key objective at the short-term planning horizon is to substantiate the declared technical and economic characteristics of VVER-S power unit (2021) and to develop the plans for long-term R&D works in the area of VVER-SKD.

25.2 Spectral regulation

Arrangements in this area: Implementation of the spectral regulation mechanism in the reactor, optimization of the uranium-water ratio, the mode with more frequent refueling, etc. They should ensure the breeding ratio of about 0.7–0.8 within the closed fuel cycle and natural uranium consumption below 130 t U/GW (e) per a year in the open fuel cycle.

It is proposed to implement spectral regulation in the course of fuel burnup by changes of the water-uranium ratio. In this case complete departure from liquid boron control in the operational modes is possible. According to the balance of fissionable isotopes VVER-S reactor can operate for about 30 years with uranium fuel and then for ~30 years more with its own plutonium. As a result, it will save ~50% of natural uranium (as compared with VVER-TOI).

25.3 VVER-S reactor technology

Main characteristics of VVER-S reactor technology. The long-term strategy of sustainable nuclear power in Russia is based on the concept for transition to the two-component system with thermal and fast reactors functioning within the closed nuclear fuel cycle. Implementation of the spectral reactivity regulation concept in light-water reactors enables qualitative changes in the technical and economic parameters of power units with thermal neutron spectrum and solution of the urgent industry-specific problems:

- Reduction of commercial risks due to decrease of the capital and operating costs by implementation of the process solutions enabling to depart from boron reactivity control as well as reduction of fuel costs;
- Reduction of systemic risks due to the possibility for efficient operation of the reactor plant in the closed nuclear fuel cycle with the use of 100% loading with regenerated uranium and plutonium fuel as well as reduction of natural uranium consumption in case of uranium fuel use;
- Maximum application of mature technologies and production facilities with regard to reactor building and fuel cycle arrangement;
- Creation of a new competitive proposal for the external market with due regard for the accumulated experience in implementation of VVER projects.

Pilot studies performed in Kurchatov Institute, OKB Gidropress JSC, and Atomenergoproekt JSC showed that used of spectral regulation in thermal reactors enabled to achieve the goals set with regard to refurbishment of VVER reactors for operation at the stage of transition to the closed fuel cycle. Spectral regulation is based on the neutron spectrum hardening mechanisms at the beginning of the reactor campaign with increase of resonance trapping in raw isotopes and subsequent spectrum softening in the core by the end of the reactor campaign. Different spectrum regulation techniques are considered: From the coolant density variation in the course of the reactor operation to changing of the water-uranium ratio with the use of mechanical displacers. Use of spectrum regulation enables to enhance fuel utilization due to increase of plutonium contribution into fission and to provide excessive reactivity compensation in the course of the reactor operation.

The concept of evolutionary development of VVER-TOI technology with the increased vessel size (VVER-1500 vessel is the counterpart) and maintenance of the power unit capacity at the level of 1300 ME (e) is considered as the basic design for VVER-S reactor plant. This approach enables to achieve the required parameters of safety and criticality control in the course of fuel burnup with low linear loads on the fuel rods. Neutron spectrum regulation is arranged with the use of mechanical displacers. Guiding channels where displacers are located are provided in the fuel assemblies. The water-uranium ratio measurement range in the fuel assemblies varies from 1.5 to 2.0 with fully inserted and fully withdrawn displacers in the reactor core. At the beginning of the reactor operation all displacers are in the core. The number of displacers is selected in such a way so that the reactor is in the critical state. Sequential withdrawal of displacers enables to compensate for reactivity losses in the course of the reactor operation. This approach enables to depart from liquid boron control.

The core of VVER-S consists of 241 fuel assemblies with rod-shaped fuel rods with the standard diameter of 9.1 mm. Fuel assemblies contain the positions for installation of displacers and CPS control rods. The list of source data for the basic variant of VVER-S core is given below. Values of the parameters may be

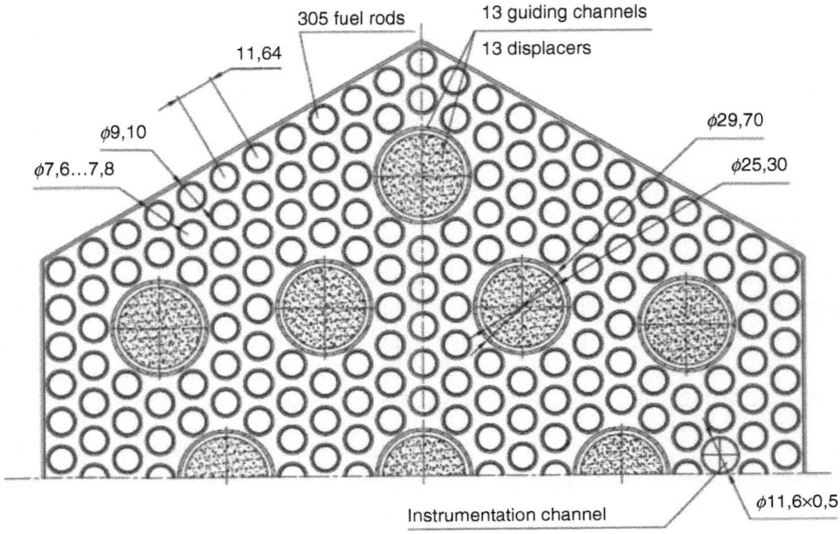

FIGURE 25.3 Map of VVER-S fuel assembly with displacers.

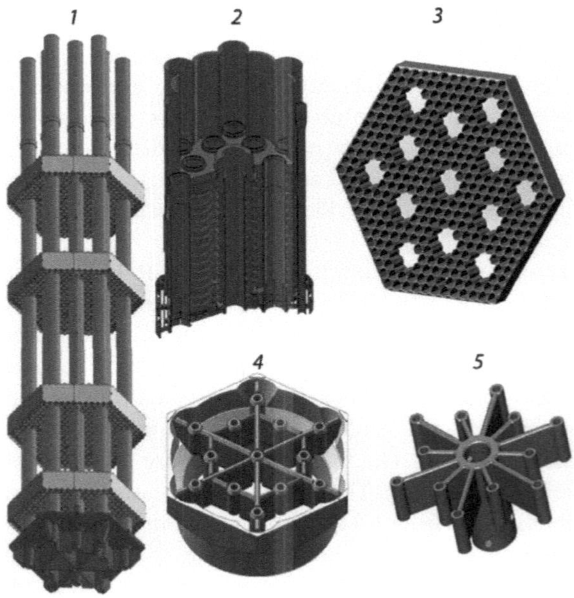

FIGURE 25.4 VVER-S FA components: 1 - frame; 2 - FA cap; 3 - spacer; 4 - tail piece; 5 - cross-bar.

adjusted in the course of development. The schematic diagram of a fuel assembly with displacers is shown in Fig. 25.3. Details of the VVER-S FA components are shown in Fig. 25.4.

Characteristics of VVER-S core

Service life, years	60
Electric power of the power unit (gross, guaranteed mode), MW	At least 1200
Gross efficiency factor, %	37.0
Capacity utilization factor, at least, relative units	0.9
Auxiliary power consumption, %	Not more than 6.4
Number of fuel assemblies, pcs	241
Ready-to-operate FA size, cm	23.4
Interval between fuel assemblies in the core, cm	23.6
Number of fuel rods in a fuel assembly, pcs	305
Interval between fuel rods in a fuel assembly, cm	1.164
Fuel rod diameter/ cladding thickness, mm	9.1/0.065
Pellet/central hole diameter, mm	7.6/1.2
Fuel pellet stack height, mm	4200
Uranium (metal) mass in the fuel rod, g	1700
Uranium (metal) mass in the core, t	125.4
Average power density in the core, W/cm	106.5
Number of fuel assemblies with displacers, pcs	133
Number of fuel assemblies with CPS ARs, pcs	54
Number of guiding channels (GCs) in a fuel assembly, pcs	13
GC outer diameter/ thickness, mm	29.7/0.8
Displacer diameter, mm	25.3
Outer diameter of CPS absorber rod (AR), mm	24.0
Inner diameter of CPS AR, mm	12.8
Main coolant parameters:	
Temperature at the core inlet, °C	297
Temperature at the core outlet, °C	330
Pressure at the core outlet, MPa	16.2

Required amount of R&D works. The designed development concept for VVER-S technology includes implementation of the integrated engineering solutions for the power unit and NPP providing for fundamental improvement of VVER light-water reactor characteristics.

Principally new solutions require additional amounts of R&D works:

- For the reactor: spectral regulation, vessel enlargement, the CPS drive with increased traction, the electrical connection block combined with the upper unit;
- For the steam generator - horizontal headers fully welded into its casing;
- For the reactor coolant pump set (RCPS): The flow channel (the coolant inlet and outlet are counter-changed similar to VVER-640 RP design), the drive section of the RCPS with the shaft, bearings, sealings and the motor (similar to RCPS 1732, 1753 designs);
- For the RP main circulation circuit (MCC): Four cold DN500 legs with four RCPSs, two hot DN850 legs, two steam generators, absence of any hydraulic locks in the legs;
- For the power unit: The possibility to locate the spent fuel pool outside the containment.

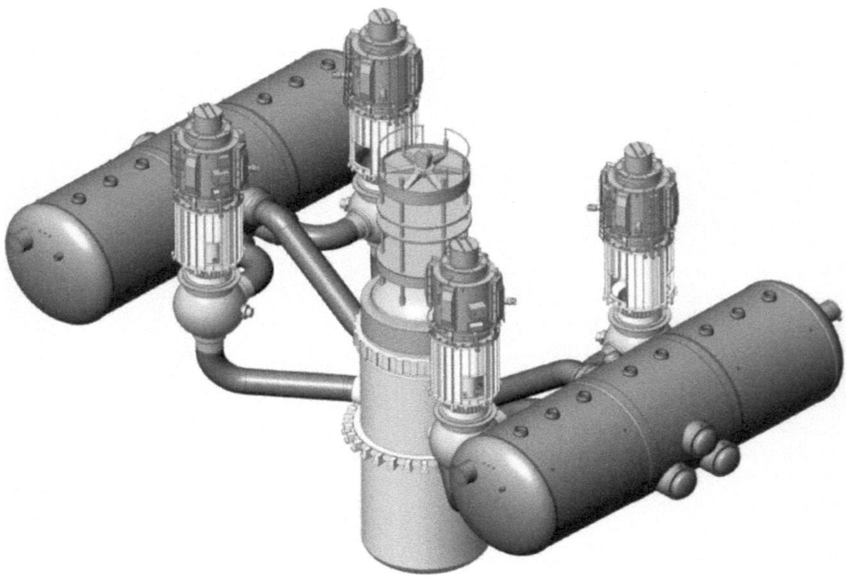

FIGURE 25.5 Reactor plant with VVER-S reactor.

Implementation of these engineering solutions will enable, in particular, to reduce power density in the core by 50% in comparison with NPP-2006 and VVER-TOI, to exclude the hydraulic locks in the MCC legs thus providing for:

- Development of the internal RP self-protection properties (assurance of stable natural circulation of the coolant);
- Exclusion of the external core catcher;
- Relaxation of the requirements for characteristics of the safety systems with regard to heat removal from the core;
- Prevention of steam-zirconium reaction in case of design basis accidents.

The new RP arrangement with two steam generators (Fig. 25.5) will increase the power unit technical utilization factor, reduce metal consumption and the number of systems connected to the primary circuit equipment (with subsequent reduction in the number of valves, penetrations, pipelines and cables).

Implementation of the above-mentioned engineering solutions for the equipment will enable to depart from the boron control system due to introduction of spectral regulation with reactivity margin compensation in the course of burnup, thus reducing the capital and operating costs (particularly due to decrease of liquid radioactive waste amounts).

Application of the new steam generator with increased capacity (one steam generator will replace the two existing SGs) with horizontal primary circuit headers. The reoriented tube bundle will reduce nonuniformity of the load on the evaporation surface and enable to decrease specific metal consumption for the steam generator.

Application of the double-loop arrangement of the MCC with two steam generators and four RCPSs enables to reduce metal consumption for the RP.

Elimination of the hydraulic lock in the MCC will considerably mitigate development of emergency conditions with the primary coolant loss.

Higher coolant flow velocity in the main circulation pipeline due to application of the pipelines with smaller diameter (two DN850 for "hot" legs and four DN500 for "cold" legs of the main circulation pipeline) enables to reduce metal consumption for the RP.

Substantiation of the corium confinement within the reactor vessel in case of any severe beyond design basis accident and, consequently, departure from the external core catcher in the design enables to reduce capital costs.

Relocation of the spent fuel pool beyond the reactor building may facilitate implementation of the scheme with one pool for two power units.

Implementation of the following works (engineering solutions) is also necessary for guaranteed assurance of compliance with the target parameters of the power unit with VVER-S:

- Development of the RP safety system tanks made of composite materials in order to reduce the loads on the civil structures and their material consumption;
- Optimization of the power unit thermal circuit, engineering solutions for the turbine generator design, the turbine hall equipment in order to ensure the maximum turbine generator efficiency factor;
- Development of integrated seismic isolation for the main nuclear island buildings (for the sites with high seismicity);
- Development of the means for long-term (up to 72 h) maintenance of the steam generator condensing capacity in case of any accidents with the primary circuit leakages in order to ensure operation of the passive heat removal system (PHRS) via the steam generator;
- Development of the PHRS for the containment (this will enable to eliminate any need for emergency releases from the containment in case of severe beyond design basis accidents and mitigation of such release consequences).

Technical and economic parameters for the VVER-S technology. Preliminary computational and analytical evaluation demonstrated that application of spectral regulation enabled to achieve important system parameters:

- The possibility for operation within the open fuel cycle with the natural uranium consumption not exceeding 130 t/GW (e) per a year;
- The possibility for operation within the closed fuel cycle with 100% loading of the core with MOX fuel (BR ~ (0.70–0.80));
- Saving of natural uranium up to 30% with the fuel cycle closing in comparison with VVER-TOI operation within the open fuel cycle;

- Assurance of the required safety level for the reactor control and protection without any reactivity compensation with boric acid in all operation modes including outage modes (reduction of the LRW amounts by 50%);
- The safety level compliant with the requirements for 3+ generation NPPs or better;
- The power unit efficiency factor may be increased at least by 10% in relative terms as compared with the parameters of VVER-TOI due to optimization of the power unit thermal circuit, including engineering solutions for the turbine generator design and the turbine hall equipment;
- Reduction of specific capital and operating costs by 10–15% as compared with the parameters of VVER-TOI;
- Reduction of the construction duration by 10% as compared with the parameters of VVER-TOI;
- The possibility for operation in the load-following modes according to the EUR requirements.

25.4 VVER-SKD reactor technology

One of the possible innovative paths for development of light-water reactors is related to usage of water coolant with supercritical parameters enabling to switch to the fast neutron spectrum in the reactor core and to improve the secondary nuclear fuel breeding characteristics of the reactor considerably with simultaneous significant enhancement of thermal energy conversion efficiency in the nuclear power installation. This is a logical trend in innovative development of light-water reactors, but due to the scale of the existing problems it requires international cooperation to solve these problems for efficient and timely implementation.

The main aim of the VVER-SKD reactor technology development is to create the new generation of reactor plants complying with the requirements for sustainable development of the nuclear power system with low capital costs and high nuclear fuel breeding parameters in the course of operation within the closed nuclear fuel cycle which is to be in demand in the second half of the XXI century.

Attractiveness of light-water SCP coolant usage in pressurized water-cooled reactors manifested particularly in the fact that the international experts included the development area for SCP reactors (SCWR) into the set of six promising reactor systems of Generation IV International Forum. Russia made the decision to join the works on supercritical reactors within the framework of Generation IV international cooperation. The set of crucial technologies impossible to be implemented on the basis of the existing R&D headstart is increasingly revealed in the course of development and deepening of the insight into the processes defining the characteristics of this area. Research and development works for expansion of databases and computational models on neutron physics, thermal and hydraulic characteristics of the coolant, corrosion and mass transport, stability of the reactor operation and development of structural materials for the reactor and the core are necessary.

In 2008–2011, Kurchatov Institute, IPPE JSC, OKB GIDROPRESS JSC and Atomenergoproekt JSC performed preliminary R&D works to substantiate the Technical Proposals on the NPP design for three options of innovative Super-VVER reactor with supercritical parameters of the coolant (VVER-SKD). The main systemic requirements VVER-SKD reactors shall comply with were determined. The following target parameters were also defined for VVER-SKD:xs

- Efficient use of fissionable materials (AC ~1.0–1.02) for self-production of fuel in the reactor within the closed NFC;
- Thermal and dynamic efficiency enhancement; NPP efficiency factor of ~40–45%;
- Acceptable specific capital costs and safety characteristics according to the requirements of Generation IV International Forum.

Current aims of the R&D works for development of VVER-SKD technology development:

- Development of the concept, design and also arrangement of the multipurpose ex-vessel SCP loop for investigation of the fluid flow processes, heat and mass exchange, physical and chemical processes in nonisothermal water SCP circuits;
- Development and implementation of the program of experiments with the multipurpose ex-vessel SCP loop.
- Successful implementation of the VVER-SKD reactor technology greatly depends on the joint efforts of all developers within the framework of Generation IV project.

Chapter 26

Expansion of the nuclear power application scope

26.1 Prospects for medium-capacity NPPs

Solution of regional administrative and economic problems, limited intersystem connections at the periphery of the Russian national grid require establishment of the regional nuclear power sector with the capacity range of 100–700 MW (e) within the common fuel and energy complex of the country. Increase of the specific capital costs of a nuclear power unit with reduction of its installed capacity as well as increase of specific operating costs are the main restricting factors. The objectives include maximum possible simplification of the design and engineering solutions and reduction of the NPP arrangement price with simultaneous safety enhancement.

External market may become a significant target for arrangement of medium-capacity NPP power units. It stands to reason that reference power units shall be constructed in Russia.

Modular arrangement of nuclear power plants may facilitate construction of medium-capacity power units with the economic parameters acceptable for this power engineering sector subject to full-scale production.

Reactor plants of VBER type that may be used as the basis for on-shore and floating nuclear power plants of modular design within the capacity range of 170–600 MW (e) are developed based on the nuclear shipbuilding technologies, full-scale manufacturing and operation of more than 460 nuclear power installations with pressurized water reactors [9].

Works for optimization of the design and engineering solutions can be continued in order to achieve economically attractive parameters of medium-capacity NPPs, substantiate their competitiveness in comparison with other power generation types and develop the integrated innovative proposal for the internal and external energy markets.

Closed Nuclear Fuel Cycle with Fast Reactors: Handbook of Russian Nuclear Power.
DOI: https://doi.org/10.1016/B978-0-323-99308-1.00008-8

26.2 Prospects for low-capacity NPPs

Arrangement of low-capacity nuclear power plants (LCNPPs) may be considered as the important factor for maintenance and development of the economic potential in the remote regions without any well-developed power transmission lines and local sources of organic fuel, power supply of mining, enrichment and other energy-intensive production facilities as well as for independent emergency power sources in large-scale grids in order to enhance energy security [9].

Arrangement of LCNPPs shall be also considered in the context of the power supply objectives for development of the Arctic and the Far North regions. Reference power units for these LCNPPs are developed for the benefit of the Ministry of Defense and for the icebreaker fleet and may become the prototypes for new developments (for example, Shelf, RITM-200).

The requirements for the characteristics of these reactor plants are the following: maximum possible factory manufacturing of all plant components, implementation of the serial production, standardization and unification effects, minimization of the on-site operations with fuel, the long core burnup cycle matching with the period of centralized factory maintenance and repair.

Innovative characteristics of floating power units open the prospects for arrangement of the power generation fleet for cooperation with large industrial companies and compensation for energy shortages as well as for the benefit of island countries. The first floating NPP (Academician Lomonosov) is planned to be commissioned for commercial operation in Russia (Chukotka) in 2020.

26.3 Role of nuclear-powered heat supply

The main demand for medium- and low-capacity power units in Russia is manifested in the area of heat supply. Heat supply for the regional needs is principally the concern of the regions. At the same time nuclear industry can propose the optimal variant of nuclear energy source.

It is necessary to analyze economic feasibility of nuclear energy sources under the peculiar conditions of location in comparison with organic fuel (primarily gas) taking into account considerably lower capital costs of steam and gas plants for medium- and low-capacity power units.

26.4 Opportunities for nuclear power installations in power-intensive industry sectors

Implementation of this area together with enhancement of power generation may provide for replacement of organic fuel in power-intensive industry sectors with achievement of the significant share of atomic energy in the energy consumption of these sectors by 2050.

Such use of atomic energy is oriented towards development of high-temperature gas-cooled reactors (HTGRs) and hydrogen production technologies, synthetic fuel production and coal gasification.

HTGRs have certain peculiarities stimulating their application in nuclear power. High temperatures increase power generation efficiency (~50%) and open the possibility to use HTGRs for the process operations.

Implementation of HTGRs will enable to solve the problems related to saving of hydrocarbons for their processing into products with high added value, particularly due to extended use of "nuclear" hydrogen and enhancement of petroleum refinery yield as well as development of high-tech production in the nuclear industry and the allied sectors and implementation of high-efficiency heat and power cogeneration cycles [9].

Application of the HTGR technology may provide the possibility to arrange industrial complexes for commercial-grade hydrogen production for oil refining, petroleum chemistry, etc. and also production of synthetic liquid fuel as coal derivative product.

Chapter 27

Alternative reactor technologies

27.1 Molten salt reactors

Development of reactors with the core in the form of molten fluorides has a long history. In 1965, MSRE-10 thermal reactor (Molten Salt Reactor Experiment) with molten Li_7F-BeF_2-ThF_4-UF_4 having operated for nearly five years reached criticality in Oakridge National Laboratory.

The pilot studies of different options for implementation of molten salt reactors are carried out around the world (Table 27.1). In some cases, for example in China attempts to arrange experimental plants are made. Start-up of the reactor reproducing MSRE is planned in China as early as 2020. Similar to MSRE this project is oriented towards usage of thorium and transmutation of minor actinides.

In Russia the research program for molten salt reactors (MSR) was implemented in 1970s and then continued in the projects of the International Science and Technology Center. Proposals on the design of MOSART reactor based on LiF-NaF-BeF_2 salt with intermediate neutron spectrum were among the results of these programs.

On the assumption of high solubility of UF_4, PuF_3, AmF_3 and lanthanide fluorides in LiF-NaF-KF salt the proposals on the fast MSR design with fast neutron spectrum for burning of minor actinides were considered in Russia. Both MOSART and fast MSRs demonstrate high level of minor actinide burning (~260 kg/year).

The main unsolved problems of this program: Lack of corrosion-resistant materials with long life span, safety analysis for the reactor plant with due regard for reduced safety barriers, MSR economy as well as absence of any regulatory framework for MSRs. High solubility of actinides in LiF-NaF-KF salt used as the basis for development of molten salt fast reactors should be also recognized disputable.

27.2 Fast reactors with the open NFC and TerraPower project

In 1958, Soviet scientists S.M. Feinberg and Ye.P. Kunegin [162] proposed the reactor operating in the depleted uranium make-up mode. It required only the first loading with enriched uranium or plutonium, and then it bred the necessary

Closed Nuclear Fuel Cycle with Fast Reactors: Handbook of Russian Nuclear Power.
DOI: https://doi.org/10.1016/B978-0-323-99308-1.00026-X

TABLE 27.1 Molten salt reactors developed worldwide.[1]

Developer	Project	Support source – purpose
European Union, CNRS, CEA, EDF, IRSN	Molten Salt Fast Reactor – MSFR	Governmental – technology development
Russia, Kurchatov Institute, Rosatom	MSR-S (MOSART)	Governmental – technology development
China, Shanghai Institute of Applied Physics	Thorium Molten Salt Reactor Liquid Fuel #1 – TMSR-LF1	Governmental – technology development
India, Reactor Engineering Division, Bhabha Atomic Research Centre	Indian Molten Salt Breeder Reactor – IMSBR	Governmental – technology development
USA, Flibe Energy Inc.	Liquid-Fluoride Thorium Reactor – LFTR	Private
USA, Thoreact LLC	Thorium Molten Salt Reactor – TMSR	Private
USA, Alpha Tech Research Corp.		Private
USA, ThorCon Power	ThorCon Reactor	Private
Canada Terrestrial Energy	Integral Molten Salt Reactor – IMSR	Private
USA, TerraPower LLC	Molten Chloride Fast Reactor – MCFR	Private
USA, Elysium Industries	Molten Chloride Salt Fast Reactor – MCSFR	Private
USA, Trans Atomic Power	Waste Annihilating Molten Salt Reactor – WAMSR	Private
Denmark, Seaborg Technologies		Private
Canada, Moltex Energy Ltd.	Stable Salt Reactors (SSRs)	Private
USA, Institute for Solid-State Nuclear Physics	Dual Fluid Reactor	Private
USA, Kairos Power LLC		Private

[1] *Molten Salt Reactor and Thorium Energy, Ed. By T. Dolan, WP, Elsevier, 2017.*

amount of plutonium from ^{238}U and burnt it. Such reactors were conventionally called "breed-and-burn." The main difference of the idea from other concepts of breeding reactors resided in the fact that the chain fission reaction took place within the limited reactor zone moving inside the core over time.

L.P. Feoktistov [104] proposing the "travelling wave" concept and H. Sekimoto whose concept was called "Sekimoto candle" [137] further developed the idea.

In 2006, Intellectual Ventures Corporation established TerraPower Venture Company for development and licensing of the technology on the basis of these

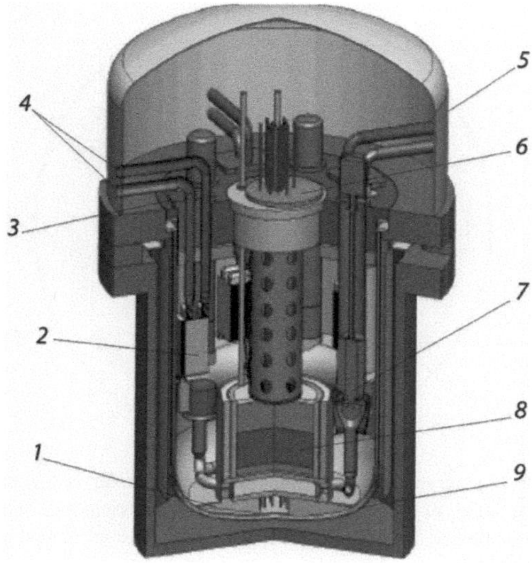

FIGURE 27.1 General diagram of TR-1 RP: 1 – reactor vessel; 2 – intermediate Na heat exchanger (1 of 4); 3 – reactor head; 4 – intermediate Na pipes to the steam generator; 5 – upper containment; 6 – CPS rotating plugs; 7 – main Na pump (1 of 4); 8 – core; 9 – lower containment.

ideas. But it should be noted that the basic principles of the technology with regard to efficient use of natural uranium and accumulated stock of depleted uranium, the need for high safety of the RP and the nonproliferation regime subject to competitiveness correlate with the requirements adopted in Proryv Project Framework. But major difference is related to prohibition of spent fuel processing adopted in TerraPower Project.

Implementation of "Sekimoto candle" in the form of the high cylindrical core with the start-up loading of enriched uranium or uranium-plutonium fuel with the ^{238}U breeding blanket abutting with it from one end was considered in the first option. Gradual accumulation of plutonium takes place in the blanket, and the chain reaction is shifted towards the blanket leaving behind the subcritical area with burnup fuel.

But quite obvious technical problems associated with the necessity to achieve extra-high furl burnup (up to 40%) with the damaging dose of up to 400–500 dpa have resulted in departure from the initial idea by now.

At present the options with refueling and fuel rejacketing or just unloading of burnup fuel assemblies and their replacement by fuel assemblies with depleted uranium are considered. Information on development of the quite realistic TR-1 reactor arrangement (Figs. 27.1 and 27.2) with the electric power of 500 MW and sodium coolant is available; this arrangement mostly repeats the existing (particularly in Russia) experience in development of BN reactors.

The possibility to achieve high fuel burnup determining efficiency of uranium use in the absence of its recycling remains the main technical problem. In this case relatively high radiation environmental hazard of the disposed spent fuel

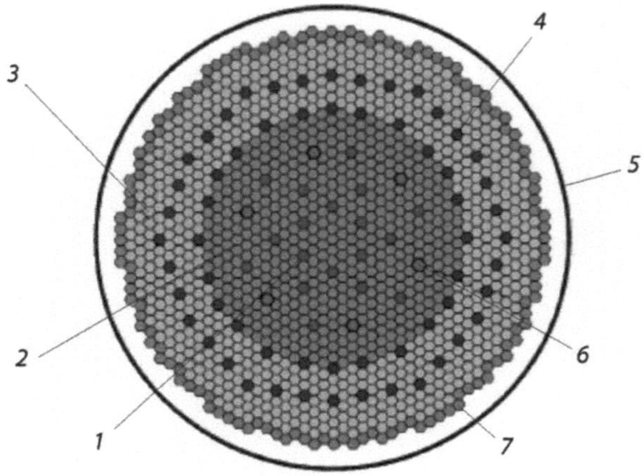

FIGURE 27.2 General diagram of TR-1 core (the core diameter is 6.4 m): 1 – main control assemblies; 2- fuel assemblies of the active control area; 3 – fuel assemblies of the fixed control area; 4 – fixed absorbing assemblies; 5 – reactor vessel; 6 – auxiliary control assemblies; 7 – redundant control/absorbing assemblies.

without its fractionation and transmutation of long-lived actinides and fission products is the main disadvantage.

27.3 Subcritical accelerator-driven systems

The concept of a subcritical nuclear reactor with an additional external neutron source (a proton accelerator is usually proposed as this source) [113] has been mentioned in the publications for quite a long time in discussion of the safety problems for nuclear energy sources. This neutron production technique was called "accelerator-driven technology" (ADS – Accelerator Driven System). The general information presented in this document was taken from the review provided in [75]. The accelerator-driven technology is integrated and more complicated from the technical and physical viewpoint than the traditional reactor technology as it combines both high energy physics which has been purely fundamental up to the present time and reactor physics.

Two main concepts are considered within the framework of the accelerator-driven technology: an accelerator-driven breeder and a subcritical energy amplifier. The former involved availability of a heavy-current proton accelerator with the energy of 1 GeV and the quasi-infinite core made of depleted or natural uranium with a narrow proton beam inserted into it. Emission of heat removed by the coolant and production of plutonium which is regularly unloaded take place in the quasi-infinite core. The generated heat is converted into electric power supplying the accelerator in the secondary circuit. Excess of generated electric power is supplied to the grid. Due to high and extremely uneven power density in the area of the proton beam insertion into the core an intermediate element in the core center – a liquid-metal (usually lead) target for conversion of the proton beam into neutrons – was considered in some studies for the purpose of heat removal.

The concept of accelerator driven breeding based on the approach when an accelerator-driven plant shall supply nuclear fuel for several traditional thermal reactors results in the necessity to arrange accelerators with the current of ~100–300 mA with the accelerator energy of 1 GeV considered as the basic value, while the real average current in the operating accelerators with approximate energy achieved as of the end of 2018 amounts to ~1 mA. Besides, implementation of the accelerator-driven breeding system required development of a great number of related technologies for both the accelerator-driven reactor itself (including energy removal in case of high and nonuniform power density in the beam insertion area and arrangement of the narrow high-current beam insertion window) and the fuel cycle closing technologies: regular unloading of the material unevenly enriched with the produced plutonium, extraction of the produced plutonium, its purification, fabrication of high-level fuel rods, etc. All these reasons taken together resulted in reduction of intensity for the works on ADS technologies and their failure to acquire systematic character.

Carlo Rubbia (Italy), a Nobel Prize winner and a former Director General of CERN, inspired a renewed interest in the accelerator-driven technology. In spite of apparent antagonism of the "reactor community" he was able to create a quite powerful group of like-minded persons, and their works resulted in widespread fascination with Energy Amplifier Project. Rubbia's proposals are essentially the return to the previously considered (1960–80s) scheme of a subcritical fast reactor controlled with the use of an accelerator. The idea of an absolutely safe (subcritical) nuclear power reactor was very positively taken by the public against the background of Chernobyl syndrome. From the viewpoint of specialists, the main advantage of Rubbia's idea resides in replacement of the previous ADS technology paradigm – one accelerator serves several thermal reactors – with the new paradigm (one accelerator – one reactor) [22]. This modification enabled to reduce the requirements for the accelerator current by an order (to ~10–30 mA) which looked quite realistic against the background of the parameters achieved by the acceleration equipment at this time.

The energy generating complex of classic ADS is a hybrid of a subcritical fast reactor and an external neutron source. A neutron-producing target of the limited size (usually Ø20 × 60 cm) located in the center of the subcritical core where a narrow proton beam with the energy of 1 GeV is inserted from the accelerator along the proton channel via an insertion window is used as an external neutron source. The start-up criticality of the core is within the range of $k_{eff} \sim 0.97–0.98$ and is ensured by "seed" ^{235}U. In case of any minor actinides placed into the core together with ^{235}U their gradual burnup will take place. The ADS scheme (accelerator – neutron-producing target – subcritical reactor) eliminating the possibility for any supercritical accident is proposed as an alternative to fast reactors.

The following conceptual advantages of the ADS scheme are specified:

- Inherent safety due to the core subcriticality;
- Harder neutron spectrum in comparison with a fast reactor enabling to burn all minor actinides, particularly the ones with threshold nature of fission;
- The possibility for partial transmutation of long-lived uranium and plutonium fission fragments due to the radiation capture reaction (n, γ) in case of their location outside the core.

But it should be noted that the above-mentioned advantages are not self-evident. So, speaking about inherent safety it should be kept in mind that it is related not only to elimination of any risk of the supercritical reactor state occurrence and uncontrolled reactor runaway, but also to elimination of any possibility for the accumulated radioactivity mobilization due to stored energy of the coolant (for example, superheated water under pressure) or residual heat emission in the fuel in case of any cooling loss. All major accidents at NPPs (except for Chernobyl) and SNF or RW storage facilities were particularly caused by loss of cooling. The supporters of the ADS concept see its strong point in the fact that it deprives the nuclear reactor of its self-consistency property when it combines the functions of a neutron source and energy liberation (which is basically capable of causing an uncontrolled runaway). But the system with an independent neutron source without any self-consistency also fails to provide any guaranteed termination of the reactor operation at the full power level in case of any core damage for any cause not related to the neutron source if this independent source is not disabled. To the contrary, a fast reactor with inherent safety and the minimum excessive reactivity margin basically enables self-shutdown of the reactor without any core damage even in case of instant insertion of the entire reactivity margin resulting in an uncontrolled runaway.

The burnup level for minor actinides returned into the regenerated fuel of fast reactors operating within the closed NFC basically enables to implement radiation-equivalent disposal of waste with the technically feasible periods of their intermediate storage [49]. Partial transmutation of long-lived fission products is also possible in fast reactors in case of their switching to "dirty" regenerated fuel with low level of purification from the above-mentioned fission products.

It may be added to the foregoing that the considered nonevident advantages of the ADS concept are outweighed by the following apparent serious drawbacks [75]:

- The problem of heat removal from the limited neutron-producing target where a narrow proton beam with the power of ~10–30 MW is inserted, and ~60% of its power is set out in an extremely uneven manner;
- The problem of the insertion window for a narrow high-current proton beam with the power of ~10–30 MW;

- The need for ~double accelerator current margin for the nuclear fuel burnup compensation which casts doubt on the declared "inherent safety";
- Burning of highly enriched uranium as the "seed" fuel and low breeding ratios with both ^{238}U and even much lower with ^{232}Th.

Serious physical, technical and technology problems arising even at the conceptual research level prevent from discussion of classic ADS as the basis for large-scale nuclear power.

The concept of a subcritical nuclear reactor in combination with a proton accelerator gained further traction in Russia after 2000 [74,75]. The new trend called "relativistic nuclear technology" (RNT) by its developers involves usage of the deeply subcritical core (made of natural or waste uranium or thorium) and transition to the proton energy of ~10–20 GeV enabling reduction of the required accelerator current by an order with the same beam power and considerable increase of the beam energy fraction for generation of the hard neutron field within the core volume. It is supposed that the RNT reactor core should be multizone with the time-dependent variable composition (and power) and will be shaped in the form of the structured pebble bed of encapsulated fuel rods manufactured in accordance with the micro-FE technology. Heat emitted in the core in the course of the RNT reactor operation should be removed by high-temperature helium coolant. This system does not require any initial loading or make-up with fissionable isotopes and basically provides complete utilization of uranium and thorium resources.

According to the opinion of the concept authors, extremely complicated technical problems associated with the high-current beam insertion window and heat removal from the limited neutron-producing target and basically unavoidable in the classic ADS scheme are almost absent in this scheme. The expected neutron and physical characteristics of the RNT reactor are evaluated through extrapolation of the results of some experimental, computational and theoretical studies. Thus it is stated that the primary proton energy fraction for neutron production may be increased from ~40% with the energy of 1 GeV in the classic ADS to ~66% with the energy of 10 GeV. Subsequent to the reactor start-up its power will increase during accumulation of fissionable ^{239}Pu isotopes in the uranium core or ^{233}U isotopes in the thorium core up to their equilibrium concentration; in this case the system will remain deeply subcritical. Usage of partially processed (only by the mechanical methods, without any radiochemical separation processes) SNF from the operating thermal reactors within the fuel composition is also possible.

Based on the above-mentioned peculiarities of the RNT system the authors of this concept assess it as "the only real opportunity to comply with the fundamental requirements for large-scale nuclear power" and specify the following key advantages:

- Solution of the problem related to fuel resources through direct involvement of ^{238}U and ^{232}Th into the NFC;

- Inherent safety of the relativistic nuclear power system due to deep subcriticality of the RNT reactor;
- Solution of the problem related to the external neutron source in the form of an accelerator by transition to proton energy of ~10 GeV with reduction of the required accelerator current to the implementable level of ~1 mA, reduction of the dimensions and simplification of the beam insertion;
- Solution of the problem related to accumulation of SNF from thermal reactors by its involvement into the NFC of RNT reactors without any radiochemical processing;
- Solution of the problem related to HLW by burn-out of all radioactive fission products and minor actinides in the hard neutron spectrum of RNT reactors with final production of short-lived neutron-deficit isotopes;
- Support of the nonproliferation regime with regard to nuclear materials by elimination of the plutonium extraction process from the NFC.

It should be mentioned here that the above-mentioned provisions are correct only partially and to a different extent. They are discussed in detail below.

27.4 Peculiarities of accelerator-driven systems

Fuel resources. The stated possibility for direct involvement of ^{238}U and ^{232}Th into the NFC is certainly attractive. But it cannot be used as the basis to declare any alternative techniques for achievement of the same goal as inadequate. The available uranium resources are quite sufficient, for example to deploy the large-scale system of fast reactors with BR ~1 on the basis of enriched uranium which will become self-sustained with regard to fuel supply after start-up and the first uploading and actually burn ^{238}U. Upon completion of deployment this system will be able to operate without any use of enriched uranium for almost unlimited period of time.

Inherent safety. As mentioned above, inherent safety is a wider concept than reactor subcriticality. Its main requirement – impossibility of any uncontrolled reactor runaway resulting in a severe accident – may be also fulfilled in a critical reactor with adequate physical properties. At the same time in this point the RNT concept authors miss the risk related to mobilization of accumulated reactivity due to stored energy of helium coolant under high pressure and residual heat emission in the fuel in case of any cooling loss, particularly in the absence of any internal physical links between the RNT reactor core and the accelerator as an external neutron source (situation: an accident at the reactor with loss of cooling, and disabling of the accelerator depends on actuation of the active engineering system with or without intervention of the operator).

Peculiarities of the proposed accelerator seem promising in comparison with the classic ADS scheme but this advantage evidently necessary for practical implementation of the relativistic nuclear power system is not sufficient to solve the issue of the reasonable scale of implementation.

Disposal of SNF from thermal reactors. The possibility to include SNF from thermal reactors into the NFC through simplified processing without any radiochemical technologies is, of course, interesting. But the issues of its relative importance on the scale of the entire nuclear power NFC and compatibility with the processing technology for own SNF from RNT reactors remain open. The latter is totally omitted by the concept authors.

HLW. The statement that the final product of RNT reactors is only in the form of "short-lived neutron-deficit isotopes" seems incorrect. There are several dozens of nuclear reactions per a high-energy proton from the accelerator involving the neutrons generated by it up to their final absorption. In this case the aggregate number of neutrons and protons in the affected nuclei shall be maintained, and their relative content in each nucleus will depend on the particular reaction taking place in it. In case of a fission reaction, the fraction of neutrons in the fragment nuclei will be lower taking into account emitted fission neutrons, but in general this corresponds to the equilibrium composition of lighter nuclei, thus they cannot be considered as knowingly neutron-deficit. In case of (n, xn) reactions the neutron fraction is really reduced in the remaining nucleus, but these reactions rather take place in neutron-excess nuclei and not in neutron-deficit ones. Finally (disregarding any leakage) all generated neutrons should be absorbed in some nuclei after losing of their initial energy and increase the neutron component in these nuclei. Taking into account that the fraction of fissionable heavy nuclei in the system is reduced (with emission of excessive neutrons) and the fraction of fission product nuclei increases the composition of the latter shall shift towards increase of the neutron fraction. New nuclei generated in the course of the above-mentioned reactions may turn out to be stable, but are highly likely to be radioactive. Their further decay depends on the neutron and proton ratio: neutron-deficit nuclei are usually subject to $\beta+$ decay or electron capture thus increasing the neutron content in a new nucleus, and neutron-excess nuclei tend to β- decay reducing the neutron content. In all cases the tendency towards approximation to the equilibrium proton and neutron fractions in stable isotopes is observed, but the above-mentioned statement on the total balance shifting towards neutron-excess nuclei remains valid. There are no reasons why the composition of fission products in the hard neutron spectrum of RNT reactors should turn out to be "better" than in ordinary reactors. Two fundamental conclusions follow from the foregoing: (1) the thesis on generation of the final product consisting mainly of neutron-deficit isotopes in the RNT reactor core is incorrect, and (2) rapid decay of the above-mentioned neutron-deficit isotopes is not the end of the HLW generation process. The composition of SNF from RNT reactors should be investigated through experiments, and at present there are no grounds for the conclusion on solution of the RW problem.

Nonproliferation regime. Departure from plutonium extraction in the course of SNF processing is the common tendency in modern developments with regard to different NFCs both for thermal reactors (REMIX technology) and fast reactors (equilibrium fuel). This statement should be considered as a mandatory requirement and not as unique peculiarity of the proposed concept. But the issue of potential abuses in usage of a powerful neutron source for production of weapon-grade plutonium in the specially arranged sections within the core or near it remains unclear. The system with an independent neutron source may be quite flexible from the viewpoint of its application options.

In conclusion it may be said that the RNT concept provides the basis for continuation of the studies in the ADS area up to development of an experimental unit. However, it cannot be considered as an alternative for nuclear power based on fast reactors with inherent safety without investigation of the RNT reactor NFC and performance of substantial technical and economic assessment.

Chapter 28

Superconducting power transmission technologies

28.1 Prospects for superconducting technologies

Peculiarities in location of NPPs far away from large cities and power consumers result in the necessity for transportation of massive energy flows over great distances. In this case connection of NPPs to the grid involves usage of high-voltage transmission lines (220–750 kV) due to commitment to minimization of energy losses in the course of transportation. This results in the necessity to arrange high-voltage step-up substations at the NPP sites, considerable energy losses (6–8%) in the course of transportation and alienation of significant land plots.

Development of the superconducting technology and creation of a great number of prototypes and experimental sections of superconducting cable lines open new opportunities in solution of the problems related to transmission of massive energy flows over great distances. The achieved level of critical currents in superconducting materials (approximately 200 A/mm^2) and the permanent tendency towards their cost reduction make their use in power engineering (particularly for power transmission) very attractive. At present cable samples with the length of 30–1000 m are developed in some countries (the USA, Japan, the Republic of Korea, China, the European Union countries, Russia); some of these cables are undergoing pilot testing in the grids. Cable lines on the basis of high-temperature superconducting (HTSC) materials with the length of several kilometers and the transmitted power of up to 1000 MW are designed. The main advantage of power HTSC cables resides in the possibility to increase the transmitted power up to the level of 0.5–2.0 GVA with reduction of energy losses and without any significant increase of cable size. They also enable to reduce the cable size without any change of power and to eliminate the intermediate transformation stages, as the power transmitted along a HTSC cable is increased due to the cable current and not the voltage. Thus, the current-carrying capability of a conductor, the energy loss level, and the possibility to transmit high power with reduced voltage are the most important working parameters of a HTSC cable determining its technical and economic parameters. Besides, such advantages of superconducting cables as minimization of the

Closed Nuclear Fuel Cycle with Fast Reactors: Handbook of Russian Nuclear Power.
DOI: https://doi.org/10.1016/B978-0-323-99308-1.00027-1

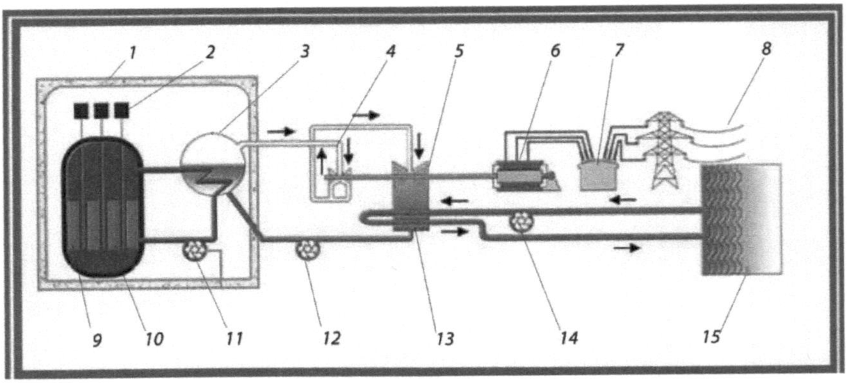

FIGURE 28.1 NPP flow diagram: 1 – containment (leak-tight box); 2 – CPS; 3 – steam generator; 4 – steam; 5 – turbine; 6 – generator; 7 – transformer; 8 – power transmission line to the consumer; 9 – fuel; 10 – reactor vessel; 11 – RCP; 12 – feed pump; 13 – condenser; 14 – circulation pump; 15 – cooling reservoir.

land alienation areas, absence of thermal and electromagnetic leakage fields, fire safety and practical elimination of the insulation ageing processes are also obvious. Based on the above-mentioned advantages we assess the possibility to use HTSC cable lines for connection of nuclear power plants to the grid.

The block diagram of a NPP is presented as an example in Fig. 28.1, and the sections where use of HTSC cable lines is possible are shown. These include short transmission lines from the generator buses to the step-up transformer, from the step-up transformer to the switchgear and the power transmission line to the consumers or the existing mains.

Let us perform technical assessment of the modern capabilities for arrangement of long transmission lines based on HTSC cables according to the following parameters:

- Possible transmitted power levels at different voltage classes;
- Energy losses in the line;
- The line cooling with determination of the maximum distance between cryogenic stations along the line length.

In assessment of long lines, we'll consider direct current lines, as any alternating current cable lines have length limitations due to occurrence of charge currents resulting in power reduction at the far end of the line.

$$I_{CH} = U\omega' C_0 L$$

Where U is the phase voltage, $\acute{\omega}$ is the cyclic frequency, C_0 is the capacity per a length unit, and L is the line length.

As a result, the length of alternating current cable lines does not exceed several dozen kilometers.

With respect to NPPs an alternating current line may be of some interest for connection of the turbine generator buses with the primary winding of the step-up transformer and further with the switchgear (see Fig. 28.1). The distance between the connection points of these lines at the operating plants does not exceed 100–200 m. In this case the output characteristics of turbine generators with the capacity of 220–1200 MW (the voltage of 15.75–24.0 kV, the current of 8.1–28.9 kA) shall be taken into account. The specified voltage values are not critical, and high current values may require arrangement of two- and even three-circuit alternating current lines. This is due to the fact that energy losses in an alternating current cable grow sharply in case of current increase and subsequent increase of the self-magnetic field intensity on the superconductor surface. The length of 1–2 km, the current of 3–5 kA, and the voltage of 220–500 kV are typical for the line connecting the transformer with the switchgear. In this case one high-voltage HTSC cable line is sufficient.

In operation of small-scale independent NPPS with their own grids power can be delivered to the grid directly with the generator voltage via superconducting cables. A generator switchgear shall be arranged at the power plant for this purpose. In some cases, ring circuits based on superconducting cables may be arranged in independent grids. These circuits enhance reliability of power supply considerably.

Apart from the above-mentioned absence of length limitations due to the reasons of electromagnetic nature, direct current superconducting lines have some significant advantages in comparison with alternating current lines, namely:

- No energy losses in the superconductor;
- No need to synchronize the generator operation with the grid;
- Lower load on the cryogenic system;
- Lower consumption of superconducting materials and, consequently, lower cost of the cable system;
- No voltage drops along the HTSC line;
- No undesirable impact on the turbine generator on the grid side.

The direct current transmission scheme in the general form is presented in Fig. 28.2.

An additional component – an integrated valve converter unit performing the functions of a rectifier on the NPP side and an inverter on the grid side – appears in direct current transmission. Manufacturing of these units is well-known for the national and international industry, and they are quite reliable.

The choice between superconducting direct current and alternating current lines in each particular case will depend on the line length, transmitted power, the infrastructure of the adjacent networks and economic considerations.

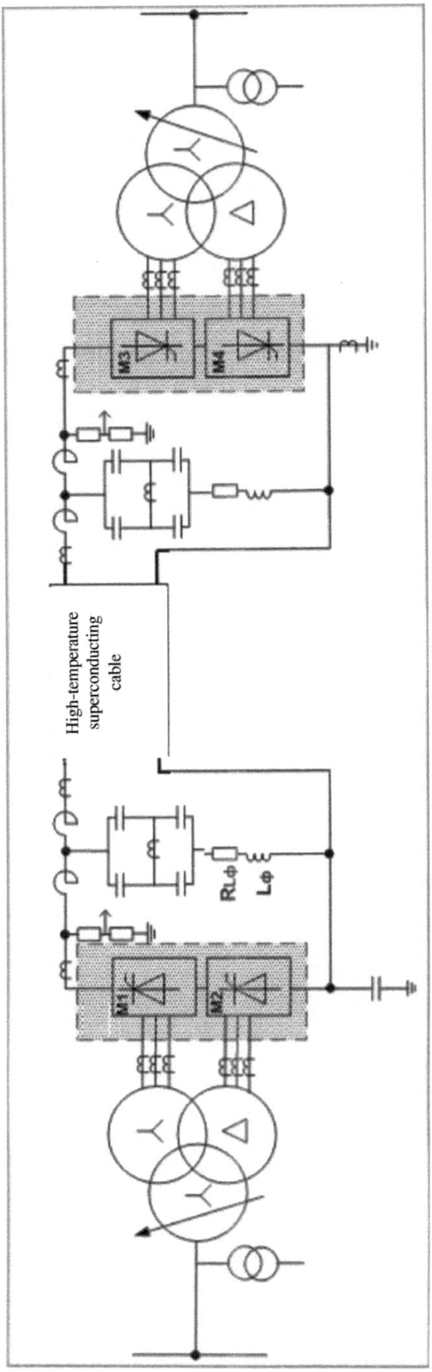

FIGURE 28.2 Basic diagram of bipolar transmission between two substations.

TABLE 28.1 Power transmitted along a direct current HTSC line at different voltage, MW.

Transmission voltage, kV	Monopolar transmission		Bipolar transmission	
	Single line	Two lines	Single line	Two lines
20	300	600	600	1200
50	750	1500	1500	3000
100	1500	3000	3000	6000
200	3000	6000	6000	12,000

28.2 Possible levels of power transmitted along the long direct current line

Let us assess the possible level of power transmitted along a HTSC line at different voltage based on the achieved characteristics of superconducting materials. Subject to the structural critical current density equal to 200 A/mm^2 arrangement of direct current cables with the rated current of 10.0–20.0 kA is real. Cables with the working current of 10.0 kA already exist. Let us assume the working current to be equal to 15.0 kA for the purpose of evaluations presented below in Table 28.1.

It is reasonable to consider transmission along two parallel lines in order to enhance transmission reliability. As can be seen from the table, power of any operating NPP may be transmitted to the grid at the voltage of as little as 50–100 kV. In this case only two or four cables will be necessary for transmission. Energy transmission from a power unit with the capacity of 300–600 MW is possible with the generator voltage without any step-up substation, or with alternating current for a short distance, or with direct current for significant distances.

28.3 Energy losses in the line

Energy losses in a direct current HTSC line consist of:

- Energy losses in the converters – about 2% of the line power;
- Heat inleakage via current leads (only a few kW);
- Losses due to heat in leakage to the cold zone via the cryostat multiplied by the refrigeration coefficient.

The first two values do not depend on the length. Heat inleakage via current leads may be neglected for long lines. Let us assume the energy loss component independent from the length and equal to 2% of the transmitted power for the purpose of calculations.

Heat inleakage via the shell of modern cryostats amounts to 1.0–1.5 W/m. The refrigeration coefficient is equal to 12–18, and thus the specific power losses per a meter of the line length will amount to 12–27 W/m. Let us assxsume the

TABLE 28.2 Length of a direct current HTSC line with the energy losses equal to 3% of the transmitted power.

Power, MW	100	300	500	1000	3000	6000
Length, km	50	150	250	500	1500	3000

average value to be equal to 20 W/m. Let us limit the aggregate losses in the course of energy transmission to 3%, that is, the losses in the superconducting line shall not exceed 1% taking into account the energy consumption for cooling. The calculation results with regard to the maximum length of a HTSC line with the losses not exceeding 3% of its power are presented in Table 28.2.

It follows from the table above that usage of a HTSC line will enable considerable reduction of energy losses in the course of its transportation over long distances.

28.4 Cooling of the line with determination of the maximum distance between cryogenic stations

It is necessary to rely on the following source data in calculation of the line cooling conditions:

- The maximum temperature of the superconducting cable cooled with liquid nitrogen shall not exceed 78 K thus resulting in permissible temperature differential of about 10 K along the length;
- The permissible pressure drop along the length is determined by the cryostat characteristics and amounts to 10–15 atm for flexible cryostats made of corrugated tubes;
- The minimum nitrogen pressure and the maximum nitrogen temperature in the high-voltage zone shall provide for absence of any possibility for gas bubble formation significantly reducing electric strength. In the majority of cases it corresponds to the following conditions: pressure of at least 1.0 atm and temperature not exceeding 78 K.

The first condition results from the fact that the liquid nitrogen phase existence range is limited by the freezing temperature as the lower limit and the boiling temperature as the upper limit and amounts to only 77.4 K–63.2 K = 14.2 K at 1 atm. Although it may be expanded by pressure increase in the system (amounting, for example, to 20.6 K at 2 atm), but the lower temperature limit (the freezing temperature) remains almost unchangeable. Consequently, expansion of the range results in temperature increase at the cryostat outlet and further to reduction of critical current in the superconducting material. Thus it can be seen that a strict physical limitation for ΔT which may be evaluated at the level of 10 K for the sake of definiteness and with certain margin exists for long lines.

TABLE 28.3 Temperature and pressure differential along the HTSC line length.

Cryostat inner diameter, mm	Cable length, km	Liquid nitrogen flow rate, L/min.	Temperature differential, K	Pressure drop, atm
60	2.5	40	4.0	4.8
	5.0	40	7.5	14.0
66	2.5	40	4.0	3.0
84	2.5	40	4.0	0.4
	5.0	40	8.0	1.1
	10.0	60	11.0	4.5
		80	8.4	8.0

In order to provide the specified temperature differential along the cable line it is necessary to pump certain amount of cooling agent for removal of heat supplied to the cryostat. With neglect of the concentrated thermal load at the line ends (current leads) the mass flow rate of liquid nitrogen required for removal of incoming heat and friction-generated heat will amount to:

$$m = \frac{L(q + q_f)}{C_p \Delta T}$$

where: m is the liquid nitrogen flow rate, kg/s; L is the cryostat length, m; q is the specific thermal load via heat insulation, W/m; q_f is the specific heat emission due to friction, W/m; C_p is the specific heat capacity of liquid nitrogen, J/kg·K.

The mass flow rate is related to the flow velocity in the following way:

$$\mathrm{m} = \rho \nu \mathrm{A}$$

where ρ is liquid nitrogen density, kg/m^3; ν is the flow velocity, m/s; A is the channel cross-section area, m^2.

There is no heat emission in the HTSC cables of the direct current transmission systems. With neglect of heat emission due to friction in the cryogenic channel which is usually much lower than heat inleakage via the cryostat shell in the operating modes the pressure drop may be evaluated according to the modified Darcy-Weisbach formula:

$$\Delta p = f \frac{L^3}{2D_h \rho} \left(\frac{q}{C_\rho \Delta NA} \right)^2$$

where Δp is the pressure drop, Pa; f is the friction loss coefficient; D_h is the hydraulic diameter, m.

Typical values of external heat inleakage for modern corrugated cryostats amount to 1.0–1.5 W/m. Based on the above-mentioned relations we assess the heat and mass exchange characteristics for a HTSC cable with the outer diameter of 39-40 mm installed in the cryostats with the inner diameter of 60, 66, and 84 mm. The calculation results are consolidated in Table 28.3.

FIGURE 28.3 Appearance and scheme of the cryostat with a radiation screen: 1 – return tube; 2 – radiation screen; 3 – cable tube.

It follows from Table 28.3 that the pressure drop can be easily adjusted by increase of the cryostat diameter, but in this case external heat inleakage will also grow. Another technique for the pressure drop reduction (at least twofold) involves usage of smooth tubes with bellows junctions. In this case, external heat inleakage may be also reduced, but installation is complicated considerably due to the need for great amounts of welding works along the cable line route.

The temperature differential along the cable length is the main limiting factor for the pumping distance increase. However, as can be seen from Table 28.3 the pumping distance for a cryostat with the inner diameter of 84 mm can reach 10–15 km, and even 20 km subject to the flow rate increase. Unfortunately, the flow rate cannot be increased unlimitedly as additional heat emission due to friction appears at high flow rates. The main technique for the temperature differential reduction is to reduce heat inleakage into the "cold" zone which can be achieved by improvement of the cryostat heat insulation and also by reduction of the temperature difference between the external and internal cryostat shell. The latter may be arranged through application of additional external heat insulation for the cryostat or cooling of the external shell with reverse nitrogen flow in the three-shell cryostat.

The results obtained within the framework of Ishikari Project in Japan may serve as an encouraging example. The developers were able to reduce heat inleakage to the cold zone considerably due to usage of the cryostat made of smooth tubes and a radiation screen cooled with the reverse cooling agent flow as shown in Fig. 28.3.

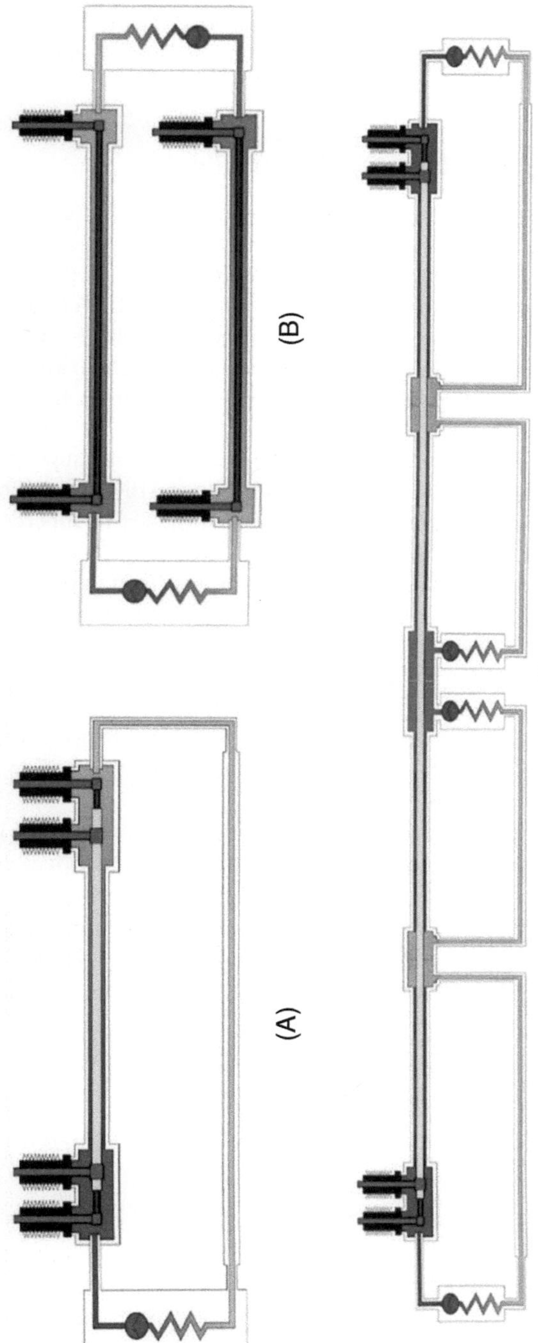

FIGURE 28.4 Possible cooling arrangement schemes for a cryogenic cable line with location of a cryogenic station at one end of the line (A), at both ends of the line (B) and along the cable line route (C).

Heat losses are extremely low due to application of the heat screen thus enabling to increase the HTSC cable length to 50 km. This project is considered in Japan at present.

28.5 Cooling schemes for HTSC cable lines

Some cooling arrangement schemes for cryogenic cable lines are presented in Fig. 28.4.

The simplest cooling arrangement schemes for single-circuit and double-circuit cable transmission are given in Figs. 28.4A and B. As follows from Table 28.3, the distance between cryogenic stations can reach 10–15 km. Subject to improvement of cryostats the pumping distance may reach 50 km. Long lines are arranged by replication of sections as per Fig. 28.4A or 28.4B as shown in Fig. 28.4C.

In conclusion, it should be noted that arrangement of long superconducting direct current cable lines for energy transmission over the distance of dozens and hundreds kilometers is possible due to the state-of-the-art in superconducting and cryogenic equipment. In this case the single line power can reach several gigawatts, and the energy losses will be twofold lower than in overhead power transmission lines. Electric voltage in the line and at the converter substation may be reduced to 100 kV and lower, but its level should be optimized in the context of efficient operation of both the cable and the converter unit. Cryogenic stations for the line cooling may be installed at both ends with the line length of 20–50 km. In case of a longer line cryogenic stations shall be located along the route with the interval of 30–80 km. The maximum line length has no technical limitations under this approach.

Combined lines including transportation of liquefied natural gas or hydrogen in combination with a HTSC cable line appear worthy of individual consideration.

Many factors including infrastructure, technical and economic aspects shall be taken into account for selection of the pilot project of long-distance transmission with the use of superconducting cable lines.

Chapter 29

Experimental facilities of nuclear power

29.1 Set of BFS test facilities

The set of BFS test facilities at IPPE JSC including two critical test facilities BFS-1 and BFS-2 represents the unique experimental facility for studies of the fast reactor physics, solution of their safety problems, optimization of the cores, solution of the fuel cycle closing problems, burnup of actinides, and plutonium disposal. A great number and variety of reactor materials enable to assemble both complex full-scale models of fast reactors and the cores of simple configuration where the experiments for adjustment of neutron and physical constants and improvement of the calculation methods are carried out.

BFS-1 critical test facility was commissioned in 1961. It is intended for studies of the neutron and physical characteristics of fuel assemblies with simple composition as well as designed research and power reactors with the capacity of up to 1000 MW (t) with various types of fuel, fertile materials, coolant, configuration of the cores and blankets. The facility may be used for experimental research on light-water and other reactor types. Availability of MI-30 electron accelerator within the facility (at present – in the temporary shutdown mode) enables to measure neutron spectra in the cores with the use of the time-of-flight method.

The vessel of BFS-1 critical test facility has the form of a vertical steel tank with the diameter of 2 m and the height of 2.2 m surrounded with concrete biological protection. The tank is completely filled with steel or aluminum tubes (~1500 pcs) with the diameter of 50 × 1 mm. Reactor materials – pellets of fuel, fertile, structural materials and coolant are loaded into the tubes in the same manner, amount and proportions as in the cores, breeding blankets and reflectors of the modeled reactors. Some tubes located in the central section of the tank perform the functions of the emergency protection devices, reactivity compensation and regulation of the chain reaction in the core. The test facility is equipped with a graphite thermal column for calibration in the methods using thermal cross-sections as reference, and a metal column for modeling

Closed Nuclear Fuel Cycle with Fast Reactors: Handbook of Russian Nuclear Power.
DOI: https://doi.org/10.1016/B978-0-323-99308-1.00004-0

of the components of in-vessel storage facilities and multilayer protection. The maximum thermal capacity is limited to 200 W. The test facility fuel in the core is cooled due to natural air convection and has the average core temperature limitation of up to 80°C.

BFS-2 critical test facility operated since 1969 is the largest operating critical test facility in the world as of the moment of this book publication. The test facility dimensions enable to perform full-scale modeling of the cores and blankets of fast reactors with the capacity of up to 3000 MW (t) as well as in-vessel protections and storage facilities with different types of fuel, fertile materials, coolant (Na, Pb, Pb-Bi, etc.). BFS-2 critical test facility is similar to BFS-1 facility in its design but has larger dimensions thus enabling to assemble the models of high-capacity fast reactors. The tank diameter is 5 m, the height is 3.3 m, and the number of tubes (fuel rods) in the tank is approximately 10,000. Eight emergency protection devices, nine reactivity compensation devices consisting of six tubes and three control rods may be used at BFS-2 facility. BFS-2 critical test facility is equipped with different experimental devices enabling to perform any studies in a fast and convenient manner, particularly through the use of a rectilinear manipulator for rearrangement of the tubes in the critical assembly tank, movement of samples and detectors within the critical assembly volume in the automated control mode and operation in the oscillator mode.

Composition of metal plutonium, and/ or uranium and/ or uranium dioxide (with ^{235}U enrichment of 36% or 90%) blocks with blocks of fertile or raw materials – metal thorium, uranium and/or depleted uranium dioxide, depleted uranium mononitride is used as nuclear fuel for both test facilities.

The main area of experimental works at the critical test facilities includes studies of the neutron and physical characteristics of designed reactors with different capacity, particularly:

- Assembling of fuel assemblies with simple composition in order to adjust the nuclear and physical constants for various reactor materials, to develop and verify task-oriented libraries of group constants for reactor and protection calculations (benchmark experiments);
- Modeling of the fast reactor cores for substantiation of designs, verification of calculation programs, particularly the calculation programs for neutron and physical characteristics of fast reactors and protection, nuclear and radiation safety parameters and the fuel cycle, dynamic calculation programs (mock-up experiments);
- Experimental studies of the neutron and physical characteristics of fast reactor models, particularly:
  - The critical configuration geometry;
  - The ratio between the average fission and capture cross-sections of various fuel and non-fuel isotopes;
  - Radial and axial distribution of the fission reaction rates for various fissionable materials;

- Sodium and lead void reactivity effect;
- Efficiency of the dummy CPS rods both individually and in groups;
- Central reactivity coefficients of the main reactor materials with the use of the small sample oscillation method;
- Neutron and gamma-quantum spectra through the use of a scintillation spectrometer with a stilbene crystal;
- Space-dependent kinetic effects;
- Effective fraction of delayed neutrons;

– Modeling of in-vessel storage facilities and protection elements as well as geological repositories for nuclear materials;
– Practicing of the measurement methods for physical start-up of the reactor including sensitivity verification for the detectors of start-up equipment, the sequence of fuel installation in the course of critical mass accumulation, measurement of the CPS rod efficiency, spatial distribution of power density, etc.

Extensive experience in operation of BFS critical test facilities and performance of experiments has been accumulated for many years of their functioning. The models of the currently operating reactors (IBR, BOR-60, BN-350, BN-600, and BN-800) and the advanced BN-1200 reactor were studied at the test facilities, compositions with uranium, plutonium and mixed fuel were investigated with both homogeneous and heterogeneous core configurations. Original methods and instruments providing for high accuracy and efficiency of experiments were developed.

The following works within the framework of international cooperation have been quite actively carried out at BFS-1 and BFS-2 critical test facilities since 1990s:

– Bilateral contacts with France, China, the USA, Japan, the Republic of Korea, and India with regard to the experiments for substantiation of the developed reactor plants, disposal of ex-weapon plutonium, the fuel cycle safety, transmutation of minor actinides, practicing and development of the accounting, control and physical protection systems for nuclear materials;
– Works via the International Science and Technology Center (more than 15 fuel assemblies);
– The international project for assessment of benchmark critical experiments (12 critical assemblies, 43 configurations);
– The international project for assessment of reactor experiments (11 critical assemblies);

international experiments for comparison of the measuring methodologies for neutron and physical characteristics with engagement of the critical test facilities of foreign participants – MASURCA (France), FCA (Japan) – in addition to BFS.

Refurbishment of the set of BFS critical test facilities was completed in 2016 within the framework of the Federal Special Program "New Generation Nuclear Power Technologies." The main objectives are accomplished in the course of refurbishment:

- The conditions for timely licensing and obtaining of the required permits for operation of BFS-1 and BFS-2 critical test facilities are established for many years ahead;
- The opportunities for experimental research are expanded considerably due to manufacturing of new materials for modeling of nuclear reactor cores and the nuclear fuel cycle facilities;
- The technical infrastructure enhancing credibility and processing rate of the experimental data is upgraded.

Thus, the set of BFS critical test facilities complies with all modern requirements for nuclear facilities and is ready for accomplishment of the assigned tasks.

29.2 Refurbishment of BOR-60 reactor

Works for retrofitting of BOR-60 reactor plant and its service life extension up to 2020 are carried out in accordance with the R&D programs and design documentation for Arrangement 2.2 "Refurbishment of the pilot fast reactor with the thermal capacity of 60 MW" of the Federal Special Program "New Generation Nuclear Power Technologies for the Period of 2010–2015 with the Extension up to 2020." The aim of the works is to substantiate feasibility of the service life extension, equipment refurbishment and replacement of the worn-out and obsolete equipment in order to ensure safe operation and performance of experimental research within the framework of innovative projects and support of operation of the existing nuclear power installations. The necessity to perform the works within the scheduled preventive maintenance period without any additional shutdown of BOR-60 RP is one of the main conditions for the refurbishment and equipment retrofitting works.

BOR-60 reactor plant has been in power operation for ~274,000 hours by the beginning of 2019. It takes the leading position in the world according to this parameter and continues to demonstrate the potential possibility to extend the service life of sodium-cooled fast reactors.

Description of BOR-60 reactor plant. BOR-60 pilot sodium-cooled fast reactor was constructed for practicing of technologies and wide-scale testing of fuel and materials for power fast reactors with sodium coolant. The basic flow diagram of the reactor is given in Fig. 29.1. Cooling of BOR-60 reactor is arranged in two circuits with sodium coolant in the primary and secondary circuit. The third circuit is steam and water one containing a turbine generator and a heating unit.

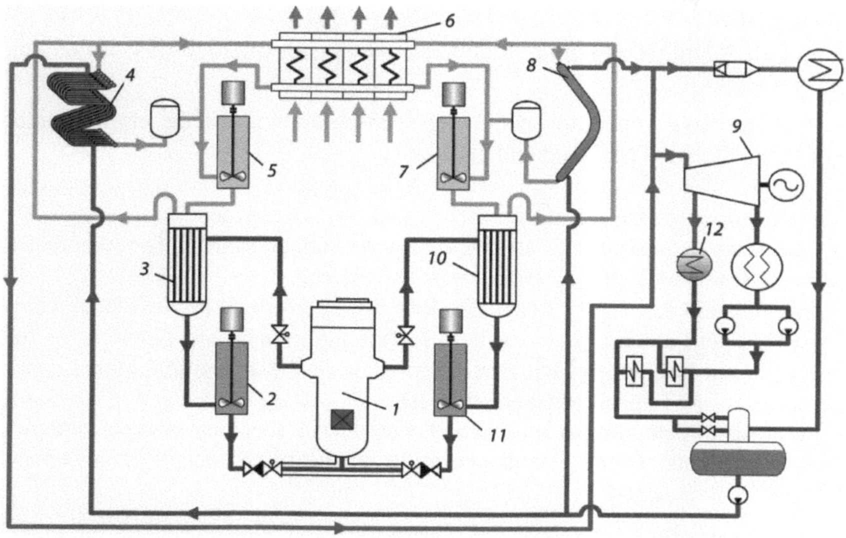

FIGURE 29.1 Flow diagram of BOR-60 RP: 1 – reactor; 2, 5, 7, 11 – primary and secondary circuit pumps; 3, 10 – intermediate heat exchangers; 4, 8 – steam generators; 6 – air heat exchanger; 9 – turbine; 12 – heating unit.

Main characteristics of BOR-60 reactor plant

Rated thermal power, MW	Up to 60
Rated electric power, MW	12
Neutron flux density, $n/cm^2 \cdot s^1$	Up to 3.6×10^{15}
Primary and secondary coolant	Sodium
Sodium flow rate through the reactor, m^3/h	Up to 1100
Maximum sodium temperature:	
At the reactor inlet, °C	310–340
At the reactor outlet, °C	Up to 530
Sodium velocity in the core, m/s	Up to 8
Sodium flow rate in two loops of the secondary circuit, m^3/h	~1400
Third circuit coolant	Steam-water
Steam pressure in the third circuit, MPa	10
Superheated steam temperature, °C	Up to 480
Power operation duration per a year, days	240
Micro-campaign duration, days	Up to 90
Interval between micro-campaigns, days	(9–45)

BOR-60 reactor design was developed on the basis of the service life assurance for 20 years from the moment of commissioning. But the reactor equipment operating experience and results of the experimental studies with regard to behavior of structural materials obtained both at BOR-60 reactor and other sodium-cooled fast reactors demonstrated the possibility and feasibility of further service life extension for the reactor plant. Thus the works for stage-by-stage extension of the reactor plant service life to 30 years at first, then to 40, and then to 50 years (i.e., by 2020) have been carried out since 1988.

Basic works for BOR-60 RP refurbishment. The package of R&D works for substantiation of the possibility to operate the reactor till 2020 was carried out, particularly:

- The equipment operation parameters and conditions for the entire reactor service life were analyzed and clarified;
- Additional studies of the structural material samples cut out of the components operated within the reactor for 3040 years were performed;
- Integrated examination of all reactor equipment and systems for compliance with the technical characteristics was carried out;
- Strength and damageability calculations were performed for the reactor components and assemblies as well as for the equipment and pipelines of the reactor cooling circuits with the use of state-of-the-art computational codes on the basis of the clarified characteristics;
- Analysis of design basis and beyond design basis accidents was performed;
- Arrangements for the reactor operation reliability and safety enhancement were implemented.

Refurbishment of the CPS, the release radiation control system, the emergency control room was carried out in order to bring it into compliance with the requirements of the regulatory documents, the emergency power supply system was retrofitted with replacement of uninterruptable power supply units, storage batteries and installation of diesel generators, refurbishment of the in-process fuel failure detection system, the batch sodium coolant flow rate measuring device for fuel assemblies, the side blanket assemblies and experimental assemblies at the shut-down reactor and the information and measurement system was performed, and revamping of the water supply and sewage systems was also carried out.

Besides, worn-out and obsolete measuring devices of the process control system and components of the core and the reactor internals were replaced.

The results of the performed works enabled to extend the operation license for BOR-60 reactor till 2020. In the long run BOR-60 reactor will be replaced by the multi-purpose fast research reactor MBIR constructed at the same site.

29.3 Multipurpose research reactor MBIR

The multipurpose fast research reactor MBIR is arranged for development of the Russian and foreign promising fast reactor designs as well as for substantiation of the nuclear fuel cycle closing technologies. It is planned to complete MBIR reactor by 2024 and proceed to the physical start-up stage, and then to commence regular material research studies from 2025.

The main purpose of the multi-purpose fast research reactor MBIR is to perform wide-scale reactor testing of innovative materials and dummy core components for the nuclear power systems of generation 4 including fast reactors with the closed fuel cycle as well as thermal reactors of low and medium capacity. It is planned that the potential experimental capabilities of the reactor will be sufficient for experimental research under the national nuclear industry development programs as well as performance of any works for foreign customers including joint international projects within the framework of the international resource sharing center planned to be arranged on the basis of MBIR.

MBIR RP including MBIR reactor is intended to ensure performance of a wide range of research and experimental works with the use of reactor radiation in the following areas:

- Radiation testing of promising fuel types, absorbing and structural materials for various types of nuclear reactors under the conditions of intensive neutron radiation with the flux density of $(2–5)\cdot 10^{15}/cm^2s$;
- Reactor testing of fuel rods, fuel assemblies, absorbing elements, other core components in steady-state, transient and emergency operation modes for the next generation innovative reactors with promising coolant types (sodium, heavy metal, gas, molten salt coolant, etc.);
- Reactor studies for substantiation of the new and modified liquid-metal coolant technologies;
- Integrated computational and experimental research in order to obtain the required information for development and verification of the design codes;

testing of the new equipment types for various process systems (components of the emergency cooldown systems, steam generators, cold oxide traps, etc.), control devices and systems, monitoring and diagnostics for different types of power reactors;

- Practicing of the technologies for production of radionuclide products for various purposes, production of modified materials;
- Applied research with the use of reactor radiation including research with the use of extracted neutron beams.

MBIR reactor is equipped with the turbine providing for electric power generation.

Main characteristics of MBIR reactor

Thermal reactor power, MW	150
Arrangement	Looped
Number of cooling circuits	Two
Primary and secondary coolant	Sodium
Third circuit working medium	Steam-water
Protective medium in the gas cavity of the main reactor vessel and the contingency case	Argon
Coolant flow direction in the reactor	From the bottom upwards
Coolant temperature:	
At the reactor inlet, °C	330
At the reactor outlet, °C	512
Pressure in the primary circuit, MPa, not exceeding	0.7

(continued on next page)

Fuel type in the working (regular) fuel assemblies	Mixed oxide uranium-plutonium
Ready-to-operate FA size, mm	72
Number of fuel assemblies in the core	93
Fuel portion height of the fuel rod, mm	550
Number of cells for location of the side blanket assemblies	Up to 350
In-vessel storage facility for spent fuel assemblies	36 cells
Reactor operation duration between refuelings, ef. days	100
Design service life, years	50

The following experimental equipment installed in the reactor vessel is provided in MBIR (Fig. 29.2):

- Loop channels for modeling of the NPI core operating conditions with various coolants;

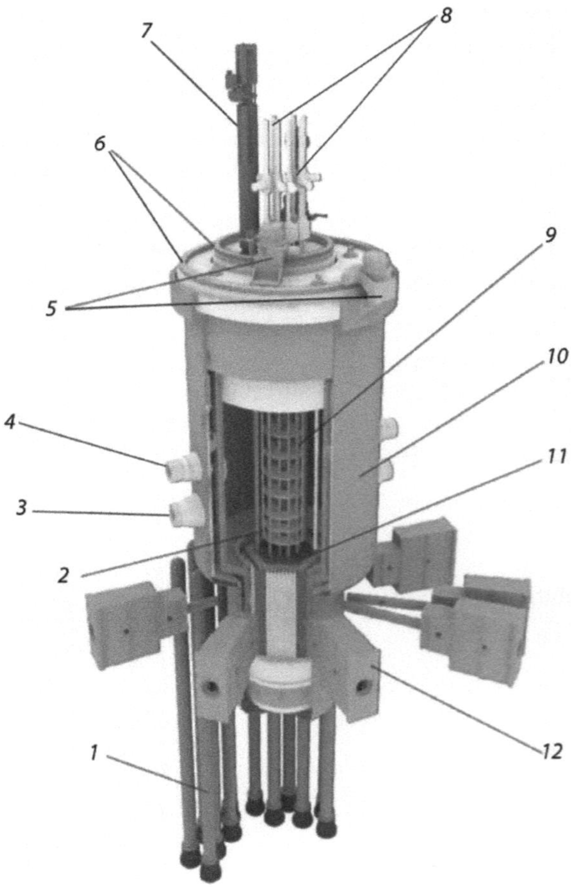

FIGURE 29.2 General view of MBIR reactor: 1 – vertical experimental channel; 2 – loop channel; 3 – inlet pipeline; 4 – outlet pipeline; 5 – plug drives; 6 – rotating plugs; 7 – refueling mechanism; 8 – drives of CPS actuators; 9 – experimental channel; 10 – vessel with the contingency case; 11 – core; 12 – horizontal experimental channel.

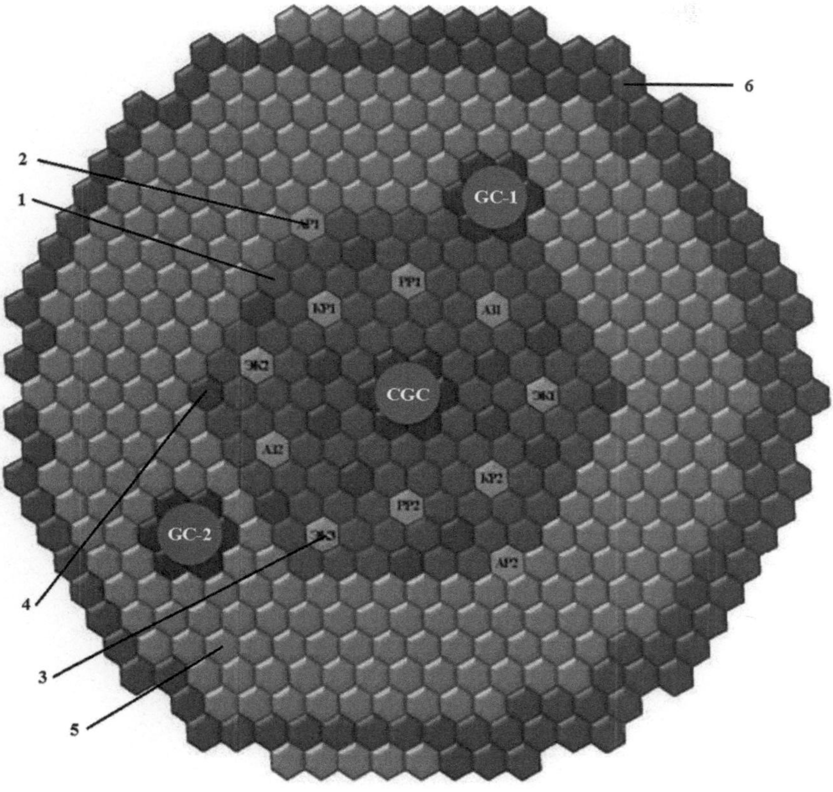

FIGURE 29.3 Core diagram: ● – fuel assemblies; ● – CR (control rods); ● – EC (experimental channel (instrumented)); ● – material research assemblies; ● – SB (side blanket) assemblies; ● – assemblies of the in-vessel storage facility. 1 - fuel assemblies; 2 - CR (control rods); 3 - EC (experimental channel (instrumented)); 4 - material research assemblies; 5 - SB (side blanket) assemblies; 6 - assemblies of the in-vessel storage facility.

- "Channel – loop" type units with various coolants – independent channels with assurance and maintenance of the specified thermal and hydraulic parameters of the coolant and cooling of fuel rods due to natural or forced circulation of the coolant arranged within the channel, with terminations of measuring cables and reactor service lines;
- Instrumented material research assemblies for testing of fuel, absorbing, and structural materials (Fig. 29.3);
- Noninstrumented material research assemblies;
- Noninstrumented assemblies for isotope production.

MBIR research reactor is unique not only due to small number of such installations, but also due to its ideology and design. It is planned to arrange the International Research Center on the basis of MBIR reactor (MBIR IRC).

Chapter 30

Digitalization in nuclear power

30.1 Digital technologies for modeling of NPE facilities

Various modeling technologies – information, 3D- and 4D-, computational, and others have been applied for a long time and are traditionally well-developed in the nuclear industry due to high cost of the facilities, the existing nuclear and radiation risks in the course of their operation and consequently extremely high price of any potential error, whether in operating control or long-term planning of design, construction, and operation. Some operation modes of nuclear facilities, first of all emergency ones, are investigated and substantiated mainly or only through the use of computational methods.

Specialization of various models even within the framework of the same area historically existed and is still preserved to a great extent due to peculiarities of the applied tasks solved through the use of these models. For example, computational models used for safety analysis of NPP power units may be quite considerably and even fundamentally different from the models used for development of full-scale and analytical simulators for the same prototype power units in their properties and characteristics – almost polar opposite sets of requirements for the modeling scope, rate, and accuracy are applied to these models. Consequently, models similar in their functions are developed and applied according to quite different technologies.

Similar situation is observed in 3D modeling – comprehensive highly detailed engineering models of any equipment and general models of the facility sites.

Nevertheless, gradual emergence of complex integrated models should be recognized as an obvious general trend in the recent 10 years. Integrated models describe the modeled facility as a whole with due regard for all constituent systems, their functions and interfaces and enable to solve various modeling tasks (i.e., they are multifunctional). They have permanent regular possibility for one-way link to the prototype facility in order to obtain relevant information on its state which can be used for prompt preparation of the calculation scenarios for modeling or verification of correct operation (both the model in the course of its adjustment and the prototype facility in this or that mode).

Closed Nuclear Fuel Cycle with Fast Reactors: Handbook of Russian Nuclear Power.
DOI: https://doi.org/10.1016/B978-0-323-99308-1.00002-7

Such integrated multitask complexes – so-called digital twins – are getting widespread as they have the following apparent advantages:

- In integrated modeling of the entire facility it is possible to investigate operation of all its process systems and equipment more thoroughly with due regard for the maximum number of interfaces, thus increasing the chances for detection of any potential errors as practice shows that the majority of possible collisions appear at the borders between systems/facilities;
- One model requires less financial and human resources for its development and use in the design approach to management than a set of different models duplicating each other partially or completely thus reducing the total costs. The approach when the integrated model is made of a set of individual models with a separate team of authors in charge of their development and practical application behind each of them is also applicable;
- The scope of the tasks possible to be solved with the use of the model with sufficient credibility and predictive performance grows considerably with the modeling scope increase.

Objective difficulties in development of digital twins include high labor intensity of their development requiring application of the advanced work planning and control systems. Besides, significant increase of the scope and complexity of modeling often require expensive engineering solutions – application of supercomputers and grid technologies for performance of the most resource-intensive calculations.

Nevertheless, in spite of any difficulties examples of multi-task computational complexes exist in the national nuclear industry:

- NPP power unit simulators used not only for the personnel training but also for development/ validation of the manuals as well as for testing of APCSs and functional analysis of the control systems;
- The integrated model of VVER-TOI had the interface with the information model of the project implemented in Intergraph SmartPlant, as well as with specialized software enabling automated generation of the design diagrams of mathematical models and their saturation with the source data based on the data import from the information model. As a result, the possibility for end-to-end in-process computational verification of the design solutions was provided:
- A virtual and digital NPP with VVER and a virtual power unit with VVER - sophisticated integrated design complexes modeling different operation modes of their prototype power units with due regard for additional failures of equipment and enabling to perform extensive computational studies. They are distinguished for highly detailed models of the primary circuit performed with the use of precision design codes and requiring application of supercomputers;

- The integrated computational mathematical model of PDEC pilot demonstration power supply complex with dynamic calculations of not only operation of the power unit with BREST-OD-300 RP but also the equipment of the near-plant nuclear fuel cycle. It contains the set of unique models, such as the model of personnel actions, the model for rapid evaluation of the economic consequences of any operational occurrences, etc.

It should be noted that the crucial issue of further development of digital twins resides not only and not so much in development of any computational and technical means used thereof, but also in arrangement of the business processes for formation and application of digital twins. The following conditions shall be also taken into account in planning of development and application of digital twins in a certain project. It is necessary to start formation of the digital twin at the earliest possible stage of development in order to provide for proactive comprehensive analysis of the facility in different virtual representations. The projects with large amounts of scientific research and R&D works require not only the digital twin of the future facility (NPP) but also availability of all scientific research and R&D work results and verified databases in the digital environment as well as initial arrangement of their performance with the use of proper digital solutions. Flexible opportunities for participation in development and application of digital twins (common information space) shall be provided for the projects with many participating organizations (some of them may not belong to the industry).

30.2 Digital technologies for nuclear facility development and life cycle management

The life cycle of nuclear facilities is quite specific with regard to both composition of stages and their number and duration. The typical life cycle model for a product or system usually has four to six stages and the duration of up to 10–20 years. The life cycle of nuclear facilities is fundamentally different – Federal Law No. 347-FZ dated 30.11.2011 defines it in the following way: "... depending on the category of the target facility for the present Federal Law the term "full life cycle of a nuclear facility" means siting, design (including surveys), engineering, production, erection or construction (including installation, adjustment, commissioning), operation, refurbishment, overhaul, decommissioning (closure), transportation, handling, storage, burial and disposal of the nuclear facilities". The life cycle duration of nuclear facilities with due regard for the design and disposal stages can exceed 75 years.

The concept of a nuclear facility life cycle has developed since 2011, and at present this life cycle is considered as the set of life cycles of NPPs (Fig. 30.1), nuclear fuel and RW overlapping at the NPP operation stage.

The facility life cycle consists of the following stages: the idea of construction, site selection, design and engineering, equipment manufacturing and

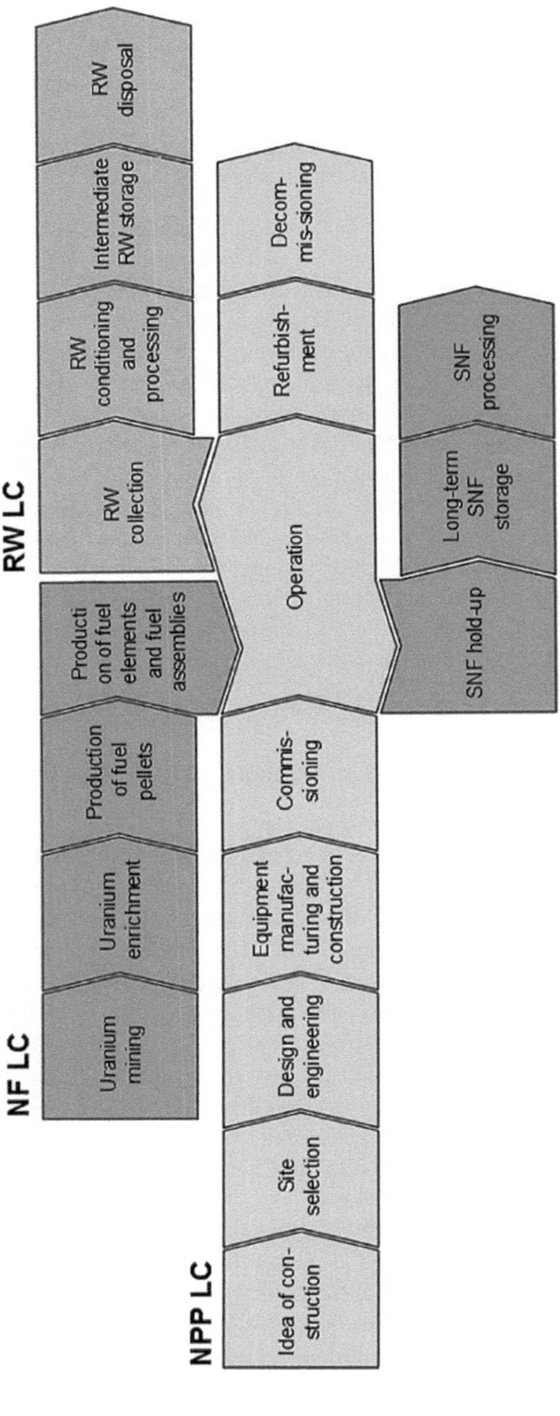

FIGURE 30.1 Integrated life cycle (LC) of a nuclear facility.

construction of the NPP, commissioning, operation and NF use, refurbishment, decommissioning. The nuclear fuel life cycle consists of the following stages: uranium mining, enrichment, manufacturing of fuel pellets, production of fuel assemblies and fuel rods, NF use, SNF hold-up, long-term storage and processing. The RW life cycle is usually represented by the stages of collection, conditioning and processing, intermediate storage and disposal. Although these three life cycles overlap at the NPP in case of the classic technology different stages are implemented at different geographically dispersed facilities. PDEC Complex within the framework of Proryv Project Framework where almost all stages of all three life cycles are implemented at the same site is an exception.

Apart from the above-mentioned life cycles the software life cycles are inevitable present at the nuclear facilities – both the design codes for confirmation of the design, engineering and process solutions and the facility APCS. The problem related to support of the software life cycle is usually solved locally by the software developers with the use of specialized products – ALM systems (Application Lifecycle Management) and is not always integrated with the facility life cycle. The Design Code Development Management System (DCDMS) developed within the framework of Proryv Project Framework and providing unified approach to management of the software life cycle process, available and timely support for the code users (due to their inclusion into the common information space), prompt control due to informational transparency of the project implementation progress, alienability of the design code from the developer can be presented as an example of successful software life cycle management.

Complexity and specificity of the nuclear facility life cycle determine the peculiarities of arrangement for the life cycle management systems. The main peculiarity resides in impossibility to manage the nuclear facility life cycle with the use of one system as in case of construction or machine building tasks. In spite of the well-developed market of life cycle data management systems (PLM – product lifecycle management), there are no ready-made systems capable of supporting all stages of nuclear facility life cycles. The only thing that can become the basis for all life cycle stages for the entire period of its existence is the set of structured data on the facility and technologies supplemented at all life cycle stages and sufficient for completion of the tasks at each life cycle stage, particularly scientific and technical substantiation of the adopted solutions, verification of the solution competitiveness criteria, project licensing, verification of the nuclear and radiation safety assurance criteria and product quality parameters, identification of intellectual property. This set is called the information model of the facility. The information model structure shall provide the following opportunities:

- Development and fixation of the information model at different life cycle stages, recording of the facility state "as designed," "as built," "as accepted for operation," "as operated," etc.;

- Structuring of the data and informational links between the data of different life cycle stages;
- Data management (versions, states, life cycle progress);
- Data visualization (3D and 4D);
- Control of access to the data for different groups of users and life cycle stages;
- Integration with the specialized software specific for particular life cycle stages.

The basic functions of the information model are supported with the use of the life cycle data management information system. At present there are no standards for such systems in the industry, and the systems of almost all engineering data management market leaders are used: ENOVIA (Dassault Systemes), Windchill (PTC), Teamcenter (Siemens), Smart Plant (Intergraph), Vault and Fusion (Autodesk).

Joint work of users with information models is supported by the common information space – the information environment enabling the participants of facility and product life cycle processes to exchange information and to use common informational resources with clear identification and understanding of the information received from each other including engineering and other data related to the facility. The common information space solves three main problems:

1) Consolidation of all life cycle support software tools into one secure software system;
2) Access for remote users via protected communication channels;
3) Reliable long-term storage of information.

Common data storage with the use of specialized data processing systems is usually arranged when it is necessary to solve the task for storage of a great amount of related heterogeneous data and their processing in accordance with various algorithms. A federated model is the most widespread option; it is arranged by cooperative efforts of different specialists in different programs with their own file formats, and assembly of the common model is carried out in special-purpose "assembly" programs (Autodesk Navis Works, Bentley Navigator and the like). In this case the model parts maintain their specific structure, and changes are implemented only via their parent programs. This approach enables more efficient working with the model parts through the use of specialized software, but also requires operability maintenance for all applied programs within the entire project life cycle amounting to more than 50 years at the lowest estimate.

An integrated model is another approach. Such model is assembled from the parts produced by conversion into universal data presentation formats. This approach requires availability of the relevant converters, can deteriorate the functional capability, but considerably reduces the amount of software to be maintained within the project life cycle.

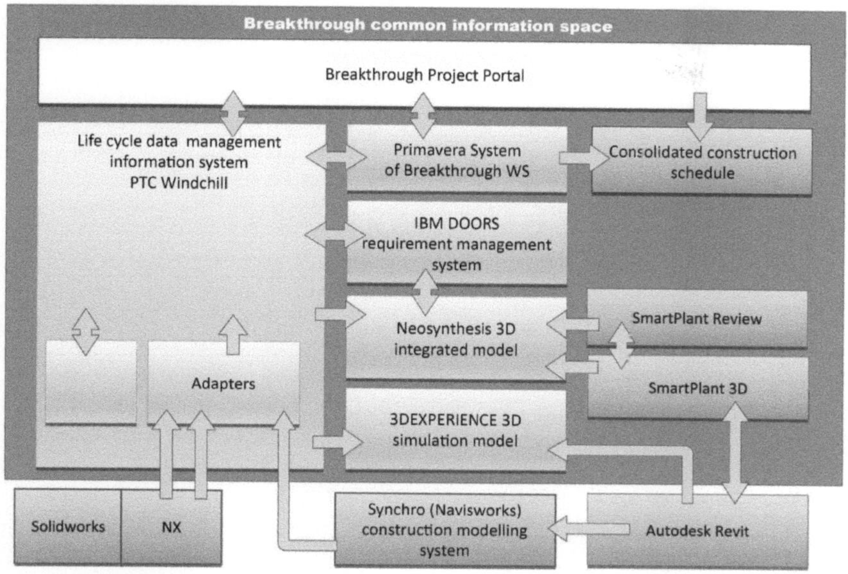

FIGURE 30.2 Structure of Proryv Project Framework common information space.

This particular approach was used in development of the common information space for Proryv Project Framework. The life cycle data management information system developed on the basis of PTC Windchill PLM system is used as the system core. Data and models developed in other systems are converted for storage through the use of the relevant converters or accepted in the universal data presentation formats: STEP, XML, PDF, and DOC. In order to ensure impossibility of any non-recoverable data loss in the course of conversion the source data are also stored in the as-developed format but used only in contingency and exceptional cases. A great amount of specialized software is applied at the development stages but its availability and functioning are not mandatory at any subsequent life cycle stages.

The main software tools of Proryv Project Framework common information space and their informational interaction are shown in Fig. 30.2. After the facility commissioning support within the remaining life cycle stages will be necessary only for the life cycle data management information system – the part of the common information space highlighted in green color. Other programs may be removed from maintenance or replaced with any counterparts or specialized software for the relevant life cycle stages.

This approach to implementation of Proryv Project Framework common information space not only provided for efficient working of many work performers at the stages of R&D, design and construction but also gave reasons to expect efficient application of the common information space at all subsequent life cycle stages of Proryv Project Framework facilities up to their decommissioning.

Chapter 31

Regulatory framework for the modern and future nuclear power

31.1 Regulatory framework for nuclear power in the Russian Federation

Legislation of the Russian Federation in the area of atomic energy use is based on the Constitution of the Russian Federation, the generally accepted principles and norms of the international law as well as on the international treaties of the Russian Federation in the area of atomic energy use and consists of federal laws of the Russian Federation and other regulations of the Russian Federation adopted pursuant to these federal laws.

In this case the federal regulations and rules (FRR) in the area of atomic energy use establishing the requirements for safe atomic energy use including the safety requirements for nuclear facilities, safety of any activities in the area of atomic energy use, particularly safety goals, principles and criteria constitute the regulatory framework for safety in atomic energy use.

At present the FRR system including more than 100 documents is established and effective. Besides, safety guidelines in the area of atomic energy use containing the recommendations, particularly on the work performance methods, methodologies, expert appraisal and safety assessment as well as clarifications and other recommendations on compliance with the safety requirements in atomic energy use are developed, approved and implemented to facilitate compliance with the FRR requirements. As of 2019 more than 130 safety guidelines in the area of atomic energy use are effective in the Russian Federation.

Development of science, technology and production in the area of atomic energy use results in the possibility to develop and implement new design and engineering solutions, technologies and structural materials improving the technical characteristics, quality and economic attractiveness of nuclear facilities. Of course, the adopted innovative engineering solutions shall ensure protection of the nuclear facility workers (personnel), the public and the environment against any unacceptable radiation and other exposure.

Closed Nuclear Fuel Cycle with Fast Reactors: Handbook of Russian Nuclear Power.
DOI: https://doi.org/10.1016/B978-0-323-99308-1.00037-4

31.2 Peculiarities of the new nuclear power technology platform projects from the viewpoint of legal regulation

Active works within the framework of the new nuclear power technology platform have been carried out in the global nuclear power since the early XXI century. The international Generation IV nuclear reactor development project and the international project on innovative nuclear reactors and fuel cycles (INPRO) were launched.

In the Russian Federation innovative reactor plant designs are developed particularly within the framework of Proryv Project Framework the main aim of which is to develop new generation nuclear power technologies on the basis of fast reactors with the closed nuclear fuel cycle for nuclear power plant satisfying the country's needs for energy resources and to enhance efficiency of natural uranium and spent nuclear fuel use with the balance of safety requirements and economic feasibility.

Proryv Project provides for development and arrangement of BRESt-OD-300 pilot demonstration lead-cooled fast reactor (chief designer – NIKIET JSC).

The main peculiarity of innovative projects resides in the fact that apart from application of the technical and organizational solutions tried in the preceding operation experience according to the requirements for implementation specified in the effective FRR these projects contain the solutions that have never been used at nuclear facilities before and consequently are not regulated in the regulatory documents.

For example, for the design of BREST-OD-300 reactor plant these solutions may include:

- Use of lead as coolant;
- Metal and concrete casing with in-built metal containments;
- Structural design of some equipment and pipelines.

Safety analysis of the solutions adopted in the design is performed through implementation of the integrated approach including:

- Reliance on the existing experience in design of reactors with liquid-metal coolant (particularly heavy-metal one);
- Analysis of the existing and proposed patented solutions;
- Design of the structures (the final stage is the basic design);
- Computational substantiation;
- Research aimed to substantiate application of new materials;
- Testing of equipment models and components.

The results of the above-mentioned works represent the basis for development of the new requirements that may be established in the FRR with regard to the nuclear facilities under consideration at the relevant life cycle stage.

Safety analysis in its turn is performed through detailed computerized modeling with the use of the design codes verified for individual phenomena on the representative basis.

Application of the codes subjected to the relevant expert appraisal is legally permitted for safety analysis. New generation precision computational complexes describing a wide range of physical processes and phenomena in the interrelated manner are developed at the RAS Institute for Safe Development of Nuclear power in cooperation with scientific, design and engineering organizations within the framework of the new technology platform. Multiphysical nature and multi-scale modeling are the main features of the new generation software packages.

Thus, innovative development of the nuclear industry shall go on simultaneously in several interrelated areas: Development of innovative nuclear power plant designs and development of the arrangements for safety assurance – compilation of FRR and other regulatory documents as well as advanced software tools (new generation codes).

33.3 Regulatory framework analysis and improvement

It is very important to ensure establishment of the legal and regulatory framework for safe atomic energy use enabling to achieve the required safety level at each stage of nuclear power development.

Analysis of FRR application aimed to develop the proposals for improvement of the regulatory system, particularly for establishment of new requirements, elimination of any omissions and collisions of the regulatory requirements and detection of any outdated requirements is carried out on the continuous basis.

In this case, it should be noted that development of any regulatory documents for nuclear facility safety assurance, particularly with due regard for the peculiarities of innovative reactor plant designs is impossible without consideration of the practical experience obtained at each project life cycle stage (for example, determination of the structural material properties) and the computational substantiation results from the very beginning of development.

The stages of interaction between the developers and the supervision bodies for development of innovative equipment may be traced through the example of Proryv Project. So, close cooperation of the specialists and experts of the Federal Environmental, Industrial and Nuclear Supervision Service, including the scientific and technical support organization (the Research Center for Nuclear and Radiation Safety), Rosatom, industry-specific research organizations and the RAS as well as the innovative project developing organizations was arranged for implementation of Proryv Project. Applicability of the effective FRR for safety analysis of a nuclear power plant with BREST-OD-300 reactor plant was analyzed with due regard for novelty of the engineering solutions, materials, and technologies adopted in the design. Subsequent to its results recommendations for establishment of the regulatory framework in the area of safe atomic energy use within the scope sufficient to obtain the construction license were developed.

Based on the performed works Rosatom approved the Plan for development (review and updating) and enactment of the regulatory documents with regard to the safety issues in accomplishment of the objective for development of fast reactors with the closed nuclear fuel cycle by the organizations of Rosatom establishing the priorities and the time limits for their implementation with the consent of Rostechnadzor.

New regulatory documents are developed on the basis of consolidated national and foreign experience in regulation of the requirements for assurance of integrity of the high-temperature nuclear installation components, particularly with lead coolant. In this case knowledge of the design, manufacturing, installation, construction and operation peculiarities of reactor plants with lead coolant at the current life cycle stage as well as the substantiating R&D results for the adopted engineering solutions in accordance with the state of the art in science, technology, and production is taken into account.

In order to maintain the leading positions of the Russian nuclear power and support competitiveness over the medium- and long-term horizon it is necessary to keep the reasonable balance of the improvement rates for the regulations in the area of atomic energy use and implementation of innovative and other promising projects of the Russian Federation in the area of atomic energy use with the engineering solutions adopted with due regard for the state of the art in science, technology and production.

In this case the regulatory framework in the area of atomic energy use shall be updated in the course of accumulation of experience in arrangement of nuclear facilities at all stages of their life cycle.

Part V

Strategic guidelines for establishment of two-component nuclear power engineering

Chapter 32

Optimal development scenarios for the Russian nuclear power

32.1 Basic provisions of scenario analysis

The scenario analysis is based on the set of documents containing the goals, fundamental provisions, and paths for implementation of the Russian nuclear power development strategy:

- State Program of the Russian Federation "Development of Nuclear power and Industrial Complex" up to 2027;
- Innovative Development and Technology Upgrade Program of Rosatom for the period up to 2030 (in the civil sector);
- Federal Special Programs "New Generation Nuclear Power Technologies for the Period of 2010-2015 with the Extension up to 2020" and "Nuclear and Radiation Safety Assurance for the Period of 2016-2020 and up to 2030";
- Basic provisions of the Russian Nuclear power Development Strategy up to 2050 with further extension for the period up to 2100.

The existing competitive selection model for new power capacities prevents from determination of competitiveness for NPP power units as it is focused on the commissioning period (4–6 years) and capital costs of TPPs. A different model is necessary for fair competition assurance where the levelized cost of electricity (LCOE) and the commissioning period increased to 7–10 years shall be the main parameter of competitiveness. In this model competitiveness of NPP power units will be primarily determined by the internal gas prices. It is obvious that the discount rate (the cost of money) will be one more pay-back factor for NPPs due to high capital costs.

NPPs of new design have greater competitiveness margin due to their optimized characteristics. The project of Kursk NPP-2 construction implemented as of the moment of this book publication loses its competitive advantages only for the variant with extreme conditions - the maximum possible capacity utilization factor of thermal generation and low gas price growth rate. So, it is necessary to rely on the economic parameters of NPP projects in selection of the strategic development variant.

Closed Nuclear Fuel Cycle with Fast Reactors: Handbook of Russian Nuclear Power.
DOI: https://doi.org/10.1016/B978-0-323-99308-1.00019-2

In cases when pay-back of new NPP power units is not ensured subsequent to the competitive selection results (for example, in construction of the first of a kind power units of new design in order to gain reference for export) additional support arrangement will be required for their construction in the form of capacity delivery agreements, cofinancing from the corporate or federal budget, etc.

Pressurized light-water reactors VVER with high level of reference supported by well-developed technology basis are the backbone of the nuclear power in Russia and foreign export in 2020. Rosatom has the predictable portfolio of contracts for construction of NPPs with VVER reactors till 2035 [9]. Nevertheless, the current development level of this reactor technology fails to guarantee maintenance of competitiveness at the internal and external markets over the long-term horizon, and the need for a new commercial product will arise as early as 2030.

Taking into account the development level of the competitive proposals particular relevance of the promising NPP designs with pressurized water reactors with the capacity of ~1200 MW (e) is notable. Due to their new properties such NPPs can participate in load regulation within the ecosystems. Application of spectral regulation in the safety system enables to decrease material consumption and to reduce the operating and capital costs and duration of construction. The efficiency factor of these NPPs is higher due to optimization of the power unit thermal circuit and engineering solutions for the turbine generator design and the turbine hall equipment. The fuel cycle is improved with accident-tolerant fuel and increased breeding ratio. The amount of generated operating RW is reduced.

In this case the advanced NPPs with pressurized water reactors with the capacity of ~1200 MW (e) provide for compliance with the systemic requirements for the safety level eliminating any need for evacuation, let alone relocation of the public from the adjacent territories of the NPP in case of any emergency situations and are competitive in comparison with alternative generation sources.

Arrangement of marketable fast reactors (hereinafter - FNRs) with liquid-metal coolant is also important for sustainable development of nuclear power, including removal of any limitations with regard to the resource base due to breeding of nuclear materials with the ratio of more than 1 and also for solution of SNF-related problems with regard to transmutation of long-lived minor actinides.

32.2 Source data for the scenario analysis

The source data for prediction of the demand for power generation at the internal market for nuclear power are based on the general layout of power generating facilities till 2035 and the predictive research performed in the RAS Institute for Energy Studies. Two development variants with due regard for accelerated and slower growth of the integral installed capacities at the

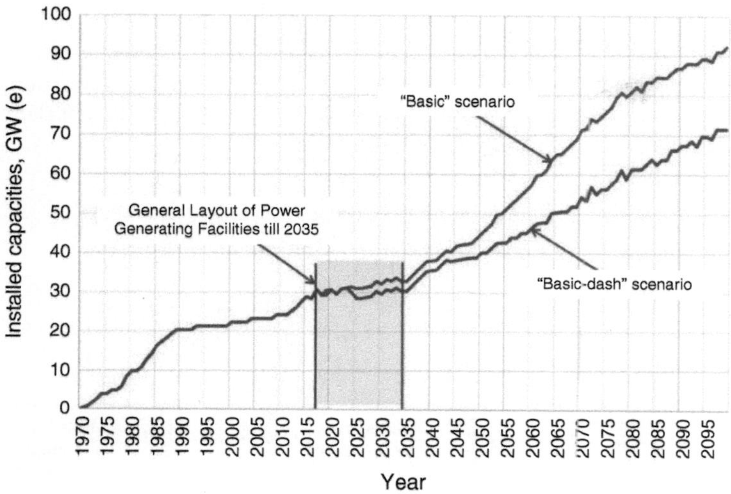

FIGURE 32.1 Scenario conditions for development of the Russian nuclear power in the XXI century.

Russian NPPs developed in the Strategic Management Department of Rosatom (Fig. 32.1) were considered in the scenario analysis. The basic scenario relies on the initial provisions established in development of the General Layout of Power Generating Facilities till 2035 approved by the Russian Government in 2017 and also on the assumed economy recovery with moderate energy efficiency growth rates. The "basic-dash" scenario is adjusted with due regard for the current predictions of electric power demand in Russia and modifications of the General Layout for commissioning and decommissioning of NPP power till 2035 approved by the Government of the Russian Federation. The "basic-dash" scenario is determined by moderate technology development rates and intensive increase of energy efficiency.

In accordance with the basic scenario achievement of the installed NPP capacity of 33 GW (e) by 2035, 44 GW (e) by 2050, and 92 GW (e) by 2100 is planned. In accordance with the "basic-dash" scenario achievement of the installed capacity of 30 GW (e) by 2035, 39 GW (e) by 2050 and 71 GW (e) by 2100 is planned. Commissioning of new power units in the scenario system studies shall ensure the required capacity level.

Evaluations of ARMZ JSC specified in Table 32.1 for natural uranium from the deposits located in Russia are used as the resource base for the Russian nuclear power.

The total domestic resources of natural uranium with due regard for the extraction possibility amount to 512.4 thous. tons. Within the framework of the calculations the total amount of domestic resources is used only as the conditional criterion of the domestic resource base sufficiency for development of the Russian nuclear power.

TABLE 32.1 Natural uranium reserves in Russia, tons.

Reserves	Price, USD/kg U				Extraction coefficient, %
	<40	40–80	80–130	130–260	
Proven	0	24,535	190,009	45,424	77
Inferred	0	15,293	255,749	125,848	79
Total	0	39,828	445,758	171,272	78

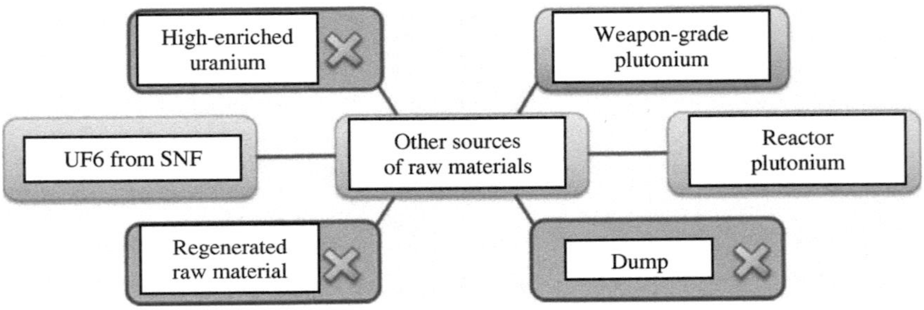

FIGURE 32.2 Other sources of raw materials: UF6 from SNF – uranium hexafluoride produced from irradiated fuel of NPP reactors, regenerated raw material – uranium hexafluoride produced from regenerated fuel of NPP reactors.

The natural uranium market in Russia is determined by both internal and external supplies. The source data for changes of the natural uranium price over time are defined on the basis of compliance with the projected prices of fuel assemblies for Russian NPPs within the period of up to 2030 in accordance with the current estimates of the NFC LC Management Department of Rosatom and the assumption on further growth with the rate of 2% per a year in accordance with the estimates of UxC Consulting Company. This prediction is conservative in relation to the natural uranium market evaluations performed by Techsnabexport JSC and reflects moderate growth of natural resource prices. Nevertheless, it should be noted that the approach selected in the scenario analysis and describing smooth change of the natural uranium price may contain significant uncertainty under the market conditions.

Other sources of raw materials include regenerated nuclear materials, uranium and plutonium produced in the course of processing (Fig. 32.2).

Stock reserves of reactor-quality plutonium extracted from SNF (with due regard for the possibility to involve ex-weapon plutonium) are assumed to the equal to ~100 tons by 2019. As of the beginning of 2018 the amount of accumulated (non-processed) NPP SNF in Russia is approximately 22.5 thous. tons (including ~ 5 thous. t_{hm} from VVER and 14.8 thous. t_{hm} from RBMK).

Fuel with addition of regenerated uranium is regularly used in RBMK reactors. The existing headstart at the level of industrial technologies enables

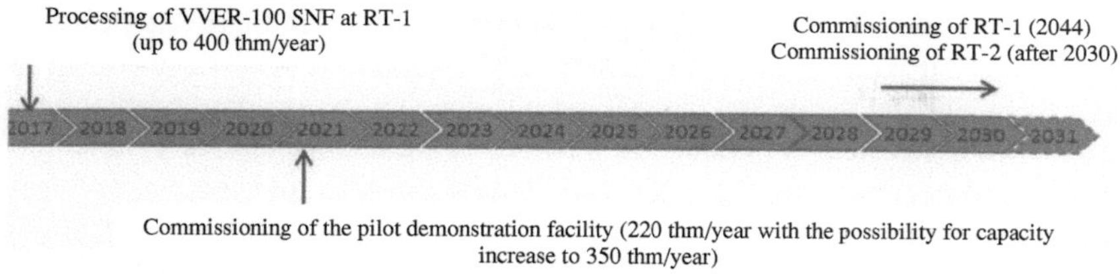

FIGURE 32.3 Development of the SNF handling infrastructure in Russia till 2030.

to solve the problem related to adjustment of the regenerated uranium isotope composition within several years. Thus, there are no limitations for utilization of all regenerated uranium produced by processing of NPP SNF in Russia.

Development of the SNF handling infrastructure is determined by the current plans defined till 2030 (Fig. 32.3). The capacity utilization for RT-1 plant of Mayak PA is determined on the basis of the current production program for the nearest 4 years.

For the scenarios with the open fuel cycle increase of the capacity of RT-1 plant at Mayak PA to the level of 200 t_{hm}/year is assumed, the pilot demonstration facility at the MCP site has the capacity of 220 t_{hm}/year. For the scenarios with the closed NFC increase of RT-1 capacity above 200 t/year and the capacity of the pilot demonstration facility above 250 t/year is not required. Commissioning of RT-2 with the capacity of ~800 t/year may be necessary on the cusp of 2060 to complete disposal of accumulated VVER SNF based on the approved roadmap for commissioning/ decommissioning of power units.

The pilot demonstration facility technology as the basic one for processing of VVER and FNR SNF may be scaled up to increase the capacity of SNF processing plants in Russia with centralized arrangement of the SNF processing operations with due regard for intermediate storage and transportation. Along with that the near-plant NFC technology for industrial power supply complexes can be developed starting with the constructed pilot demonstration power supply complex with lead-cooled FNR and enabling significant reduction of the external NFC part duration, accumulation of non-processed SNF and the amount of SNF and NF transportation.

The possibility to use water-based processing of FNR SNF together with SNF from thermal reactors including processing of RBMK SNF is considered in the scenario studies. Disposal of all accumulated SNF by 2100 is one of the conditions for development of the scenarios for the closed NFC with FNRs. It is assumed for calculations that processing of SNF from fast reactors in the near-plant NFC is more expensive than processing of SNF from thermal reactors due to limitations related to the short external fuel cycle and safety assurance at the production site. At the same time, it should be taken into account that the amount of processed SNF from thermal reactors for production of the same quantity of FNR fuel is ~15 times more than the amount of processed FNR SNF. Evaluation of the SNF processing costs and RW amounts is provided by the specialists of SNF Handling System Formation PA of Rosatom based on the current parameters of the enterprises, predictions with regard to their refurbishment and the design data taking into account the standard profitability.

Russia has certain experience in manufacturing and irradiation of uranium and plutonium fuel in research and commercial reactors with thermal and fast neutron spectra.

The current level of the uranium and plutonium fuel manufacturing technology maturity enables to aim at the possibility for commercial production of MOX fuel for BN reactors.

The core of operating BN-800 commercial power unit is partially loaded with fuel assemblies containing MOX fuel. Pilot operation of the automated plant for production of pelletized MOX fuel for BN-800 at MCP JSC commenced in 2015. The technology practicing and achievement of the design MOX FA production capacity parameters should be completed by 2021.

The program for manufacturing and irradiation of dense uranium and plutonium MNIT fuel in BN-600 commercial reactor aimed at improvement of the characteristics of lead- and sodium-cooled FNRs is implemented.

The main area of the medium-term development of reactor technologies in Russia shall include FNRs with lead or sodium coolant operated within the closed NFC with simultaneous continued improvement of pressurized thermal VVER type reactors.

In accordance with the principles for selection of technology solutions specified in the basic provisions of Strategy-2018 (the principles of integrity, phasing and staging) the time intervals required for implementation of new projects shall be taken into account in review of the Russian nuclear power development scenarios. It is assumed for the calculations that at least 4 years shall pass from the commissioning moment of the first power unit based on any fundamentally new engineering solutions till the second and third power units may be commissioned. Large-scale commercial construction of power units may be commenced in 8 years after the first power unit start-up.

Chapter 33

Comparative analysis of the Russian nuclear power development scenarios

33.1 The initial Russian nuclear power development scenario based on the existing technologies (Variant 0)

The scenario of the Russian power engineering development under the conditions of limited application of the closed NFC technologies on the basis of the existing technologies is considered as the initial one (Fig. 33.1):

- Increase of the installed capacities (Fig. 33.2) due to commissioning of power units with VVER-TOI;
- Total domestic natural uranium resources (see Table 30.1) that should provide for guaranteed operation of the reactor fleet (within the entire service life) are the limiting factor for its growth;
- SNF from VVER-440 and BN-600 is processed to the full extent, SNF from VVER-1000/1200 – to the extent of the functioning of RT-1 plant with the maximum capacity of 200 t_{hm}/year and the pilot demonstration facility – 220 t_{hm}/year, commercial processing of SNG from RBMK is not arranged (due to absence of any prospects for significant involvement of plutonium produced in the course of processing into the nuclear fuel cycle and considerable cost of its storage);
- Regenerated uranium is completely recycled, plutonium from SNF processing is stored or used for fuel supply to BN-800;
- The series of improved light-water reactors: VVER-TOI from 2023.

The scenario calculation objective is to determine the Russian nuclear power development boundaries subject to absence of any new solutions in the area of reactor building.

The systemic analysis demonstrated that limitation with regard to the natural uranium resource base assumed in the study prevented from the nuclear power development with the integral capacity growth rates set for the "basic" and "basic dash" scenarios. The maximum integral capacity value for the "basic" scenario

Closed Nuclear Fuel Cycle with Fast Reactors: Handbook of Russian Nuclear Power.
DOI: https://doi.org/10.1016/B978-0-323-99308-1.00009-X

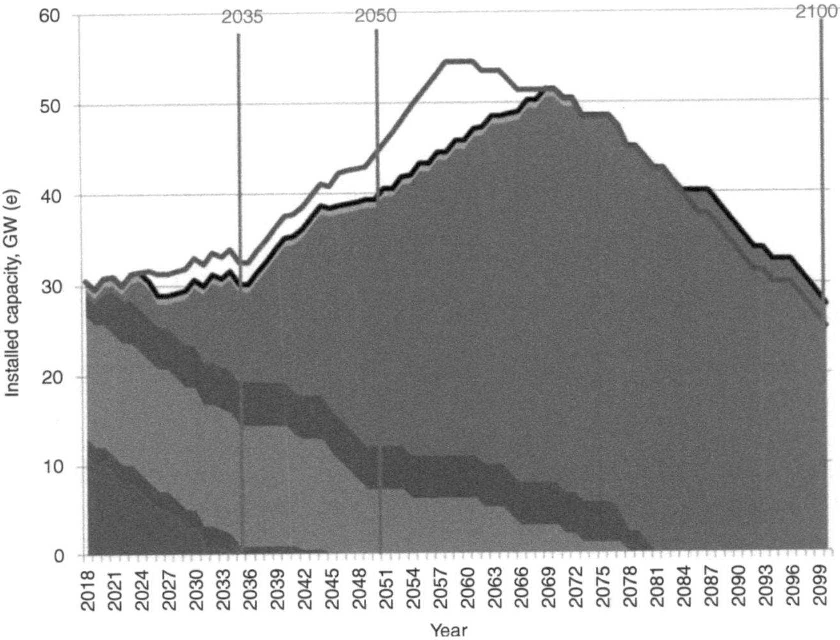

FIGURE 33.1 Russian nuclear power integral capacity increase for Variant 0 by the types of reactor plants: – BN-800; – BN-600; – VVER-1300; – VVER-1200; – VVER-1000; – VVER-440; – RBMK-1000; and scenarios: basic dash; basic.

will be reached in 2058 and for the "basic dash" scenario – in 2069. The amount of accumulated SNF stock will reach 60 thous. t_{hm} by 2100, and the cost of its processing and disposal will exceed 1 tln. RUB. Absence of any efficient solution for involvement of regenerated materials into the NFC will also affect the significant amounts (~200 t) of high-background plutonium accumulated by 2100.

According to Variant 0 the reactors built by 2100 will consume 1116 thous. tons of natural uranium in the "basic" scenario and 896 thous. tons in the "basic dash" scenario for their entire service lives.

33.2 Scenario for the Russian nuclear power establishment based on evolutionary development of the VVER technology with the open NFC (Variant 1)

The scenario for the Russian nuclear power establishment based on evolutionary development of the VVER technology with the open NFC (see Fig. 33.2) is determined by the following conditions:

- It is similar to the initial scenario with regard to the series of reactors, regenerated uranium recycling, the amount of processed SNF and RW handling technologies;

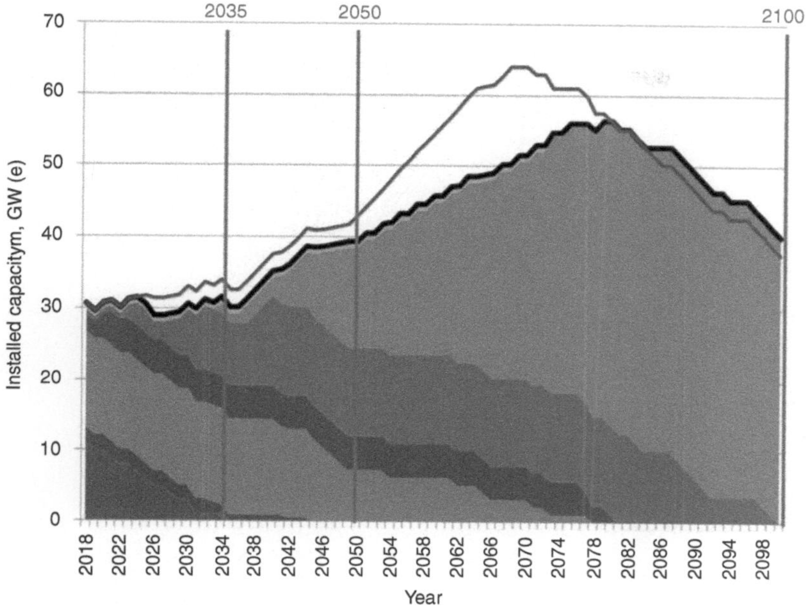

FIGURE 33.2 Russian nuclear power integral capacity increase for Variant 1 by the types of reactor plants: ▬ – BN-800; ▬ – BN-600; ▬ – VVER-S; ▬ – VVER-1300; ▬ – VVER-1200; ▬ – VVER-1000; ▬ – VVER-440; ▬ – RBMK-1000; and scenarios: ▬ basic dash; ▬ basic.

- The series of improved light-water reactors: VVER-TOI from 2023, VVER-S – the first of a king unit of a small series in 2034, two subsequent units within the period of 2038-2042, then – commercial construction of VVER-S.

The scenario calculation objective is to determine the VVER technology evolutionary development boundaries with reduction of natural uranium consumption through the use of the new spectral reactivity regulation technology in VVER-S RPs and establishment of the requirements for the NFC closing.

The systemic analysis demonstrated that for Variant 1 scenario limitation with regard to the natural uranium resource base also prevented from the nuclear power development with the integral capacity growth rates set for the "basic" and "basic dash" scenarios. The maximum integral capacity value for the "basic" scenario will be reached in 2068 and for the "basic dash" scenario – in 2079, that is, 10 years later than in the initial variant (Tables 33.1 and 33.2).

The VVER-S technology will enable to reduce annual natural uranium consumption by one third and to decrease fuel costs with due regard for the risks of its price increase over time; however, the resource base limitation is not basically eliminated in this case but only deferred. Besides, the problems related to assurance of the acceptable safety level for the growing NPP fleet, accumulation of SNF and competitiveness remain.

TABLE 33.1 Integrated material parameters for the "basic dash" scenario.

"Basic dash" scenario	Measurement unit	Variant 0	Variant 1	Variant 2	Variant 3	Variant 4	Variant 5
Maximum integral capacity (year of the maximum value achievement)	GW (e)	51.47 (2069)	56.42 (2079)	69.03 (2096)	70.02 (2100)	70.06 (2100)	70.01 (2100)
Electric power generation by 2100	thous. TW·h	25.87	28.07	30.53	30.37	30.38	30.38
Integral amount of natural uranium consumed in the nuclear power system	thous. t_{hm}	517.24	510.17	514.25	177.12	184.91	185.74
Integral amount of stored SNF by 2100	thous. t_{hm}	59.79	61.74	28.31	0.05	1.33	0.10
SNF encumbrance (without discount) by 2100	bln. RUB	1044.53	1096.12	591.77	64.81	120.78	38.76
Integral amount of stored Pu by 2100	thous. t_{hm}	199.49	199.49	5.02	17.59	34.95	24.73
Average natural uranium mining amount	t_{hm}/year	5679.20	5313.62	4646.93	2160.33	2254.25	2265.95
Average U fuel production capacity	t_{hm}/year	754.24	821.65	799.81	339.36	351.87	349.26
Average U-Pu fuel production capacity	t_{hm}/year	5.87	5.87	120.51	228.38	341.33	238.15
Average enrichment capacity	SWU/year	6852.37	6171.82	5960.67	2964.01	3024.29	3090.40
Average SNF processing capacity for the period	t_{hm}/year	272.26	272.26	709.29	1002.80	1163.55	848.96
Average LCOE (new power units) d = 5.4 %	RUB/MW·h	2043.9	2039.1	2041.6	2045.0	2098.6	1998.0
Average LCOE (new power units) d = 0 %	RUB/MW·h	1293.3	1289.1	1350.5	1293.2	1363.9	1241.3
– Operating costs		379.11	387.90	388.31	390.03	390.03	389.69
– Fuel costs + SNF liabilities (2100)		506.96	450.08	390.93	242.04	315.58	215.68
– Investment expenditures		407.23	451.08	570.83	661.60	658.33	635.87

TABLE 33.2 Integral material parameters for the "basic" scenario.

"Basic" scenario	Measurement unit	Variant 0	Variant 1	Variant 2	Variant 3	Variant 4	Variant 5
Maximum integral capacity (year of the maximum value achievement)	GW (e)	54.5 (2058)	63.9 (2068)	77.6 (2080)	91.3 (2100)	91.3 (2100)	91.3 (2100)
Electric power generation by 2100	тыс. TW·h	27.00	30.18	35.6	38.1	38.4	38.6
Integral amount of natural uranium consumed in the nuclear power system	th. t_{hm}	516	518	520	197	204	226
Integral amount of stored SNF by 2100	th. t_{hm}	62.9	66.9	28.8	0.04	0.09	0.06
SNF encumbrance (without discount) by 2100	bln. RUB	1088	1173	596	85.8	134	47.3
Stored Pu amount:							
– Initial		50	50	50	50	50	50
– Maximum		199	199	?	?	?	116
– by 2100		199	199	10.0	34.8	77.1	47.0
Average natural uranium mining amount	t_{hm}/year	5925	5715	5319	2403	2487	2708
Average U fuel production capacity	t_{hm}/year	786	878	920	372	392	380
Average U-Pu fuel production capacity	t_{hm}/year	5.9	5.9	127	265	428	312
Average enrichment capacity	kSWU/year	7147	6625	6748	3273	3356	4742
Average SNF processing capacity for the period	t_{hm}/year	272	272	837	1139	1384	936
Average LCOE (new power units) d = 5.4%	RUB/MW·h	2056	2011	2020	2041	2105	2020
Average LCOE (new power units) d = 0%	RUB/MW·h	1286	1232	1236	1265	1344	1226
– Operating costs		379	388	389	390	390	390
– Fuel costs + SNF liabilities (2100)		522	432	375	227	311	220
– Investment expenditures		385	412	474	648	643	616

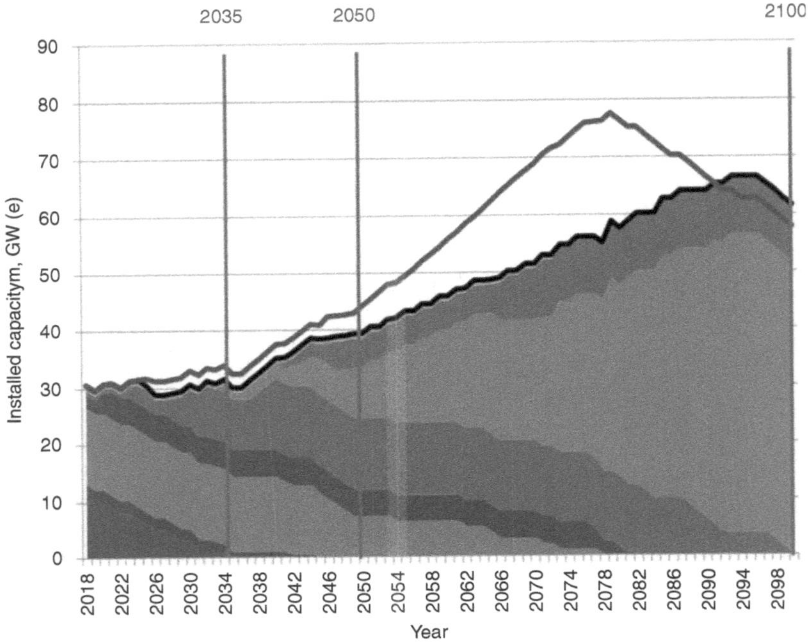

FIGURE 33.3 Russian nuclear power integral capacity increase for Variant 2 by the types of reactor plants: – BN-800; – BN-600; – VVER-S; – VVER-1300; – VVER-1200; – VVER-1000; – VVER-440; – RBMK-1000; and scenarios: basic dash; basic.

According to Variant 1 the reactors built by 2100 will consume 848 thous. tons of natural uranium in the "basic" scenario and 695 thous. tons in the "basic dash" scenario for their entire service lives.

33.3 The development scenario with SNF processing, partial closing of the NFC and usage of regenerated nuclear materials in thermal reactors (Variant 2)

Development of the Russian nuclear power system with thermal reactors, SNF processing and partially closed NFC (Fig. 33.3) is determined by the following scenario conditions:

- It is similar to the initial scenario with regard to the reactor series increase, regenerated uranium recycling and RW handling technologies;
- Use of uranium and plutonium fuel (see Table 4 of the Appendix) for the series of improved light-water reactors (particularly for removal of any limitations due to the available domestic natural uranium reserves);
- Duration of the external fuel cycle for thermal reactors is 7 years, and the possibility for MOX fuel recycling as well as involvement of RBMK SNF into the cycle are not considered due to low quality of the plutonium isotope content.

The objective of this scenario consideration is to evaluate the potential for growth of capacities for the VVER technology subject to increased SNF processing and usage of uranium and plutonium fuel at NPPs with VVER-S RPs.

The systemic analysis demonstrated that for Variant 2 scenario limitation with regard to the natural uranium resource base also prevented from the nuclear power development with the integral capacity growth rates set for the "basic" and "basic dash" scenarios. The maximum integral capacity value for the "basic" scenario will be reached in 2080 and for the "basic dash" scenario – in 2096, that is 22 years later than in the initial variant. The share of VVER-S reactors with 100% loading with MOX fuel amounts to 15–20% in the system thus enabling complete implementation of single regenerated plutonium recycling. It may be noted that partial closing of the NFC will not solve the SNF handling problem to the full extent, as RBMK SNF and MOX remain at the warehouses amounting to 26 thous. t_{hm} by 2100. As in Variant 1, the problems related to the acceptable safety level assurance for the growing NPP fleet and competitiveness also persists.

This approach to the NFC arrangement may be intermediate in transition to establishment of two-component power engineering with thermal and fast reactors as it does not solve the minor actinide transmutation problems.

The main conclusion from review of the Russian nuclear power development scenarios based on thermal reactors (Variants 0, 1, 2) resides in the necessity to develop FNR technologies for sustainable development and elimination of the risks with regard to resource availability in the long run.

Three main areas of the FNR technology and the NFC development are further considered in the scenario studies from the viewpoint of the balanced Russian nuclear power structure establishment:

- The NFC based on MOX fuel and BN-1200 RPs (high burnup characteristics for nuclear fuel, medium BR parameters);
- The NFC based on MNIT fuel and BN-1200 RPs (moderate burnup characteristics for nuclear fuel, high BR parameters for the option with breeding blankets);
- The NFC based on MNIT fuel and BN-1200/BR-1200 RPs (moderate burnup characteristics for nuclear fuel, BR ~1 for the option without breeding blankets).

33.4 FNR reactor technology development scenario with MOX fuel (Variant 3)

The Russian nuclear power system development scenario with assessment of the possibility for large-scale development of BN-1200 technology based on MOX fuel with BR>1, BRC<1 (accelerated development of the BN technology, Fig. 33.4):

- The installed capacity growth scenarios (Fig. 33.5) due to commissioning of power units with VVER-TOI RPs and fast reactors (see Table 5 of the Appendix);

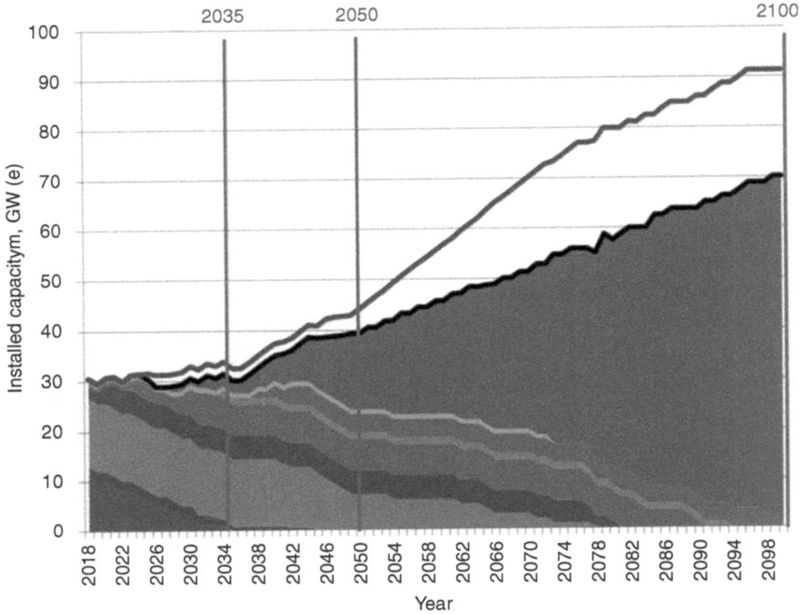

FIGURE 33.4 Russian nuclear power integral capacity increase for Variant 3 by the types of reactor plants: ▬ – BN-1200; ▬ – BN-800; ▬ – BN-600; ▬ – VVER-S with MOX ▬ – VVER-S; ▬ – VVER-1300; ▬ – VVER-1200; ▬ – VVER-1000; ▬ – VVER-440; ▬ – RBMK-1000; and scenarios: ▬ basic dash; ▬ basic.

- Processing of the entire amount of SNF including RBMK SNF with involvement of regenerated nuclear materials from thermal (regenerated uranium) and fast (regenerated plutonium) reactors;
- The series of improved light-water reactors: VVER-TOI from 2023, VVER-S – the first of a kind power unit of the small series in 2034, two subsequent power units within the period of 2038–2042, then construction of VVER-S reactors with maintenance of the positive plutonium stock balance is possible.
- BN-1200 series of reactors – the first power unit in 2028, two subsequent power units – within the period of 2032–2036, then – commercial construction;
- Duration of the external fuel cycle for fast reactors is 3 years;
- Stage-by-stage transition to the fuel cycle with enhanced burnup parameters for MOX fuel of fast reactors.

The objective of this scenario consideration is to assess the potential for development of two-component nuclear power with the priority of the BN reactor technology with the use of MOX fuel.

The NFC closing in two-component nuclear power with thermal and fast reactors enables complete solution of the SNF and RW handling problem for

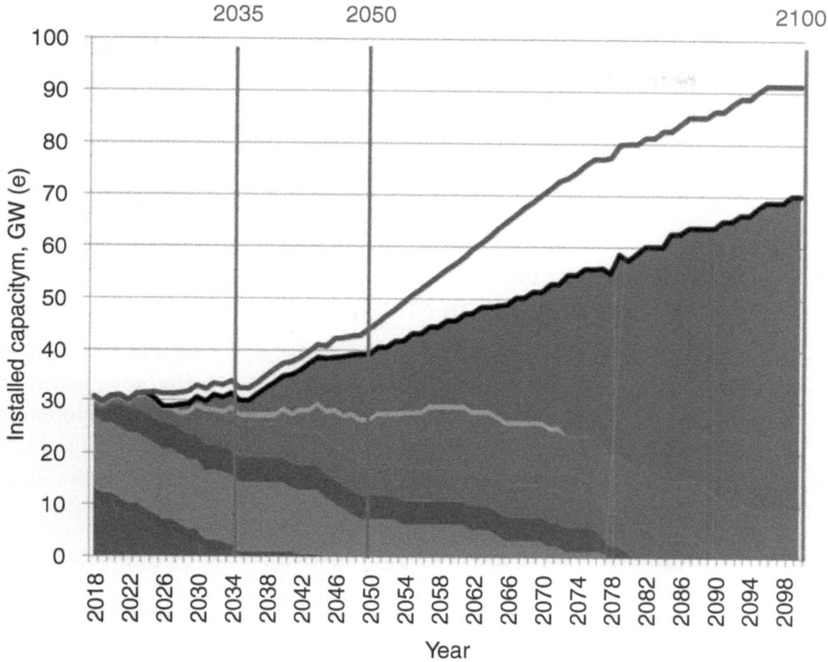

FIGURE 33.5 Russian nuclear power integral capacity increase for Variant 4 by the types of reactor plants: – BN-1200; – BN-800; – BN-600; – VVER-S with MOX; – VVER-1300; – VVER-1200; – VVER-1000; – VVER-440; – RBMK-1000; and scenarios: basic dash; basic.

all reactor types with efficient utilization of the energy potential of regenerated nuclear materials. Excessive plutonium generated in the NFC can be used in VVER-S thermal reactors; in this case their share in the system will be small due to the need for maintenance of the nuclear power growth rate specified in the scenarios with relatively low breeding parameters for MOX fuel. The share of VVER-S thermal reactors may be increased subject to processing of SNF from foreign NPPs without any return of the plutonium generated in the course of processing to the customers within uranium and plutonium fuel.

Development of nuclear power based on the BN reactor technology with MOX (BR = 1.25) enables complete self-production of fuel without any involvement of natural uranium resources with high prime cost of extraction. But this variant only partially complies with the long-term sustainable development principles, as the problems related to competitiveness and acceptable safety of the nuclear power system still persist with significant increase of the FNR fleet capacities using potentially flammable and explosion-hazardous sodium coolant.

In spite of the fact that the scenario analysis has not revealed any fuel supply problems for the variant based on the BN-1200 reactor technology with MOX fuel it should be taken into account that the growing commissioning rate of new power units may require new solutions for BR increase.

Besides, the problem of acceptable safety maintenance for the nuclear power system is not resolved due to significant increase of the FNR fleet capacities using potentially flammable and explosion-hazardous sodium coolant.

33.5 Development scenario with commissioning of fast reactors with increased BR on MNIT fuel (Variant 4)

The Russian nuclear power system development scenario with assessment of the possibility for large-scale development of the BN-1200 technology based on MOX fuel with BR>1 and fast reactors with dense fuel (accelerated development of the fast reactor technology with enhanced breeding parameters, Fig. 33.5) different from the development scenario based on the BN reactor technology with MOX fuel in the following:

- Stage-by-stage transition to the fuel cycle with enhanced burnup parameters for MNIT fuel;
- Usage of excessive plutonium in VVER-S reactors in the form of MOX fuel (the scope of VVER-S commissioning is larger than in Variant 3).

The objective of this scenario consideration is to evaluate the potential for development of two-component nuclear power with the priority of the fast reactor technology and enhanced breeding characteristics and to define the maximum resource potential for the nuclear power system which can be used to expand its application scope.

BR increase to 1.5 for BN-1200 reactors operating with dense MNIT fuel and equipped with breeding blankets enables to increase the share of thermal reactors in the system to 10–20% and to reduce the risks related to the FNR technology development. Nevertheless, the necessity to maintain the nuclear power growth rates set in the scenarios without any use of plutonium from SNF of foreign customers prevents from equalization of the balance between thermal and fast reactors in the system, and large amounts of processed SN increase fuel costs (see Tables 33.1 and 33.2).

As in Variant 3, the problems related to acceptable safety of the nuclear power system and competitiveness achievement still persist due to significant increase of the FNR fleet.

33.6 Development scenario with commissioning of fast reactors with BR ~1 based on MNIT fuel (Variant 5)

The Russian nuclear power system development scenario with assessment of the possibility for large-scale development of FNR technologies based on MNIT

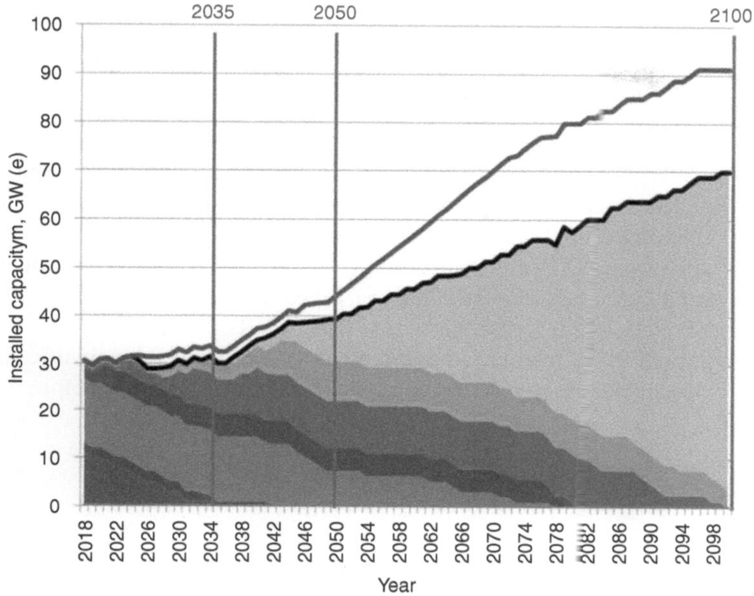

FIGURE 33.6 Russian nuclear power integral capacity increase for variant 5 by the types of reactor plants: – BR-1200; – BN-1200; – BN-800; – BN-600; – VVER-1300; – VVER-1200; – VVER-1000; – VVER-440; – RBMK 1000; and scenarios: basic dash; basic.

fuel with BR = BRC ~1 (accelerated development of the FNR technology, Fig. 33.6):

- Scenarios of the installed capacity growth (see Fig. 30.1) due to commissioning of power units with VVER-TOI RPs and fast reactors (see Table 4 of the Appendix);
- Processing of the entire amount of SNF in the centralized NFC with involvement of regenerated nuclear materials in TNRs (regenerated uranium) and FNRs (regenerated plutonium);
- The series of improved light-water reactors: VVER-TOI from 2023;
- BN-1200 power units with MNIT fuel (see Table 32.2) – the first power unit in 2028, two subsequent power units within the period of 2032-2036, then – commercial construction;
- BR-1200 power units with MNIT fuel (see Table 32.2) – the first power unit in 2037 (2034 is specified in the table as there is uncertainty), two subsequent power units within the period of 2041-2045, then – commercial construction (partial start-up with uranium loading is also considered in order to maintain the positive plutonium balance in the system).

The objective of this scenario consideration is to evaluate the potential for development of two-component nuclear power with the priority of the FNR reactor technology with inherent safety and lead coolant operating on MNIT fuel with BR = BRC~1, including determination of the uranium fuel amount necessary for FNR development subject to maintenance of the integral installed capacity growth.

The nuclear power development scenario based on FNR reactor technologies with MNIT fuel and BR = BRC~1 complies with the long-term sustainable development principles and provides for resource availability over the entire horizon considered up to 2100. Deficiency of fissionable material arising in the second half of the XXI century according to Variant 5 basic scenario may be compensated by introduction of additional breeding blankets or engagement of some enriched uranium for the start-up loads of FNRs.

The possibility to maintain the acceptable safety level with significant increase of the nuclear power generating capacities is a matter of principle.

Chapter 34

Russian nuclear power development variants for different integral capacity growth scenarios

The results of the comparative analysis of material and economic parameters for the Russian nuclear power development variants up to 2100 for the "basic dash" integral capacity growth scenario were presented in the previous chapter (Tables 31.1 and 31.2).

Let us consider the comparative analysis results for the integral characteristics.

34.1 Competitiveness according to LCOE

In all Variants 0-5 the average LCOE parameter calculated for new power units (subject to compliance with the scenario conditions for achievement of the declared technical and economic indices) is lower than the value adopted for comparison of the conventional CCGT-410 at the site in Kursk region with the capacity utilization factor of 50%. The possibilities to reduce LCOE for CCGTs due to the capacity utilization factor increase are restricted because of the need to operate under the conditions of variable network load and are highly likely to be outweighed by the risks of gas price growth over the strategic horizon due to exhaustion of the raw material base with low extraction prime cost and probable introduction of fees for CO_2 releases.

34.2 Assurance of the required share of nuclear generation in the energy balance of the Russian Federation

Only the variants with fast neutron generation (Variants 3, 4, 5) comply with the requirements of the basic development scenarios up to 2100 according to the maximum integral capacity criterion subject to fulfillment of the scenario condition with regard to the integral natural uranium consumption at the level

Closed Nuclear Fuel Cycle with Fast Reactors: Handbook of Russian Nuclear Power.
DOI: https://doi.org/10.1016/B978-0-323-99308-1.00021-0

of ~512 thous. tons (the resource base of natural uranium in the Russian Federation).

Variants with construction of only thermal reactors have high risks of competitiveness reduction in case of any natural uranium price growth and do not provide the required level of NPP power generation by 2100 without any risks of dependence on imported uranium; they also require development of the Russian uranium deposits with high extraction prime costs.

Transition to the new technology platform with FNRs and the closed NFC enables twofold reduction of natural uranium consumption in comparison with complete exhaustion of the national resources in the development variants with TNRs alone and provides for guaranteed achievement of the Russian nuclear power capacities of both 70 GW (e) in the "basic dash" scenario and 92 GW (e) in the "basic" scenario by 2100. The existing uranium enrichment facilities are also sufficient for the above-mentioned capacities.

34.3 Safety level maintenance for the growing system

The course toward reduction of nuclear power adopted in some developed countries is primarily associated with the idea of its insufficient safety. On the other side, many countries including Russia focus on preservation and further development of nuclear power, so it is fair to assume that the existing level of safety is considered acceptable in these countries. However, this fact cannot be used as the basis for implementation of the plans for large-scale nuclear power development through simple replication of the existing NPP types. The global fleet of operating NPPs based on pressurized water-cooled reactors as well as traditional fast sodium-cooled breeding reactors are characterized with irremovable hazard potential in the form of stored energy capable to mobilize the accumulated radioactivity in case of any beyond design basis accident, for example with loss of cooling and subsequent fuel melting as at EBR-1 reactor and Three Mile Island NPP (USA), Fukushima NPP (Japan), or as a result of the reactor runaway with prompt neutrons as in case of the accident at Chernobyl NPP. In case of many-fold increase of the generating capacity higher and whole new safety level shall be provided for individual power units. Transition of nuclear power to the new technology platform based on FNRs with inherent safety is necessary; it is implemented only in Variant 5 among all variants under consideration.

34.4 Solution of the problems related to SNF and its processing products

For Variants 0, 1 (thermal reactors without MOX fuel for reactors with spectral regulation) the amount of SNF accumulated at storage facilities will reach 60 thous. tons by 2100 thus requiring significant expansion of the existing infrastructure and also contradicting the approved national SNF and RW management policy. For Variants 0, 1 with the open NFC the share of SNF handling encumbrance exceeding 1 tln. RUB by 2100 (~60 thous. tons of SNF) will become significant. Absence of any possibility for large-scale plutonium involvement into the NFC of fast reactors results in increased stock of regenerated nuclear materials, including plutonium.

Partial closing of the NFC in Variant 2 (with MOX fuel for reactors with spectral regulation) enables considerable reduction of the stored SNF amount to ~25 thous. tons but does not solve the problem related to processing and handling of regenerated nuclear materials for RBMK SNF. The problems related to handling of irradiated MOX fuel and transmutation of minor actinides also remain unsolved thus resulting in a large amount of long-lived HLW.

Sound resolution of the SNF problems is provided only in the scenarios with development of fast neutron generation (Variants 3, 4, 5).

34.5 Financial and economic stability

Traditional evaluation of competitiveness in construction of new generating facilities is based on the LCOE parameter. From the strategic viewpoint it is necessary to compare the expenditures for development of generating systems based on the compared technologies for the country energy supply assurance in the sufficiently long term (exceeding the life cycle duration for individual power plants). The relevant system parameter - SLCOE shall take into account total expenditures within the period of time under consideration including the investments not only into the generating facilities but also into the fuel cycle, as well as long-term effects of fuel price increase and environmental impact.

Export potential is an essential element of financial stability. Due to increase of competition with foreign vendors the NPP product with considerably improved consumer performance will be necessary in 2030s. Implementation of improved power units on the basis of the VVER technology will enable to gain additional time for achievement of competitive FNR parameters.

Scenario calculations are based on the declared parameters for promising thermal and fast reactor designs. The technology maturity level of the projects with evolutionary development of VVER and BN series is higher than for lead-cooled FNRs. Temporal risks related to achievement of the declared parameters both with regard to the technical and economic indices of the power unit and the fuel characteristics are present for FNR technologies.

Implementation of the FNR technology with lead and sodium coolant removes the resource base limitations of the nuclear power development scenarios over the entire horizon under consideration. Nevertheless, the issues of plutonium stock management and development of the uranium and plutonium fuel manufacturing technologies become urgent for fast reactors in the closed NFC. The existing nuclear power structure with thermal reactors and the amount of SNF accumulated in the course of their operation (including RBMK SNF) are the drivers for the transient stage in establishment of two-component nuclear power

TABLE 34.1 Integral export parameters of compliance with the system requirements.

System requirements	Variant 0	Variant 1	Variant 2	Variants 3 and 4	Variant 5
Compliance with the scenario requirements for energy generation	–	–	–	+	+
Safety level maintenance for the growing system	–	–	–	–	+
Development potential after 2100	–	–	–	+	+
SNF and RW problem solution	–	–	+/–	+	+
Efficient involvement of nuclear materials into the NFC	–	–	+/–	+	+
Reduction of natural uranium consumption in the system	–	–	–	+	+

with thermal and fast reactors. Characteristics of the nuclear fuel breeding and the possibilities to use enriched uranium also have significant effect on the FNR fuel balance management.

The NFC closing and reduction of natural uranium consumption in Russia provide the possibility for its export to the external market and also extend the potential for provision of services at the final NFC stage (SNF and RW handling). Processing of SNF from foreign customers without any return of plutonium to them in the form of uranium and plutonium fuel provides the possibility to increase the nuclear generation amount/ growth rate in the Russian Federation.

Integral comparison of the Russian nuclear power development variants in the context of compliance with the modern system requirements demonstrating the advantages of the NFC closing in two-component nuclear power is presented in Table 34.1.

The main risks of long-term Russian nuclear power development with due regard for the considered development scenarios are the following:

- Resource base limitations and natural uranium price increase for the open NFC;
- Failure to maintain the acceptable safety level in the growing nuclear power system;
- Increased cost of new power units in relation to the planned parameters;
- Failure to achieve the burnup depth parameters for MOX and MNIT fuel in FNRs;
- Absence of any financial mechanism to guarantee pay-back of the investments in order to maintain the nuclear generation development rate.

Failure to achieve the declared parameters with regard to the new power unit construction costs is one of the main risks in long-term development of the Russian nuclear power. This refers both to advanced FNRs with sodium and lead coolant as the degree of their commercialization and technology availability is still very low, and new VVER designs that have not demonstrated any tendencies toward reduction of specific capital costs through history with some implemented improvements.

Maintenance of the acceptable safety level with considerable increase of the generating capacities is of crucial significance for developing nuclear power. In this context the most rapid transition of nuclear power to the NTP with lead-cooled FNRs is especially important.

The expected natural uranium price growth rates and risks related to its limited natural reserves require implementation of new technologies providing for reduction of natural uranium consumption in thermal reactors and establishment of two-component nuclear power with fast reactors and the closed NFC.

The main recommendations on establishment of balanced two-component nuclear power system with the closed NFC in Russia in relation to implementation of the VVER and FNR reactor technology development programs are the following:

- Implementation of the VVER technology evolutionary development program will enable to enhance the economic parameters and to reduce natural uranium consumption;
- Implementation of the FNR technologies will provide the possibility to demonstrate commercial efficiency of the NFC closing, ensure efficient use of regenerated nuclear materials, enable to increase the processing rate and to reduce SNF stock, solve the long-lived HLW transmutation problems, facilitate development of new products for the external markets and establish the basis for sustainable growth of nuclear power in the long run;
- Implementation of BREST-300 technologies will enable to demonstrate the main advantages of lead-cooled reactors with regard to economic feasibility, safety and the FNR NFC arrangement and to improve the FNR operating characteristics in the future;
- Commercial production of MOX fuel for BN-800 and commercial production of MNIT fuel for BREST-300 will enable to practice the uranium and plutonium fuel fabrication technologies;
- Use of uranium and plutonium fuel in the operating BN reactors will enable to substantiate the performance characteristics of MOX and MNIT fuel.

The decision on implementation of new reactor technologies in nuclear power shall be made subsequent to analysis of their current availability level and confirmation of credibility of the declared technical and economic parameters. Passing of technology bifurcation points opens the opportunities for nuclear power development in accordance with this or that scenario. The key dates

TABLE 34.2 Description of technological forks for key reactor technologies, that define Russian Strategy for Nuclear Power development.

Technology	Measure	Fork passage time	Further stages of development
VVER-S	Design selection, feasibility study, R&D program development. 2018–2020	2021	Preparation of design documentation for the reactor plant – 2025. Construction and commissioning of the reactor plant – 2033.
BN-1200	Substantiation of design characteristics of NPP with BN-1200M reactor plant, substantiation of design technical and economic indicators for MOX fuel. 2018–2021	2020	Construction and commissioning of the reactor plant – 2028.
BR-1200	Commissioning and successful operating experience of PDEC with BREST-OD-300 reactor plant, substantiation of design technical and economic indicators for MNIT fuel/ 2025–2030	2029	Preparation of design documentation for the reactor plant – 2028. Construction and commissioning of the reactor plant – 2034.

of bifurcation point passing for the reactor technologies with high degree of availability are given in Table 34.2.

Successful passing of technology bifurcation points will enable to state the reasonable claim for large-scale development of nuclear power in the Russian Federation with governmental support in formation of the financial mechanism to guarantee pay-back of the investments providing for the competitive level of electric power prices for the Russian consumers and to create the conditions for strengthening of the positions of Russian nuclear technologies at the global market.

Conclusion

The safety factor is determined not only by real physical properties of the facilities but also by their common understanding. The recognized safety level of nuclear power is the prerequisite for achievement of its potential level of development.

Six major accidents that have taken place for 60 years of the global nuclear power development cannot be disregarded. Probabilistic approaches efficient in search of poor design solutions and detection of any drawbacks of the operating regulations cannot be considered as the main safety criterion, let alone the indicator of any particular accident occurrence probability.

The prospects of large-scale nuclear power should not be associated with the technologies providing for classification of accidents into design basis and beyond design basis ones as well as for possible urgent evacuation of the public in case of any accident and the scale of economic losses not covered by the operating organization or insurance companies. It should be recognized that almost the entire global fleet of operating NPPs based on water-cooled reactors and also the proposed sodium-cooled fast breeding reactors are characterized with irremovable hazard potential in the form of stored energy capable to mobilize the accumulated radioactivity in case of any beyond design basis accident, for example with loss of cooling and subsequent fuel melting as at EBR-1 reactor and Three Mile Island NPP (USA), Fukushima NPP (Japan), or as a result of the reactor runaway with prompt neutrons as in case of the accident at Chernobyl NPP. Reorientation of both NPP reactors and the NFC facilities toward the inherent safety concept adopted in Strategy-2000 is necessary; the principles of this concept are well-developed and substantiated theoretically.

At present it may be considered theoretically proven and substantiated through calculations and experiments that three additional requirements for fast reactors such as the integrated design, the BRC close to one, lead coolant and dense nitride uranium-plutonium fuel (MNIT fuel) provide the possibility for significant increase of the safety level for fast reactors complying with these requirements as compared to fast reactors with the BRC of much lower than one (which is higher only in case of blanket use), with sodium coolant and MOX fuel. Thus, a fast reactor of integrated design with the BRC closes to one, lead coolant and dense uranium and plutonium fuel is necessary for large-scale nuclear power component.

Taking into account the requirement for the safety level enhancement seven main engineering solutions ensuring compliance with the requirements for inherent safety of large-scale nuclear power may be specified for NPP reactor plants:

1) The equilibrium core of a fast reactor enabling to minimize the reactivity margin for nuclear fuel burnup and basically eliminate any runaway with prompt neutrons;
2) Dense fuel enabling to implement the equilibrium core with the BRC close to one without any uranium blanket;
3) A wide core lattice enabling to provide the natural circulation level sufficient for residual heat removal and to reduce power consumption for the coolant pumping;
4) Integral arrangement of the reactor plant enabling to prevent any core cooling loss and to create the conditions for efficient natural circulation;
5) Heavy liquid-metal coolant enabling to implement the wide lattice and to prevent any positive void reactivity effect;
6) The lead coolant quality maintenance system enabling to use heavy coolant in high-capacity fast reactors;
7) Use of the atmospheric air as the ultimate heat sink in the course of natural circulation residual heat removal in high-capacity reactor plants.

Apart from the NPP safety requirements there is a set of requirements for the closed NFC technologies:

- Low-waste processing of SNF from fast reactors;
- Involvement of the thermal reactor SNF processing products into the fuel cycle of fast reactors;
- Reduction of the storage period for SNF from fast reactors prior to its processing to one-two years;
- Assurance of radiation balance between the extracted fuel raw materials and disposed radioactive waste (RW);
- Technology support of the nonproliferation regime.

In the context of the requirements for the closed NFC technologies three engineering solutions may be specified:

1) "Dry" processing of SNF from fast reactors in order to reduce the SNF storage duration prior to its processing and to prevent extraction of pure plutonium in the course of its processing;
2) Transmutation of minor actinides in fast reactors in order to ensure balance between the extracted fuel raw materials and disposed RW;
3) Departure from the blanket in fast reactors in order to prevent weapon-grade plutonium production (for export of technologies).

The projects of Russian fast reactors described in the present monograph include technology peculiarities that may be useful to solve the problems related to nonproliferation of nuclear weapons (departure from uranium enrichment, absence of any blanket, the near-plant fuel cycle, processing technologies without any separation of uranium and plutonium).

The scenarios for implementation of fast reactors with the closed NFC shall be based on the existing structure of nuclear power. Development of nuclear power based on fast reactors with inherent safety replacing potentially dangerous thermal reactors will require mutual coordination of their fuel balance, processing of the accumulated irradiated fuel and reasonable use of fuel resources. In this regard clear identification of the difference between perception of two-component nuclear power as a stage in transition to nuclear power with predominant fast reactors and the well-known concept of two-component nuclear power with permanently coexisting thermal reactors and fast breeding reactors for their fuel make-up is necessary.

As far as competitiveness is concerned the possibility to implement the above-mentioned preconditions for cost reduction will arise in the course of transition to NPPs with complete set of inherent safety features. In general, the conclusion may be made that the concept of nuclear power structure based on fast reactors with inherent safety complies with the main requirements for development of really large-scale power system.

After 2001, the works within the framework of the strategic objectives and approaches stated in Strategy-2000 were suspended for a long time. All basic principles of safe nuclear power growth specified in Strategy-2000 are waiting for their engineering solution and implementation in new plants. The top-priority objective of the industry - implementation of Proryv Project uniting the projects for strategic accomplishment of the targets with regard to development of nuclear power technologies with inherent safety based on fast reactors and the closed NFC - is defined.

Zero-carbon power generation in which nuclear power is the most well-developed section is gaining increasing significance in the global power engineering but it starts to experience competitive pressure on the part of renewable energy sources in the context of both safety and economic attractiveness, especially in case of the cost growth tendency persistence for modern NPPs. From the strategic point of view, the position of Russia stated in Strategy-2000 and oriented toward acquisition of the closed NFC remains unchanged. Use of self-sustained fast reactors with BR of $\sim$1 where excessive neutrons are used not for maximum plutonium production but for achievement of the inherent safety properties including operation with equilibrium fuel and burning of long-lived minor actinides is the fundamental provision. Departure from high BR of fast reactors in favor of new opportunities granted by reactors with BR $\sim$1 is also recognized in the recent systemic research of the US nuclear power.

Safety, raw material resources and economic competitiveness in comparison with other types of power generation remain the key factors for large-scale development of nuclear power in Russia and worldwide. The technology platform of the Russian nuclear power based on VVER reactors is sufficient for the scope of

NPP construction and export supplies predicted in ES-2030. But its potential for solution of long-term strategic problems of the country is limited due to noncompliance of the safety level with the requirements for large-scale nuclear power, limited resources of natural uranium, SNF accumulation and the issues of its transportation and storage.

The concept of the nuclear power structure based on fast reactors and the CNFC facilities with inherent safety complies with the key requirements for development of really large-scale power system.

It is possible to deploy large-scale nuclear power by the end of this century in order to achieve considerably improved fuel balance in the fuel and energy industry of Russia in general, to save valuable resources of fossil fuels and to solve the problem of CO_2 emissions completely. It means the necessity of the most rapid transition to the closed NFC and maintenance of the reasonably high growth rates for nuclear power for the major part of the century.

From the viewpoint of efficiency of the nuclear power switching to the closed NFC it is reasonable to use all available fuel resources (both plutonium and enriched uranium) primarily for accelerated commissioning of self-sustained fast reactors. Inefficient burning of plutonium in the form of MOX fuel for thermal reactors is unpractical.

The economy of transition to the closed NFC requires inclusion of the deductions for the generated SNF processing and RW handling into the prime cost of electric power at NPPs.

Investments into construction of the pilot reactor with inherent safety and the near-plant NFC (and the subsequent series of commercial ones) shall become the top priority of the industry in the area of nuclear power within the nearest 10-20 years.

The results obtained within the framework of Proryv Project Framework enable to make the following conclusions:

- The characteristics of fast reactors with inherent safety and low breeding coefficient ($\sim$1.05) are acceptable for accelerated arrangement of the self-sustained system of fast reactors subject to use of enriched uranium as additional fuel resource;
- Transition to the nuclear power with predominating fast reactors with equilibrium fuel provides the possibility for significant reduction of the processing ($\sim$1 ton of equilibrium fuel SNF per 1 ton of regenerated one instead of $\sim$15 tons of VVER SNF) and transportation amount of irradiated fuel from thermal reactors as well as the fuel stream size due to achievement of increased fuel burnup (8-12% for fast reactors as compared to 4.5–7.5% for thermal reactors); in this case technology support of the nonproliferation regime will be also provided by the final departure from uranium and plutonium separation processes as well as uranium enrichment;
- Implementation of the closed near-station NFC for fast reactors with significant reduction of the external NFC duration and "burning" of long-lived

minor actinides within regenerated fuel enables considerable reduction of the content of the latter in HLW providing the possibility for their radiation equivalent disposal;
- According to the fuel factor of NPP power the closed NFC of fast reactors in its final form with regenerated equilibrium fuel turns out to be knowingly more profitable than the open NFC of LWR even in case of expected significant price increase for its stages.

The works for arrangement of the entire set of the new nuclear power technology platform components deployed within the framework of Proryv Project Framework in accordance with the principles of Strategy-2000 and developed in Strategy-2018 ensure leadership of Russia in the following areas:

- Construction of reactor plants with inherent safety (deterministic prevention of accidents requiring evacuation of the public);
- Development of dense nitride fuel optimal for fast reactors;
- Ultimate solution of the SNF accumulation problem;
- Radiation-equivalent RW handling;
- Arrangement of the world's first pilot power supply complex with FNR and the CNFC technologies (PDEC).

It is possible to overcome crisis of the global nuclear power through arrangement of the first power units within the IPC based on fast reactors with BN-1200 RPs within the period of up to 2035 subject to confirmation of competitiveness in comparison with VVER by the conceptual design. Besides, the IPC with BN-1200 RPs ensures competitiveness in comparison with CCGTs and RES.

Appendix

TABLE 1 Potential biological hazard (PBH) of significant radionuclides in the nuclear power waste from thermal and fast reactors in 2100.

Nuclide	PBH, Sv	Nuclide	PBH, Sv	Nuclide	PBH, Sv
^{3}H	5.14E+07	^{135}Cs	2.00E+04	^{232}U	1.88E+08
^{14}C	4.70E+04	^{137}Cs	6.13E+09	^{234}U	9.00E+05
^{63}Ni	9.32E+04	^{133}Ba	9.01E+04	^{236}U	4.69E+04
^{79}Se	7.13E+05	^{144}Ce	5.62E+11	^{238}U	6.16E+04
^{85}Kr	1.22E+11	^{146}Pm	1.32E+07	^{237}Np	9.41E+05
^{90}Sr	5.81E+09	^{147}Pm	4.49E+10	^{236}Pu	1.04E+08
^{93}Zr	3.99E+06	^{151}Sm	1.33E+09	^{238}Pu	4.08E+10
^{93m}Nb	1.25E+06	^{152}Eu	1.89E+08	^{239}Pu	2.67E+09
^{99}Tc	1.01E+05	^{154}Eu	2.15E+10	^{240}Pu	4.94E+09
^{106}Ru	9.92E+11	^{155}Eu	3.93E+09	^{241}Pu	2.36E+09
^{102m}Rh	1.83E+06	^{153}Gd	5.04E+05	^{242}Pu	1.04E+07
^{107}Pd	5.95E+04	^{157}Tb	2.37E+04	^{241}Am	1.86E+10
^{108m}Ag	5.21E+04	^{158}Tb	1.18E+05	^{242m}Am	2.22E+09
^{109}Cd	7.33E+04	^{166m}Ho	1.78E+05	^{243}Am	1.16E+08
^{113m}Cd	9.81E+09	^{171}Tm	5.13E+04	^{243}Cm	2.87E+08
^{119m}Sn	1.29E+07	^{210}Pb	2.92E+04	^{244}Cm	4.79E+09
^{121m}Sn	1.09E+08	^{227}Ac	8.04E+04	^{245}Cm	5.48E+06
^{126}Sn	4.39E+07	^{228}Th	7.12E+08	^{246}Cm	6.91E+05
^{125}Sb	1.63E+10	^{229}Th	7.27E+05	^{249}Bk	2.93E+06
^{129}I	3.31E+04	^{230}Th	4.92E+04	^{249}Cf	1.60E+05
^{134}Cs	6.56E+08	^{231}Pa	6.16E+04	^{250}Cf	2.94E+05

TABLE 2 PBH of significant radionuclides in the nuclear power waste from thermal and fast reactors subsequent to storage for 100 to 1000 years starting with 2100.

	Storage time, years						
Nuclide	100	200	300	400	600	800	1000
^{3}H	1.97E+05	–	–	–	–	–	–
^{14}C	4.65E+04	4.59E+04	4.54E+04	4.48E+04	4.37E+04	4.27E+04	4.17E+04
^{63}Ni	4.70E+04	2.35E+04	1.18E+04	–	–	–	–
^{79}Se	7.13E+05	7.13E+05	7.13E+05	7.13E+05	7.13E+05	7.13E+05	7.12E+05
^{85}Kr	2.09E+08	3.35E+05	–	–	–	–	–
^{90}Sr	5.36E+08	4.82E+07	4.33E+06	3.90E+05	–	–	–
^{93}Zr	3.99E+06	3.99E+06	3.99E+06	3.99E+06	3.99E+06	3.99E+06	3.99E+06
^{93m}Nb	1.70E+06	1.71E+06	1.71E+06	1.71E+06	1.71E+06	1.71E+06	1.71E+06
^{99}Tc	6.14E+05	1.01E+05	1.01E+05	1.01E+05	1.01E+05	1.00E+05	1.00E+05
^{107}Pd	4.50E+05	5.95E+04	5.95E+04	5.95E+04	5.95E+04	5.95E+04	5.95E+04
^{113m}Cd	1.56E+08	5.53E+05	–	–	–	–	–
^{121m}Sn	2.22E+08	8.88E+06	2.52E+06	7.14E+05	5.74E+04	–	–
^{126}Sn	2.80E+08	4.38E+07	4.38E+07	4.38E+07	4.37E+07	4.37E+07	4.36E+07
^{129}I	3.31E+04	3.31E+04	3.31E+04	3.31E+04	3.31E+04	3.31E+04	3.31E+04
^{135}Cs	2.00E+04	2.00E+04	2.00E+04	2.00E+04	2.00E+04	2.00E+04	2.00E+04
^{137}Cs	6.22E+08	6.24E+07	6.22E+06	6.20E+05	–	–	–
^{151}Sm	6.19E+08	1.87E+09	1.33E+08	6.14E+07	1.32E+07	2.82E+06	6.04E+05
^{152}Eu	1.19E+06	3.14E+04	–	–	–	–	–
^{154}Eu	7.32E+06	2.01E+04	–	–	–	–	–
^{157}Tb	–	3.33E+04	–	–	–	–	–
^{158}Tb	8.09E+04	5.50E+04	3.74E+04	2.55E+04	1.18E+04	–	–
^{166m}Ho	1.68E+05	7.36E+05	1.49E+05	1.41E+05	1.26E+05	1.12E+05	9.97E+04
^{210}Pb	1.11E+04	1.85E+04	2.86E+04	4.07E+04	7.06E+04	1.08E+05	1.53E+05
^{226}Ra	–	–	1.21E+04	1.69E+04	2.87E+04	4.33E+04	6.05E+04
^{227}Ac	9.64E+04	9.70E+04	9.70E+04	9.70E+04	9.69E+04	9.68E+04	9.68E+04
^{228}Tb	1.68E+07	6.16E+06	2.25E+06	8.24E+05	1.10E+05	1.47E+04	–
^{229}Th	7.20E+05	7.14E+05	7.08E+05	7.02E+05	6.90E+05	6.79E+05	6.67E+05
^{230}Th	5.66E+04	6.91E+04	8.40E+04	1.00E+05	1.34E+05	1.68E+05	2.02E+05
^{231}Pa	6.16E+04	6.16E+04	6.15E+04	6.15E+04	6.15E+04	6.14E+04	6.14E+04
^{232}U	7.59E+07	2.78E+07	1.01E+07	3.71E+06	4.96E+05	6.64E+04	–

(*continued on next page*)

TABLE 2 PBH of significant radionuclides in the nuclear power waste from thermal and fast reactors subsequent to storage for 100 to 1000 years starting with 2100—cont'd

Nuclide	Storage time, years						
	100	200	300	400	600	800	1000
^{234}U	2.65E+06	3.50E+06	3.91E+06	4.11E+06	4.27E+06	4.31E+06	4.32E+06
^{236}U	4.96E+04	5.23E+04	5.50E+04	5.76E+04	6.29E+04	6.80E+04	7.30E+04
^{238}U	6.16E+04	6.16E+04	6.16E+04	6.16E+04	6.16E+04	6.16E+04	6.16E+04
^{237}Np	1.29E+06	1.61E+06	1.87E+06	2.10E+06	2.46E+06	2.72E+06	2.91E+06
^{238}Pu	1.96E+10	9.46E+09	4.651409	2.33E+09	6.20E+08	1.81E+08	5.70E+07
^{239}Pu	2.67E+09	2.66E+09	2.65E+09	2.65E+09	2.63E+09	2.62E+09	2.60E+09
^{240}Pu	4.92E+09	4.86E+09	4.81E+09	4.76E+09	4.66E+09	4.57E+09	4.47E+09
^{241}Pu	1.99E+07	2.83E+05	1.26E+05	1.23E+05	1.22E+05	1.20E+05	1.18E+05
^{242}Pu	1.05E+07	1.05E+07	1.06E+07	1.06E+07	1.06E+07	1.06E+07	1.06E+07
^{241}Am	1.87E+10	1.60E+10	1.36E+10	1.16E+10	8.41E+09	6.11E+09	4.43E+09
^{242m}Am	1.36E+09	8.34E+08	5.10E+08	3.12E+08	1.17E+08	4.37E+07	1.63E+07
^{243}Am	1.15E+08	1.14E+08	1.13E+08	1.12E+08	1.10E+08	1.08E+08	1.06E+08
^{243}Cm	2.71E+07	2.51E+06	2.31E+05	2.14E+04	–	–	–
^{244}Cm	1.08E+08	2.35E+06	5.10E+04	–	–	–	–
^{245}Cm	5.43E+06	5.39E+06	5.35E+06	5.31E+06	5.22E+06	5.14E+06	5.05E+06
^{246}Cm	6.82E+05	6.72E+05	6.63E+05	6.53E+05	6.34E+05	6.16E+05	5.98E+05
^{249}Cf	1.32E+05	1.08E+05	8.88E+04	7.29E+04	4.91E+04	3.31E+04	2.23E+04

Note: the PBH values marked with "-" were not calculated for the relevant isotopes.

TABLE 3 PBH of natural uranium with the total mass of 541.7 thous. Tons.

Nuclide	PBH, Sv	Nuclide	PBH, Sv	Nuclide	PBH, Sv
^{222}Rn	1.03E+10	^{238}U	3.01E+08	^{227}Th	2.71E+06
^{210}Po	8.23E+09	^{231}Pa	2.21E+08	^{214}Pb	9.59E+05
^{210}Pb	4.73E+09	^{223}Ra	3.12E+07	^{214}Bi	7.53E+05
^{226}Ra	1.92E+09	^{234}Th	2.27E+07	^{231}Th	1.06E+05
^{230}Th	1.41E+09	^{235}U	1.46E+07	^{211}Pb	5.63E+04
^{227}Ac	3.43E+08	^{218}Po	9.59E+06	^{211}Bi	1.99E+04
^{234}U	3.29E+08	^{210}Bi	8.94E+06		

TABLE 4 Characteristics of fuel campaigns for light-water reactors in the closed NFC.

Parameter	VVER TOI	VVER-S	VVER-SKD
Electric/ thermal power, MW	1255/3300	1255/3300	1200/2800
Availability factor	0.93	0.90	0.90
Cycle duration, ef. days	500	330	330
Initial loading, t_{hm}	77	125	40/60
Initial loading enrichment/ Pu content*, %	3.9/–	2.9/–	0.2/14.0
Annual loading with UOX/U-PuOX, t_{hm}	13.7/9.7	–/21	6.2/13.8
Annual loading enrichment/ Pu content*, %	4.86/9	–/8.2	0.2/16.5
UOX/U-PuOX burnup, MW*day/kg_{hm}	49/46	–/50	15/62
Accumulation coefficient for fissionable materials	0.44/0.72	0.66	0.04/0.92

**To be adjusted.*

TABLE 5 Basic technical characteristics of power units with fast reactors.

Reactor	BN-600	BN-800	BN-1200
Rated thermal power, MW	1470	2100	2800
Gross electric power, MW	600	880	1220
Number of heat-generating loops	3	3	4
Primary coolant temperature, °C (at the intermediate heat exchanger inlet/ outlet)	535/368	547/354	550/410
Secondary coolant temperature, °C (at the SG inlet/ outlet)	505/318	505/309	527/355
Third circuit parameters:			
Main steam temperature, °C	505	490	510
Main steam pressure, MPa	14	14	17
Feedwater temperature, °C	240	210	275
Gross/ net efficiency factor, %	42.5/40	41.9/38.8	43.5/40.7

References

[1] Federal Law "On RW handling and introduction of changes into individual legal acts of the Russian Federation" dated July 11, 2011 No. FZ-190.

[2] Fundamentals of the State Policy in the Area of Ecological Development of the Russian Federation up to 2030, approved by the RF President on 30.04.2012.

[3] Energy Strategy of Russia up to 2030 [approved by Order of the Russian Government No. 1715-p dated November 13, 2009].

[4] Energy Strategy of the Russian Federation up to 2035: draft. Developed in accordance with Federal Law No. 172-FZ dated June 28, 2014 "On strategic planning in the Russian Federation", rev. 01.02.2017.

[5] New Generation Nuclear Power Technologies for the Period of 2010-2015 with the Extension up to 2020: Federal Special Program [approved by Resolution of the Russian Government No. 5 dated 03.02.3010].

[6] Criteria for Classification of Solid, Liquid and Gaseous Waste as Radioactive Waste, Criteria for Classification of Radioactive Waste as Special Radioactive Waste and Disposable Radioactive Waste and Classification Criteria for Disposable Radioactive Waste. Resolution of the Russian Government No. 1069 dated October 19, 2012.

[7] Land-Use Planning Scheme of the Russian Federation in the Area of Power Generation [approved by Resolution of the Government of the Russian Federation No. 1634-p dated 01.08.2016].

[8] Russian Nuclear power Development Strategy for the First Half of the XXI Century. Basic Provisions. - M.: Russian Ministry of Atomic Energy, 2000.

[9] Basic provisions of the Russian Nuclear power Development Strategy up to 2050 with further extension for the period up to 2100. M. Rosatom, 2019.

[10] PBTVR-73. Safety Rules in Transportation of Radioactive Substances//Electronic library of legal, regulatory and technical documentation [Electronic source]: official site. - M., 2000. - URL: http://docs.cntd.ru/document/1200035576.

[11] OPBZ-83. Basic Safety and Physical Protection Rules for Transportation of Nuclear Materials//Electronic library of legal, regulatory and technical documentation [Electronic source]: official site. - M., 2000. - URL: http://docs.cntd.ru/document/901864810.

[12] NPP SR 03. Sanitary Rules for Design and Operation of Nuclear Power Plants (approved on 22.04.2003)//Electronic library of legal, regulatory and technical documentation [Electronic source]: official site. - M., 2000. - URL: http://docs.cntd.ru/document/901862274.

[13] NP-053-04 Safety Rules for Transportation of Radioactive Materials (approved by Order of Rostechnadzor No. 5 dated 04.10.2004)//Russian Gazette 2004.

[14] Rules for Safe Transportation of Radioactive Materials. IAEA, Vienna, 2009. Section I. P.2.

[15] NRB-99/2009. Radiation Safety Standards: Hygienic Norms SP 2.6.1.2523-09. - M.: Federal Hygiene and Epidemiology Center of the Russian Agency for Health and Consumer Rights, 2009. - 100 p.

[16] NP-093-14. Criteria of RW Acceptability for Disposal: Federal Rules and Regulations in the Area of Atomic Energy Use.

[17] NP-019-15. Collection, Processing, Storage and Conditioning of Liquid Radioactive Waste. Safety Requirements: Federal Rules and Regulations in the Area of Atomic Energy Use.

[18] Amendments to the FRR in the area of atomic energy use "Collection, Processing, Storage and Conditioning of Liquid Radioactive Waste. Safety Requirements" (NP-019-15)//Nuclear and Radiation Safety. 2016. No. 3.

[19] A.A. Goverdovsky, S.G. Kalyakin, V.I. Rachkov, Alternative strategies of nuclear power development in the XXI century, Heat Power Eng. (5) (2014) 3.

[20] V.A. Tsykanov, A.A. Maershin, A.A. Petukhov, et al., Operability analysis for fuel rods of BOR-60 reactor with vibro-compacted uranium-plutonium oxide fuel, Atomic Energy 66 (5) (1989) 299–302.

[21] A.A. Arakelyan, M.V. Vedernikova, A.A. Kiselev, S.V. Panchenko, S.V. Strizhova, V.A. Chukharev, Analysis of risks for the health of the population of Obninsk from harmful releases into the atmosphere, Risk Anal. Probl. 15 (5) (2018) 26–37.

[22] V.S. Barashenkov, Problems of ADS Technology, JINR preprint P2-94-56, Dubna, 1994.

[23] White Book of Nuclear power/under general editorship of E.O. Adamov. - M.: NIKIET, 2001. - 269 p.

[24] Introduction to Application of INPRO Methodology for Assessment of Nuclear Power Systems. IAEA Nuclear Energy Series No. NP-T-1.12. [Electronic source]. Vienna: IAEA, 2011. 49 p. URL: https://www-pub.iaea.org/MTCD/Publications/PDF/Publ478R_web.pdf (date of reference: 21.01.2019).

[25] Introduction to Nuclear Power Technologies M.V. Rachkov, L.A. Lebedev, V.V. Ivanov; ed. V.I. Rachkov, RAS A.M. - M.: Science, 2015. - 214 p.

[26] A.Yu. Smirnov, V.A. Nevinnitsa, et al., Inclusion of regenerated uranium into the fuel for light-water reactors as the technique for protection against proliferation, Univ. News. Nucl. Power (4) (2011) 93–103.

[27] A.V. Lopatkin, V.V. Orlov, Effect of neutron spectrum on the characteristics of Np, Am and Cm transmutation chains, Atomic Energy. 100 (6) (2006) 452–458.

[28] Annual Report of Rosenergoatom JSC for 2010. - M.: Rosenergoatom JSC, 2011.

[29] V.A. Grachev, Interrelation of global environmental problems of the public health and nuclear power development, Hum. Ecol. (2) (2018) 9–15.

[30] Exposure Doses for the Population of the Russian Federation in 2015: Information Digest. - St. Petersburg: NIIRG, 2016. - 73 p.

[31] I.H. Ganev, V.V. Orlov, E.O. Adamov, Achievement of radiation equivalence in handling of radioactive waste from nuclear power, Atomic Energy 73 (1) (1992) 44–50.

[32] V.I. Chissov, V.V. Starinsky, G.V. Petrova, Malignant Tumors in Russia in 2007 (incidence and death rate), Moscow: Herzen Moscow Oncology Research Institute of the Russian Federal Agency for Medical Technologies, 2009, p. 244.

[33] A.D. Kaprin, V.V. Starinsky, G.V. Petrova, Malignant Tumors in Russia in 2016 (Incidence and Death Rate), M.: Herzen Moscow Oncology Research Institute - Branch of Radiology National Medical Research Center of the Russian Ministry of Health, 2018, p. 250.

[34] Manufacturing of the Pilot Batch of the Master Mixture of REMIX Fuel Powders, REMIX Fuel Pellets and Pre-Reactor Studies of REMIX Fuel Characteristics (including the master mixture delivery to VNIINM JSC): R&D Report D.V. Ryabkov et al.

[35] V. Kapyrin, V.A. Ivanov, G.V. Kopytov, et al., GeRa Integral Code for RW Disposal Safety Analysis/I, Mining J. (10) (2015) 44–50.

[36] Yu.S. Fedorov, B.A. Bibichev, B.Ya. Zilberman, et al., Use of Regenerated Uranium and Plutonium in Thermal Reactors, Atomic Energy 99 (2) (2005) 136–141.

[37] V.A. Astafiev, A.Ye. Glushenkov, M.V. Skupov, Investigation of MOX fuel powder pulverization and mixing processes, Nucl. Sci. Technol. Issues. Ser. Mater. Res. New Mater. 1 (59) (2002).

[38] Sh.Sh. Ibragimov, T.S. Menshikova, A.Ya. Ladygin, Studies of fuel rods from the work packages of BR-5 reactor with uranium monocarbide fuel, Atomic Energy 30 (1) (1971) 3–7.

[39] Results of Rosatom State Atomic Energy Corporation Activities. Public Annual Report. 2013. [Electronic source]. URL: www.rosatom.ru/resources/4691a78045da51aaab41ftd 203d7eel8/GO+ROSATOM+2013.pdf.

[40] E.O. Adamov, P.N. Martynov, V.I. Rachkov, A.P. Sorokin, On selection of coolant for nuclear power installations with inherent safety, RAS News. Power Eng. (6) (2015) 3–14.

[41] R. Fletcher, S. Fletcher, E. Wagner, Clinical Epidemiology. Fundamentals of Evidence-Based medicine, Rev. 3. - M.: Media Sphere (2004) 352.

[42] D.N. Kolupaev, Development of SNF Processing Technologies in Russia, in: Materials of Atomeco-2015 International Exhibition, Moscow, 10.11.2015.

[43] E.O. Adamov, A.V. Jalavyan, A.V. Lopatkin, N.A. Molokanov, Ye.V. Muraviev, V.V. Orlov, S.G. Kalyakin, V.I. Rachkov, V.M. Troyanov, Ye.N. Avronin, V.B. Ivanov, R.M. Aleksakhin, Conceptual provisions of the Russian nuclear power development strategy up to 2100, Atomic Energy. 112 (6) (2012) 319–330.

[44] I.V. Kapyrin, S.S. Utkin, Yu.V. Vasilevsky, Concept of development and application of GeRa computational package for safety analysis of radioactive waste repositories, Nucl. Sci. Technol. Issues. Ser. Math. Model. Phys. Process. (4) (2014) 44–54.

[45] Large-Scale Two-Component Nuclear power with the Closed NFC Based on Fast Reactors: Analytical Note. M.: NIKIET, 2017.

[46] A.V. Lopatkin.Radiation-Equivalent RW Handling: Technical Information 01.2017 NRRE. - M., 2017. - 21 p.

[47] Ye.P. Makarov. Handling of RW from SNF Processing at Mayak PA FSUE Materials of Atomeco-2013 International Exhibitions. 31.10.2013. Moscow.

[48] V.V. Pudanov, A.Ye. Aksenova, V.A. Pervichko, et al., Mathematical modelling of fluid dynamics in the components of nuclear power installations with the use of super-computers, Atomic Energy 117 (6) (2014) 307–311.

[49] E.O. Adamov, I.H. Ganev, A.V. Lopatkin, et al., Minimization of high-level waste from functioning of large-scale nuclear power in Russia, Atomic Energy 82 (3) (1997) 209–218.

[50] Ye.S. Smotritsky, N.T. Chebotarev, V.I. Kutaitsev, et al., Multi-component alloys on the basis of U-Pu as nuclear fuel for fast reactors, in: Symposium on the use of Pu, Brussels, IAEA, Vienna, 1967 13-17 March 1967.

[51] A.I. Morozov, Introduction to plasma dynamics, M.: Science (2008) 576 pp.

[52] Ye.V. Muraviev, Concept of two-component nuclear power in Russia within the framework of development strategy, Energy Policy (3) (2017) 49–59.

[53] A.M. Pavlovichev, Yu.S. Fedorov, B.A. Bibichev, et al., Neutron and physical characteristics of VVER-1000 core with 100% loading of regenerated uranium and plutonium fuel, Atomic Energy 101 (6) (2006) 407–413.

[54] G.N. Vlaskin, V.I. Rachkov, Yu.S. Khomaykov, Neutron and physical substantiation of Np, Am, Cm transmutation, Atomic Energy 116 (5) (2014) 262–266.

[55] V.V. Orlov, Ye.N. Avrorin, E.O. Adamov, et al., Non-traditional concepts of NPPs with inherent safety (new nuclear technology for the next stage of large-scale nuclear power), Atomic Energy 72 (4) (1992) 317–329.

[56] E.O. Adamov, V.I. Rachkov, New technology platform for establishment of the national development strategy for nuclear power, RAS News. Power Eng. (2) (2017) 3–12.

[57] E.O. Adamov, V.I. Rachkov, Yu.S. Khomyakov, Yu.Ye. Shvetsov, Possible engineering solutions for emergency residual heat removal in liquid metal cooled fast reactors, RAS News. Power Eng. (2) (2019) 17–32.

[58] Ye.M. Melikhova, Ye.M. Byrkina, Yu.A. Pershina, Some mechanisms of social health risk enhancement through presentation of Fukushima accident in mass media, Med. Radiol. Radiat. Saf. 58 (4) (2013) 5–6.

[59] State of Sanitary and Epidemiological Wellbeing of the Russian Federation Population in 2016: Governmental Report. M.: Federal Service for Surveillance on Consumer Rights Protection and Human Wellbeing, 2017. - 220 p.

[60] I.V. Khonina, A.A. Lumpov, A.Yu. Shadrin, B.Ya. Zilberman, N.G. Kravchenko, Formation of molybdenum and zirconium deposits in concentrated uranyl nitrate solutions, Radiochemistry 52 (2) (2010) 151–154.

[61] a.A. Maershin, V.A. Tsykanov, Yu.M. Golovnenko, et al., Experience in testing of advanced fuel compositions in BOR-60 Reactor, Atomic Energy 91 (5) (2001) 385–389.

[62] Experience in comparative assessment of radiation and chemical risks for the public health at the regional level S.M. Novikov, T.A. Shashina, N.S. Dodina, V.A. Kislitsin, S.V. Panchenko, A.A. Arakelyan International Scientific and Practical Conference on Environmental, Industrial and Energy Safety-2018, September 24-27, 2018, Sevastopol.

[63] Pilot and Demonstration Center at MCP: Objectives, Technologies, Prospects I.A. Maslennikov, Yu.S. Fedorov, A.Yu. Shadrin et al., 2012 [Electronic source: Atomic Energy 2.0]: http://www.atomic-energy.ru/technology/33473.

[64] V.V. Orlov, What should the doubling time of fast reactor be? Atomic Energy. 64 (3) (1988) 195–197.

[65] G.G. Onishchenko, S.M. Novikov, Yu.A. Rakhmanin, S.L. Avaliani, K.A. Bushtueva, Fundamentals of the Public Health Risk Assessment under the Impact of Chemical Substances Contaminating the Environment, M.: Research Institute of Human Ecology and Environmental Hygiene, 2002, p. 408. Yu.A. Rakhmanin, G.G. Onishchenko.

[66] A.V. Belyaeva, I.F. Gilmutdinov, I.Yu. Zhemkov, et al., Peculiarities of uranium-plutonium nitride fuel swelling in the course of low-temperature irradiation in a fast reactor up to burnup of 5% ha, Atomic Energy, Moscow 122 (5) (2017) 263–266.

[67] V.I. Rachkov, S.G. Kalyakin, O.F. Kukharchuk, Yu.I. Orlov, A.P. Sorokin, From the first NPP to generation IV nuclear power installations (for the 60-th anniversary of the first NPP), Heat Power Eng. (5) (2014) 11–18.

[68] V.M. Troyanov, A.F. Grachev, L.M. Zabudko, M.V. Skupov, Opportunities to use nitride fuel for fast reactors with the closed fuel cycle, Atomic Energy. 117 (2) (2014) 69–75.

[69] A.I. Burnazyan, P.V. Ramzaev, A.A. Moiseev, Radiation hygiene prospects, Hygiene Sanit. (10) (1969) 8–14.

[70] O.V. Skiba, V.A. Kisly, Yu.P. Savochkin, S.K. Vavilov, Pyro-electrochemical Processes in the Fuel Cycle of Fast Reactors, JSC "RIIAR", 2012, p. 347.

[71] Preparation of the Device and Reactor Operability Testing for Experimental Fuel rods with REMIX Fuel in the Loop of MIR Reactor: R&D Report V.A. Ovchinnikov et al.

[72] B.F. Myasoedov, Yu.M. Kulyako, A.M. Fedoseev, et al., Production of Np, Pu and U dioxides in nitric-acid solutions in presence of hydrazine hydrate, Radiochemistry 55 (6) (2013) 487.

[73] Yu.M. Kulyako, S.A. Perevalov, T.I. Trofimov, et al., Production of uranium oxides in nitric-acid solutions through uranyl nitrate and hydrazine hydrate reaction, Radiochemistry 55 (6) (2013) 481.

[74] I.N. Ostretsov, V.V. Chilap, Post-Chernobyl Nuclear power: "Innovations" Based on the Ideas of the 50-ies, XX Century, or... Relativistic Heavy Nuclei Power Engineering? in: RAS Institute of Natural Science and Technology History. Annual Scientific Conference. - M., 2006, pp. 541–545.

[75] Problems of Large-Scale Nuclear power Arrangement and Relativistic Nuclear Technologies (RNT) ATOMENERGOMASH Center of Physical and Technical Projects CJSC. M., 2010.

[76] A.A. Makarov, L.M. Grigoriev, T.A. Mitrova, Prediction of the Global and Russian Power Engineering Development for 2016, Analytical Center of Energy Research Institute of the Russian Academy of Sciencesat the Russian Government. - Moscow, 2016, p. 196.

[77] V.M. Troyanov, A.F. Grachev, L.M. Zabudko, M.V. Skupov, G.A. Kireev, Program and some results of pre-reactor studies of nitride fuel for fast reactors, Atomic Energy. 117 (4) (2014) 192–197.

[78] Publication 103 of the International Commission for Radiation Protection (ICPR): tr. from Eng./Under the general editorship of M.F. Kiselev and N.K. Shandala. [Electronic source]. M.: Alana, 2009. 312 p. URL: http://www.icrp.org/docs/P103_Russian.pdf (date of reference: 18.01.2019).

[79] G.G. Onishchenko, A.Yu. Popova, I.K. Romanovich, A.N. Barkovsky, T.A. Kormanovskaya, I.G. Shevkun, Radiation hygiene certification and the uniform system for accounting and control of individual exposure doses as the information basis for managerial decisions on radiation safety assurance for the population of the Russian Federation, Radiation Hygiene 10 (3) (2017) 18–35.

[80] A.V. Tselishchev, V.S. Ageev, Yu.P. Budanov, A.G. Ioltukhovsky, N.M. Mitrofanova, M.V. Leontieva-Smirnova, I.A. Shkabura, L.M. Zabudko, A.V. Kozlov, V.V. Maltsev, A.V. Povstyanko, Development of structural steel for fuel rods and fuel assemblies of sodium-cooled fast reactors, Atomic Energy. 108 (4) (2010) 217–221.

[81] I.A. Shkabura, V.M. Poplavsky, B.A. Vasiliev, Development of materials and fuel rods for the core of fast reactors, Rosenergoatom (12) (2009) 6.

[82] V.V. Chudanov, A.Ye. Aksenova, A.A. Makarevich, et al., Development of the direct numerical simulation method for turbulent flows with the use of super-computers, Atomic Energy 118 (4) (2015) 197–202.

[83] F.Ye. Reshetnikov, I.S. Golovnin, Yu.K. Bibilashvili, et al., Development, Production and Operation of Fuel rods for Power Reactors, 2 books. - Moscow: Energoatomizdat, 1995.

[84] V.I. Rachkov, Fundamentals of hazardous system theory, M.: Science (2015) 165.

[85] T.K. Zhaboev, N.A. Mosunova, A.R. Arutunyan, Implementation of the design code development management system based on IBM rational Jazz platform, Comput. Inf. Technol. Bull. (1(115)) (2014) 34–38.

[86] B.D. Rogozkin, N.M. Stepennova, Yu.E. Fedorov, et al., Testing results for mixed mononitride fuel (45%PuN+55%UN and 60%PuN+40%UN) in BOR-60 reactor up to burnup of ~12% ha, Atomic Energy 110 (6) (2011) 332–346.

[87] I.G. Shevkun, V.S. Stepanov, S.V. Matyukhin, et al., Results of radiation hygiene certification for 2015 (Radiation Hygiene Certificate of the Russian Federation), - M. (2016) 125.

[88] A.Yu. Shadrin, K.N. Dvoeglazov, O.A. Ustinov, PH-PROCESS - Processing Technology for Mixed Uranium-Plutonium Fuel from BREST-OD-300 Reactor, Radiochemistry 58 (3) (2016) 234–241.

[89] Ye.P. Velikhov, A.Yu. Gagarinsky, S.A. Subbotin, V.F. Tsibulsky, Russia in the global power engineering of the XXI century, M.: IzdAt (2006) 136.

[90] Yu.A. Rakhmanin, S.M. Novikov, S.L. Avaliani, O.O. Sinitsyna, T.A. Shashina, Current problems in assessment of the risks related to the impact of environmental factors on the public health and opportunities for its improvement, Health Risk Anal. (2) (2015) 4–11.

[91] R.V. Arutunyan, L.M. Vorobieva, S.V. Panchenko, R.I. Bakin, S.M. Novikov, T.A. Shashina, N.S. Dodina, D.V. Goryaev, I.V. Tikhonova, S.V. Kurkatov, S.Ye. Skudarnov, O.Yu. Ivanova, Comparative analysis of radiation and chemical health risks for the population of the Krasnoyarsk territory, Radiat. Risks 23 (2) (2014) 123–136.

[92] F.N. Kryukov, B.N. Susev, O.N. Nikitin, et al., Condition of alloyed metal fuel subsequent to irradiation in fast reactors, Atomic Energy 113 (4) (2012) 222–227.

[93] S.V. Panchenko, A.A. Arakelyan, M.V. Vedernikova, Comparative assessment of radiation and toxic risks in angarsk, Radiat. Risk. 26 (2) (2017) 83–96.

[94] A.A. Arakelyan, M.V. Vedernikova, S.V. Panchenko, Comparative assessment of radiation and chemical risks for the population of Obninsk, in: Environmental, Industrial and Energy Safety. - 2018: collection of articles on the materials of the international scientific and practical conference, Sevastopol, 2018, pp. 103–106.

[95] , Comparative assessment of damage to health of the personnel and the public in the course of power generation at NPPs and TPPs, Nuclear Power Experience, 4, IAEA, Vienna, 1983, pp. 573–579.

[96] E.O. Adamov, L.M. Zabudko, V.I. Matveev, V.I. Rachkov, V.M. Troyanov, Yu.S. Khomyakov, V.N. Leonov, Comparative analysis of advantages and drawbacks of metal and nitride mixed uranium-plutonium fuel in fast reactors, RAS News. Power Eng. (2) (2015) 3–15.

[97] A.A. Arakelyan, S.V. Panchenko, S.V. Strizhova, T.A. Shashina, Comparative analysis of radiation and chemical risks in the location area of Leningrad NPP, in: Safety, Efficiency and Economy of Nuclear power: plenary and sectional reports of the 11-th International Scientific and Technical Conference, Moscow, 2018, pp. 410–416. May, 23-24, 2018- M.

[98] E.O. Adamov, I.H. Ganev, A.V. Lopatkin, V.T. Muratov, V.V. Orlov, Degree of approximation to radiation equivalence of high-level radioactive waste and natural uranium in the nuclear power fuel cycle of Russia, Atomic Energy 81 (6) (1996) 403–409.

[99] A.N. Dorofeev, L.A. Bolshov, I.I. Linge, S.S. Utkin, Ye.A. Savelieva, Strategic master plan of investigations for safety analysis of construction, operation and decommissioning of geological RW repositories, Radioactive Waste (1) (2017) 33–40.

[100] Solid Oxide Fuel rods: Problems, Solution Approaches, Opportunities for Development and Commercialization: Analytical Review. M.: Republican Scientific, Research and Consulting Expert Center FSBIS.

[101] O.V. Skiba, P.T. Porodnov, A.V. Bychkov, et al., Technologies of pyro-electrochemical processing and production of oxide nuclear fuel, in: Collection of NIIAR papers, Dimitrovgrad, 1994, pp. 3–13. Isib. P. 21-28.

[102] A.Yu. Shadrin, K.N. Dvoeglazov, V.A. Kashcheev, Yu.S. Mochalov, A.T. Terentiev, M.N. Gerasimenko, S.A. Cheshuyakov, Technology for Hydro-Metallurgical and Combined Processing of MNIT and MOX FNR SNF, in: Fuel Cycle Closing for Nuclear power Based on Fast Reactors: materials of the industry conference, Tomsk, 2018 October 11-12.

[103] Fuel Composition for Water-Cooled Thermal NPP Reactors. Patent RU 2 537 013 B.Ya. Zilberman, Yu.S. Fedorov, A.A. Rimsky-Korsakov, B.A. Bibichev, M.N. Chubarov, P.N. Alexeev Bul. No. 36, 2014.

[104] Second Generation Fuel Assembly TVS-2M/OKB Gidropress [Electronic source]. URL: http://www.gidropress.podolsk.ru/files/booklets/ /ru/T VS_rus.pdf.

[105] E.O. Adamov, I.H. Ganev, A.V. Lopatkin, et al., Transmutation fuel cycle in large-scale nuclear power of Russia, M.: NIKIET SUE (1999) 252.

[106] V.K. Ivanov, S.Yu. Chekin, A.N. Menyailo, M.A. Maksyutov, K.A. Tumanov, P.V. Kashcheeva, S.S. Lovachev, E.O. Adamov, A.V. Lopatkin, Radiological protection levels for the public in implementation of the radiation equivalence principle: risk-oriented approach, Radiat. Risk 27 (3) (2018) 9–23 V.

[107] R.V. Arutyunyan, L.A. Bolshov, Ye.M. Melikhova, S.V. Panchenko, Lessons of Chernobyl and Fukushima: urgent problems related to improvement of the protection system for the public and territories in case of NPP accidents, Med. Radiol. Radiat. Saf. (3) (2016) 36–51.

[108] L.P. Feoktistov, Analysis of a physically safe reactor concept: AEI preprint -4605/4, M.: Atominform (1988).

[109] A.V. Khaperskaya.Problems of SNF handling in Russia and the opportunities for their solution. - 2015. [Electronic source]. URL: http://www.atomic-energy.ru/articles/2015/04/01/55910.

[110] E.O. Adamov, I.H. Ganev, Environmentally perfect nuclear power, M.: NIKIET FSUE (2007) 145.

[111] R.V. Arutyunyan, L.A. Bolshov, L.M. Vorobieva, Ye.K. Khandogina, S.M. Novikov, T.A. Shashina, N.S. Skvortsova, M.I. Chubirko, N.M. Pichuzhkina, Environmental protection and sustainable development of the location area of Novovoronezh NPP, Atomic Energy 109 (2) (2010) 109–114 VIs.

[112] A.V. Beznosov, S.S. Pinaev, D.V. Davydov, A.A. Molodtsov, T.A. Bokova, P.N. Martynov, V.I. Rachkov, Experimental study of the contact heat exchange characteristics for lead coolant and working medium, Atomic Energy 98 (3) (2005) 182–187 VIs.

[113] V.T. Vasilkov, V.I. Goldansky, V.P. Jelepov, V.P. Dmitrievsky, Accelerator-driven method for generation of neutrons and production of fissile materials, Atomic Energy. 29 (3) (1970) 151–158 VIs.

[114] N.S. Babaev, V.F. Demin, L.A. Ilyin, V.A. Knizhnikov, I.I. Kuzmin, V.A. Legasov, Yu.V. Sivintsev, Nuclear power, human beings and environment, Rev. 2. - M.: EnergoIzdat (1984) 312.

[115] R. Bonifetto, S. Dulla, R Ravetto, et al., A full-core coupled neutronic/thermal-hydraulic code for the modeling of lead-cooled nuclear fast reactors, Nucl. Eng. Des. 261 (2013) 85–94.

[116] W. Keller, A. Modarres, A historical overview of probabilistic risk assessment development and its use in the nuclear power industry: a tribute to the late Professor Norman Carl Rasmussen, Reliab. Eng. Syst. Safety 89 (3) (2005) 271–285.

[117] OECD, in: Actinide and Fission Product Partitioning and Transmutation: Book of Abstracts 11th Information Exchange Meeting, San Francisco, California, USA, 2012, p. 406. 1-4 November 2010NEA No. 6996.

[118] Actinides Recovery from Irradiated MOX fuel by Pyrochemical Reprocessing I Kato T. et al. Global 2011 (Proc. Int. Conf., Makuhari, Japan, 2011). – P. 391320.

[119] Advanced Fuel Cycle Cost Basis //INL/EXT-17-43826 - 2017.

[120] Advances in the Development of the Code FRE-NETIC for the Coupled Dynamics of Lead-Cooled Reactors Bonifetto R., Caron D., Dulla S. et al. CERSE-POLITO RL 1572/2015. Torino, 2015. - 63 p.

[121] W.F. Lyon, R.B. Baker, R.D. Legget, R.B. Matthews, Advancing liquid metal reactor technology with nitride fuels, in: Proc. of Int. Conf. on Fast Reactors and Related Fuel Cycles, II, Kyoto, Japan, 1992, p. 14.8-1. October 28-November 1.

[122] R.K. Amirov, N.A. Vorona, A.V. Gavrikov, G.D. Lizyakin, V.P. Polishchuk, I.S. Samoilov, V.P. Smirnov, R.A. Usmanov, EM. Yartsev, Vacuum arc with a distributed cathode spot as a plasma source for plasma separation of spent nuclear fuel and radioactive waste, Plasma Phys. Rep. 41 (2015) 808–813.

[123] C.E. Ahlfeld, J.D. Wagoner, D.L. Sevier, R.L. Freeman, Application, design and project implementation of a plasma mass separator for enhanced high level waste processing, in: Fusion Engineering 2005. Twenty-First IEEE/NPS Symposium on, 2005 -Pl.

[124] R. Freeman, S. Agnew, R Anderegg, B. Cluggish, J. Gilleland, R. Isler, A. Litvak, R. Miller, R. O'Neill, T Ohkawa, S. Pronko, S. Putvinski, L. Sevier, A. Sibley, K. Umstadter, T Wade, D. Winslow, Archimedes plasma mass filter *I*, AIP Conf. Proc. 694 (2003) 403.

[125] Archimedes plasma mass filter Litvak A., Agnew S., Anderegg E, Cluggish B., Freeman R., Gilleland J., Isler R., Lee W., Miller R., Ohkawa T, Putvinski S., Sevier L., Umstadter K., Winslow D. 30th EPS Conference on Contr. Fusion and Plasma Phys. (St. Petersburg, 7-11 July 2003). EGA Vol. 27A. - P. O-1.6A.

[126] E.O. Adamov, V.V Orlov, I.H Ganev, Attainment of radiation equivalence in nuclear power radioactive product management *I*, Nucl. Technol. 104 (2) (1993) 233–241.

[127] A.L Morozov, V.V. Saveliev, Axisymmetric plasma-optic mass separators *I*, Plasma Phys. Rep. 31 (5) (2005) 417–424.

[128] T. Ohkawa, R.L. Miller, Band gap ion mass filter *I*, Phys. Plasmas 9 (2002) 5116.

[129] M. Fromont, J. Lamontagne, M. Asou, I. Aubrun, Behavior of Uranium-Plutonium mixed nitride and carbide fuels irradiated in Phenix *I*, in: Nuclear Energy Systems for Future Generation and Global Sustainability: Int. Conf. Global 2005, Tsukuba, Japan, 2005, p. 329. October 9-13.

[130] H. Blank, Specification and characterization of dense LMFBR fuels, J. Nucl. Mat. 153 (1988) 7.

[131] Blees T. Prescription for the planet. The painless remedy for our energy and environmental crises. Moscow: Center for the promotion of social and environmental initiatives of the nuclear industry, 2011. - 282 p.

[132] BloombergNEF: Clean Energy Investment Trends, 2018 [Electronic source]. URL: https://data.bloomberglp.com/professional/sites/24/BNEF-Clean-Energy-Investment-Trends-2018.pdf.

[133] N.E. Bolus, NCRP report 160 and what it means for medical imaging and nuclear medicine, J. Nucl. Med. Technol. (41) (2013) 255–260.

[134] Brueziere J. MOX Recycling: An Industrial Reality Global 2011 (Proc. Int. Conf., Makuhari, Japan, 2011). - P. 360024.

[135] D.L. Preston, S. Kusumi, M. Tomonaga, S. Izumi, E. Ron, A. Kuramoto, N. Kamada, FL Dohy, T. Matsuo, H. Nonaka, D.E. Thompson, M. Soda, K. Mabuchi, Cancer incidence in atomic bomb survivors. Part III: leukemia, lymphoma and multiple myeloma, 1950-1987, Radiat. Res. 137 (1994) 68–97.

[136] K.F. Eckerman, R.W. Leggett, C.B. Nelson, J.S. Puskin, A.C.B. Richardson, Cancer Risk Coefficients for Environmental Exposure to Radionuclides. Federal Guidance Report 13, Oak Ridge National Laboratory. Office of radiation and indoor air United States Environmental Protection Agency, Washington, DC, 1999 EPA 402-C-99-00120460.

[137] H. Sekimoto, K. Ryu, Y. Yoshimura, CANDLE: the new burnup strategy, Nucl. Sci. Eng. 139 (2001) 1–12.

[138] Characterization of dissolution residues - fuel rod cladding and fission product insolubles Jeskins I.L., Brown P.E. Radiochimica Acta. - 1984. Vol. 36. - P. 25-30.

[139] Y Korobkin, N. Lebedev, V. Paperny, Charge separation of plasma flow in curvilinear magnetic field, Tech. Phys. Lett. 38 (3) (2012) 254.

[140] Combined (Pyro+Hydro) Technology of FNR SNF Reprocessing Volk V. et al. Global 2011 (Proc. Int. Conf., Makuhari, Japan, 2011). – P. 386756.

[141] Yu.E. Shvetsov, Yu.M. Ashurko, S.L. Osipov, V.S. Gorbunov, Comparative Analysis of Effectiveness of Various Emergency Core Cooling System Design Options for Sodium Fast Reactors of High Power, in: Int. conf. on Fast Reactors and Related Fuel Cycles: Safe Technologies and Sustainable scenarios, Paris, France, 2013 4-7 March.

[142] Comparison of the decay heat removal systems in the KALIMER-600 and the DSFR Kwi-Seok Ha, Hae-Yong Jeong. Nucl. Eng. Technol. - 2012. Vol. 44(5). - P. 535-542.

[143] H. Kleykamp, J. Paschoal, R. Pejsa, E Thummler, Composition and structure of the fission products precipitates in irradiated nuclear fuels: correlation with phase studies in the Mo-Ru-Rh-Pd and BaO-UO2-ZrO2-MoO2 systems, J. Nucl. Mater. 130 (1985) 426–433.

[144] Zh. Chen, X.-N. Chen, A. Rineiski, et al., Coupling a CFD code with neutron kinetics and pin thermal models for nuclear reactor safety analyses, Ann. Nucl. Energy. 83 (2015) 41–49.

[145] V.I. Volk, T.A. Boytsova, S.N. Veselov et al. Crystallization purification full scale test of U–Pu mixture. J. Radioanal. Nucl. Chem. 318 (2018) 2501–2505. https://doi.org/10.1007/s10967-018-6336-9

[146] V.L. Vidanov, L.I. Tkachenko, A.Yu. Evsyukova, A.Yu. Shadrin, Degradation of diluent in TODGA - F-3 extraction system under a-, 0-, y- irradiation, in: 18th Radiochemical Conference, Marianskie Lazni, Czech Republic, 2018, p. 91. 13- 18.05.2018, Book of Abstracts, RadChem.

[147] Y. Arai, Y. Suzuki, T. Iwai, T. Ohmichi, Dependence of the thermal conductivity of (U, Pu) N on porosity and plutonium content, J. Nucl. Mater. 195 (1992) 37–43.

[148] E. Esbelin, P. Gareil, M. Masson, J.-L. Emin, Determination of Mo(VI)-Zr(IV) interaction parameters in highly acidic solutions by UV absorbance spectrophotometry, Analytica Chimica Acta. 442 (2001) 171–182.

[149] Grachev A.F., L.M. Zabudko, E.A. Zvir, D.V Zozulya, Ivanov Yu.A., F.N. Kryukov, Mochalov Yu.S., M.V Skupov, Development of innovative fast reactor nitride fuel in Russian Federation, in: state-of-art Proc. of Int. Conf. "Fast Reactors and Related Fuel Cycles: Next Generation Nuclear Systems for Sustainable Development, Yekaterinburg RE Book of Abstracts, IAEA, 2017, p. 57. 26-29 JuneCN-245-62.

[150] T Ishizu, H. Endo, I. Tatewaki, et al., Development of integrated core disruptive accident analysis code for FBR - ASTERIA-FBR, in: Proc. of the 2012 Intern. Congress on Advances in Nuclear Power Plants (ICAPP'12), Chicago, USA, 2012 ID 12100.

[151] O.V. Shmidt, T.V. Podymova, A.Yu. Shadrin, Development of mathematical model for balance settlement of product and waste flows from SNF reprocessing, Proc. Chem. 7 (2012) 387–391.

[152] H. Hayashi, et al., Development of pyrochemical process of spent nitride fuels for ADS; its elemental technologies and process flow diagram, in: Global 2011: Proc. Int. Conf., Makuhari, Japan, 2011, p. 39209.

[153] L. Tkachenko, E. Kenf, V. Babain, M. Alyapyshev, M. Logunov, Yu. Voroshilov, V. Vidanov, A. Shadrin, D. Zverev, Dynamic Test of Extraction Process for Americium Partitioning from the Purex Raffinate, in: International Conference on Fast Reactors and Related Fuel Cycles, Yekaterinburg, Russian Federation, 2017, pp. CN228–CN245. 26-29 June.

[154] Energy, Electricity and Nuclear Power Estimates for the Period up to 2050. Reference Data Series No. 1, 2018 Edition. IAEA-RDS-1/38 ISBN 978-92-0- 104918-6.

[155] V.A. Grachev, Environmental Policy of the State Atomic Energy Corporation ROSATOM, Center for the Promotion of Social and Environmental Initiatives of the Nuclear Industry, Moscow, 2011, p. 350.

[156] A.K. Sengupta, et al., Experience on mixed carbide fuels with high Pu content for Indian fast breeder reactor - an overview, J. Nucl. Mater. 385 (l) (2009) 161–164.

[157] P. Gavrilov, et al., Experimental demonstration center on MCC as a prototype of the 3-rd generation plant for thermal reactor SNF reprocessing, in: Global 2009: Proc. Int. Conf, Paris, France, 2009, p. 9075.

[158] A. Rodin, V. Vidanov, in: Explosion safety of Am and Cm partitioning by chromatography methodActinide and fission product partitioning and transmutation: Book of Abstracts 15th information Exchange Meeting, United Kingdom, 2018, pp. 119–120. 30th September to 3rd October 2018.

[159] R.B. Matthews, K.M. Chidester, C.W. Hoth, et al., Fabrication and testing of uranium nitride fuel for space power reactors, J. Nucl. Mat. 151 (1988) 334.

[160] Y. Arai, S. Fukushima, K. Shiozawa, M. Handa, Fabrication of uranium-plutonium mixed nitride and thermally stable carbide fuels, in: Advanced fuel for fast breeder reactors: fabrication and properties and their optimization: Proceedings of technical meeting IAEA, Vienna, 1987 IAEA-TECDOC-466.

[161] Y. Fedorov, Ecological safety of the base technology of EDC, in: Materials of Atomeco-2013 International Exhibition, Moscow, 2013 31.10[Electronic source]. URL:.

[162] S.M. Feinberg, Discussion Content, in: Record of Proceedings Session B-10, Int. Conf, on the Peaceful Uses for Atomic Energy, 9, Geneva, Switzerland: United Nations, 1958, p. 447.

[163] M. Conquerelle, C.T. Walker, Fission gas release and microscopic swelling in highly rated advanced fuels, Nucl. Techn. 48 (1) (1980) 43–54.

[164] Hippel. Frank von, Managing spent fuel in the United States: the illogic of reprocessing, IPFM Res. Rep. (3) (2007) 44 pp.

[165] T Okawa, I. Tatewaki, T. Ishizu, et al., Fuel behavior analysis code FEMAXI-FBR development and validation for core disruptive accident, Progress Nucl. Energy. 82 (2015) 80–85.

[166] D. Crawford, D. Porter, S. Hayes, Fuels for sodium cooled fast reactors: US perspective, J. Nucl. Mater. 371 (2007) 202–231.

[167] A. Hasegawa, K. Tanigawa, A. Ohtsuru, H. Yabe, M. Maeda, J. Shigemura, T. Ohira, T. Tominaga, M. Akashi, N. Hirohashi, T. Ishikawa, K. Kamiya, K. Shibuya, S. Yamashita, R.K. Chhem, Health effects of radiation and other health problems in the aftermath of nuclear accidents, with an emphasis on Fukushima, Series: From Hiroshima and Nagasaki to Fukushima 386 (2015) 2[Electronic source]. URL:. www.thelancet.com .

[168] M. Vitaliy, K. Oleg, F. Lubov, K. Evgeniy, Y. Nikolaiy, F. Sergey, L. Mikhail, V. Yuriy, V. Vitaliy, S. Andrei, Hot test of technique separation of americium and curium, Int. Conf, on Fast Reactors and Related Fuel Cycles (2017) CN237–CN245 26-29 June.

[169] Hydrogen Production Using Nuclear Energy, International Atomic Energy Agency, Vienna, 2013.

[170] IAEA Power Reactor Information System. [Electronic source]. URL: http://www.iaea.org/pris.

[171] IAEA, 2015. The Fukushima Daiichi Accident. Technical Volume 4/5. Radiological Consequences. - 262 p.

[172] ICRP. Problems Involved in Developing an Index of Harm, (ICRP Publication 27). Ann. ICRP 1 (4) (1977).

[173] ICRP, Recommendations of the international commission on radiological protection, Ann. ICRP 1 (3) (1977) 1977.

[174] ICRP, Publication 45. Quantitative bases for developing a unified index of harm, Ann. ICRP 15 (3) (1985).

[175] ICRP Database of Dose Coefficients: Workers and Members of the Public; Ver. 3.0. [Electronic source]. URL: http://www.icrp.org/page.asp?id=145 (Accessed 18.01.2019).

[176] IEA Atlas of Energy, International Energy Agency [Electronic source]. URL: http://energyatlas.iea.org.

[177] IEA. 2019 Snapshot of Global PV Markets. Report IEA-PVPS Tl-35, April 2019 [Electronic source]. URL: http://www.iea-pvps.org/fileadmin/dam/public/report/statistics/IEA-PVPS_Tl_35_Snapshot2019-Report.pdf.

[178] International Energy Outlook 2017. DOE/EIA-0484(2017). US Energy Information Administration, 2017.

[179] Y. Ito, Electrochemistry of nitrogen and nitrides in molten salts, J. Nucl. Mater. 344 (2005) 128–135.

[180] Kenneth McClellan, in: U.S. DOE 2016 Advanced Fuel Update. 17th GenIV SFR Advanced Fuel PMB Meeting, Beijing,China, 2016 March 21-23.

[181] T. Koyama, Current Status on Reprocessing Technology of Fast Reactor Fuel Cycle Technology Development (FACT) Project in Japan, Overview of Reprocessing Technology Development, in: Proc. Int. Conf. Global 2009, Paris, France, 2009 9100.

[182] Y. Kulyako, A study of nitrogen oxides released into the gas phase during uranium nitride dissolution in nitric acid, J. Radioanal. Nucl. Chem. 299 (3) (2014) 1293–1298.

[183] Ph. Dubuisson, Y. De Carlan, Main ways to increase SFR structural materials irradiation stability, in: Development of Nuclear Power Industry based enclosed fuel cycle with Fast Reactors: Innovative technologies and materials: 3rd International Scientific & Technical Conference, Moscow, 2009 November 11-12.

[184] B.A. Chin, R.J. Neuhold, J.L. Straalsund, Materials development for fast breeder reactor cores, Nucl. Techn. 57 (3) (1982) 426.

[185] P. Bardelle, D. Warin, Mechanism and kinetics of the uranium-plutonium mononitride synthesis, J. Nucl. Mater. 188 (1992) 36–42.

[186] A.A. Bauer, P. Cybulskis, J.L. Green, Mixed-nitride fuel performance in EBR-II *I*, in: Proc. of Int. Meeting on Advanced LMFBR Fuels, Tucson, 1977, p. 299. A2Oct. 10-13.

[187] New Energy Outlook 2018. BNEF's annual long-term economic analysis of the world's power sector out to 2050 [Electronic source]. URL: https://bnef.turtl.co/story/neo2018?teaser=true.

[188] L. Zabudko, A. Kamaev, L. Mamaev, A. Trufanov, Nitride fuel for advanced fast sodium reactors, in: Proc. Int. Conf. Global 2003, New Orleans, LA, 2003, p. 1679. Nov. 16-20.

[189] N. Kimura, Y. Miyake, et al., Noise Reduction Techniques for the Particle Image Velocimetry, Application to an Experimental Study of Natural Convection in Fast Reactor Core, ICONE-8405, Baltimore, MD, USA, 2000 April 2-6.

[190] Nuclear Energy Data - 2009, © OECD 2009.

[191] Nuclear Energy Data - 2018, © OECD 2018, NEA No. 7416.

[192] S.M. Rashad, F.H. Hammad, Nuclear power and the environment: Comparative assessment of environmental and health impacts of electricity-generating systems, Appl. Energy 65 (2000) 211–229.

[193] E.O. Adamov, V.N. Mihailov, Nuclear Power in Russia - Realities and Prospect, in: World Energy 15th Congress, Madrid, 1992.

[194] V.V Orlov, V.S. Smirnov, Physical Characteristics of Lead Cooled Fast Reactor, in: Proc. of Int. Conf. on Advances in Reactor Physics, 1, Knoxvill, USA, 1994, pp. 348–356. Apr. 11-15.

[195] V.A. Zhiltsov, V.M. Kulygin, N.N. Semashko, A.A. Skovoroda, V.P. Smirnov, A.V. Timofeev, E.G. Kudryavtsev, V.I. Rachkov, V.V. Orlov, Plasma separation of the elements applied to nuclear materials handling, Atomic Energy 101 (4) (2006) 755.

[196] A.M. Yegorov, V.B. Yuferov, S.V Shariy, V.A. Seroshtanov, O.S. Druy, V.V. Yegorenkov, E.V. Ribas, S.N. Khizhnyak, D.V. Vinnikov, Preliminary study of the demo plasma separator, Probl. Atomic Sci. Techn. Plasma Phys. Ser. (1 (59)) (2009) 122 (15).

[197] A.A. Bekhtenev, V.I. Volosov, Problems of a thermonuclear reactor with a rotating plasma, Nucl. Fusion. 20 (1980) 579.

[198] B.D. Rogozkin, Yu.K. Bibilaschvili, et al., Production and studies of U and Pu nitrides as nuclear fuel and forms of weapon grade plutonium storage, in: Proc. Int. Conf. Global, 2, France, 1995, p. 1359. September 11-14, 1995.

[199] Projected Costs of Generating Electricity, IEA/NEA, 2015 2015 Edition.

[200] Y. Arai, Y. Suzuki, M. Handa, Research on actinide mononitride fuel, in: Evaluation of Emerging Nuclear Fuel Cycle System: Proc. Int. Conf. Global, Paris, 1995, p. 538. Sept. 1-14, 1995.

[201] Renewables 2018 Global Status Report [Electronic source]. URL: http://www.ren21.net/gsr-2018.

[202] Review of non-aqueous nuclear fuel reprocessing and separation methods Bychkov A.V., et al. NATO Advanced Study Inst. on Chemical Separation Technologies and Related Methods of Nuclear Waste Management. Applications, Problems, and Research Needs (Proc. Int. Conf. Dubna, Russia, May 18-28, 1998), Vol. 2(53). - P. 71-98.

[203] SCALE Nuclear Systems Modeling & Simulation. [Electronic source]. URL: https://www.ornl.gov/scale.

[204] A.Yu. Shadrin, V.L. Vidanov, L.I. Tkachenko, Separation and partition of Am and Cm from closed nuclear fuel cycle high-level waste, in: Actinide and fission product partitioning and transmutation: Book of Abstracts 15th information Exchange Meeting, Manchester, UK, 2018, p. 57. 30th September to 3rd October 2018.

[205] A. Shadrin, Closing nuclear fuel cycle with fast breeder reactors: problems and prospects, in: Proc. Int. Conf. Global 2013, Salt-Lake-City, 2013, p. 7683.

[206] V.P. Smirnov, A.A. Samokhin, N.A. Vorona, A.V. Gavrikov, Study of charged particle motion in fields of different configurations for developing the concept of plasma separation of spent nuclear fuel, Plasma Phys. Rep. 39 (6) (2013) 456.

[207] D.L. Preston, E. Ron, S. Tokuoka, S. Funamoto, N. Nishi, M. Soda, K. Mabuchi, K. Kodama, Solid cancer incidence in atomic bomb survivors: 1958-1998, Radiat. Res. 168 (1) (2007) 1–64.

[208] Status and Trends of Nuclear Fuels Technology for Sodium Cooled Fast Reactors. IAEA Nuclear Energy Series, No. NF-T-4.1, 2011.

[209] F. Delage, J. Carmack, C.B. Lee, et al., Status of advanced fuel candidates for Sodium Fast Reactor within the Generation IV International Forum, J. Nucl. Mater. 441 (2013) 324–330.

[210] L. Zabudko, I. Kurina, A. Mayorshin, et al., Status of CEA-Minatom collaborative experiment BORA-BORA: fuels with high plutonium content, Backend of the Fuel Cycle: From Research to Solutions, Global, Paris, 2001, p. 292. September 9-13,2001.

[211] V.L. Vidanov, L.I. Tkachenko, M.G. Dmitriev, V.E. Davidov, E.V. Kenf, V.A. Babain, A.Yu. Shadrin, A.Yu. Evsyukova, The batch test of TODGA/F-3 solvent using gram-amounts of americium, in: Proc. Int. Conf. ISEC 2017, Miyazaki, Japan, 2017, p. 76.

[212] L. Tkachenko, V. Vidanov, E. Kenf, A. Shadrin, A. Murzin, The effect of radiolysis on the extraction properties of TODGA in the fluorinated diluent F-3, in: Actinide and fission product partitioning and transmutation: Book of Abstracts of 15th information Exchange Meeting, Manchester, UK, 2018, p. 126. 30th September to 3rd October.

[213] F. Doucet, et al., The formation of hydrated zirconium molybdate in simulated spent nuclear fuel reprocessing solutions, Phys. Chem. Chem. Phys. 4 (2002) 3491.

[214] The Future of Nuclear Power, An Interdisciplinary MIT Study, Massachusetts Institute of Technology, USA, 2003.

[215] E.O. Adamov, V.V. Orlov, A.I. Filin, et al., The next generation fast reactor, Nucl. Engng Des. 173 (1997) 143–150.

[216] Y. Suzuki, Y. Arai, Thermophysical and thermodynamic properties of actinide mononitrides and their solid solutions, J. Alloys Compd. 271-273 (1998) 577–582.

[217] V.M. Bardakov, Dan Vo Nhu, G.N. Kichigin, N.A. Strokin, Three-component plasma-optical separation of spent nuclear fuel, Nuclear Power (2) (2011) 3.

[218] A.V. Timofeev, Plasma method for processing spent nuclear fuel, Plasma Phys. Rep. 33 (11) (2007) 890.

[219] Zh. Gu, G. Wang, Wang Zh, et al., Transient analyses on loss of heat sink and overpower transient of natural circulation LBE-cooled fast reactor, 11 Ibid 81 (2015) 60–66.

[220] D. Kolupaev, M. Logunov, A. Shadrin, Used mixed oxide fuel reprocessing at RT-1 plant *I*, Procedia Chem. 10 (2016) 198–202.

[221] G. Wang, Zh. Gu, Zh. Wang, et al., Verification of neutronics and thermal hydraulics coupled simulation program NTC by the PDS-XADS transient simulation, Progress Nucl. Energy. 85 (2015) 659–667.

[222] World Energy Outlook 2018. OECD/IEA, 2018.

[223] World Wind Energy Association. Wind power capacity worldwide reaches 597 GW, 50,1 GW added in 2018 [Electronic source]. URL: https://wwindea.org/blog/2019/02/25/wind-power-capacity-worldwide-reaches-600-gw-539-gw-added-in-2018.

List of abbreviations

AC	automatic control
ADS	accelerator driven system
APCS	automated process control system
BFS	large-scale test facility
BN	sodium cooled fast reactor
BOR	test fast reactor
BR	breeding ratio
BRC	breeding ratio in the reactor core
BREST-OD-300	pilot demonstration lead-cooled fast reactor plant
BWR	boiling water reactor
CCGT	combined cycle gas turbine
CED	committed effective dose
CEFA	combined experimental fuel assemblies
CERN	European Organization for Nuclear Research
CI	capital investments
CNFC	closed nuclear fuel cycle
CPS	control and protection system
CR	control rod
CZ	central zone
DDREF	dose and dose rate effectiveness factor
DOE	U.S. Department of Energy
EAR	excessive absolute risk
ECCS	emergency core cooling system
ECS	emergency cooldown system
EFA	experimental fuel assembly
EHE	emergency heat exchangers
EHRS	emergency heat removal system
EP	emergency protection
ERR	excessive relative risk
FA	fuel assembly
FAI	federal autonomous institution
FF	fuel factor of electricity cost
FFTF	fast flux test facility (sodium-cooled fast reactor with the capacity of 400 MW (t))
FR, FNR	fast neutron reactor
FP	fission products
FRM	fabrication/ refabrication module

FRR	federal rules and regulations
GFP	gaseous fission products
GIF	Generation IV International Forum
HE	heat exchanger
HLW	high-level waste
HM	heavy metals
HPP	hydroelectric power plant
HTGR	high-temperature gas-cooled reactor
HTSC	high-temperature super-conducting
IAEA	International Atomic Energy Agency
IBR	pulsed fast reactor
ICRP	International Commission on Radiological Protection
IEA	International Energy Agency
ILW	intermediate level waste
INPRO	International Project on Innovative Nuclear Reactors and Fuel Cycles
IPC	industrial power supply complex
IRS	ionizing radiation sources
ITER	International Thermonuclear Experimental Reactor
KEU	set of experimental facilities
LCNPP	low-capacity nuclear power plants
LCOE	levelized cost of energy
LLW	low-level waste
LRW	liquid radioactive waste
MA	minor actinides
MBIR	multi-functional fast research reactor
MCC	main coolant circuit
MNIT	mixed nitride uranium-plutonium
MOX	mixed oxide uranium-plutonium
MSR	molten salt reactor
MSRE	molten salt reactor experiment
NF	nuclear fuel
NFC	nuclear fuel cycle
NPE	nuclear power
NPP	nuclear power plant
NRES	nonrenewable energy sources
NRS	nuclear and radiation safety
NTP	new technology platform
PDEC	pilot demonstration power supply complex
ODS	oxide dispersion strengthened
OECD	Organization for Economic Cooperation and Development
OPBZ	basic rules of safety and physical protection in transportation of nuclear materials
PBH	potential biological hazard
PBTRV	safety rules for transportation of radioactive materials
PFS	passive feedback system
PHRS	passive heat removal system
PM	processing module

PWR	pressurized water reactor (see VVER)
PZ	periphery zone
R&D	research and development works
RBMK	high-power channel-type reactor
RC	reactivity compensation
RCP	reactor coolant pump
RCPS	reactor coolant pump set
REE	rare-earth element
RES	renewable energy sources
RNT	relativistic nuclear technology
RP	reactor plant
RW	radioactive waste
SC	State Corporation
SCP	supercritical pressure
SCWR	supercritical water-cooled reactor
SFA	spent fuel assembly
SG	steam generator
SLCS	steam generator leakage confinement system
SNF	spent nuclear fuel
SPP	solar power plant
SRW	scientific and research work
SSC	stress and strain condition
SSS	spent sealed sources
SW	software
TEI	technical and economic indices
TNR	thermal neutron reactor
TPP	thermal power plant
TPS	transportation package set
TR	thermal reactors
TUE	transuranium elements
VVER	pressurized water reactor
VVER-TOI	standard optimized reference pressurized water reactor
WPP	wind power plant
WS	work stream

Index

Page numbers followed by "*f*" and "*t*" indicate, figures and tables respectively.

S

T

U

V

W

MIX
Papier | Fördert
gute Waldnutzung
FSC® C083411

Kolibri 360 GmbH
Ferdinand-Jühlke-Straße 7
99095 Erfurt, Deutschland
produktsicherheit@kolibri360.de